Photographs

We have included these photographs as a separate section to illustrate some of the ideas and provide portraits of some of the people discussed. These photographs were not a part of the original document and are not the complete set that would be required to appropriately represent the manuscript; rather, they are a sample of those available from the time period and history discussed.

Figure 1. The Aneroid barometer was used to measure differences in elevation. It was more convenient than the mercurial or cistern barometer but less reliable.

Figure 2. The Odometer was used to measure distance traveled by counting the revolutions of a wheel (1871).

Figure 3. The Berger theodolite was a precision instrument used for measuring horizontal and vertical angles. Manufactured by C.L. Berger & Sons, Boston (circa 1901).

Figure 4. Clarence King, the first Director of the U.S. Geological Survey (1879–81).

Figure 5. John Wesley Powell, the second Director of the U.S. Geological Survey (1881–94).

Figure 7. John Karl Hillers was the photographer with the Powell Surveys and later Chief Photographer of the U.S. Geological Survey (circa 1890s).

Figure 6. A U.S. Geological Survey pack train carries men and equipment up a steep slope while mapping the Mount Goddard, California, Quadrangle (circa 1907).

Figure 8. Copper plate engraving of topographic maps provided a permanent record.

Figure 9. U.S. Geological Survey printing presses for use of lithostones in printing topographic maps.

Figure 10. Topographic information was transferred from copper plates to a lithographic stone for printing.

Figure 12. A 300-foot steel tape used for accurately measuring distances.

Figure 11. Plane table mounted on a Johnson tripod. Inset shows Johnson tripod head.

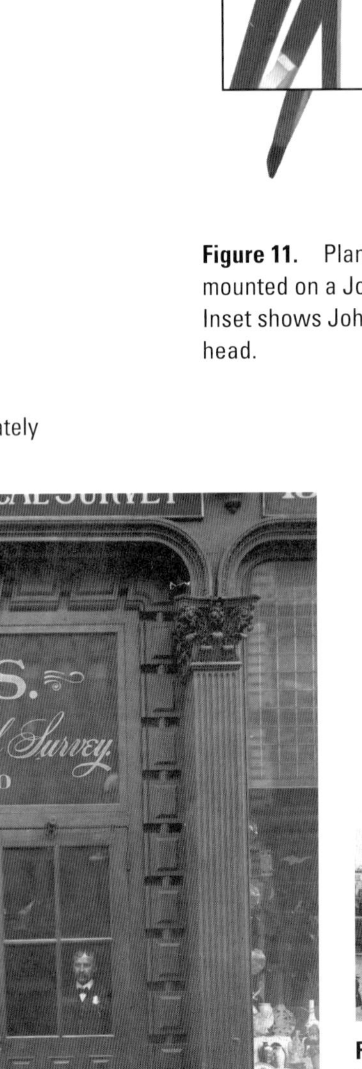

Figure 13A. Entrance to the Hooe Iron Building (1917).

Figure 13B. View of the Hooe Iron Building, U.S. Geological Survey headquarters, 1330 F Street, NW., Washington, D.C. (1917).

Figure 14. Bench marks were established with bronze-stemmed disks (left and middle) and caps (right) were stamped with a rounded elevation and placed by U.S. Geological Survey survey crews.

Figure 15. When a structure was not available for cementing a stemmed disk, a 4-foot long wrought iron post with a bronze bench mark cap riveted to the top was planted by survey crews.

Figure 16. Henry Gannett, U.S. Geological Survey Chief Geographer in 1882. He is often referred to as the father of American Topographic Mapping (1899).

Figure 17. U.S. Geological Survey flag with emblem, a design consisting of a white triangle and crossed hammers encircled by 13 stars on a blue field.

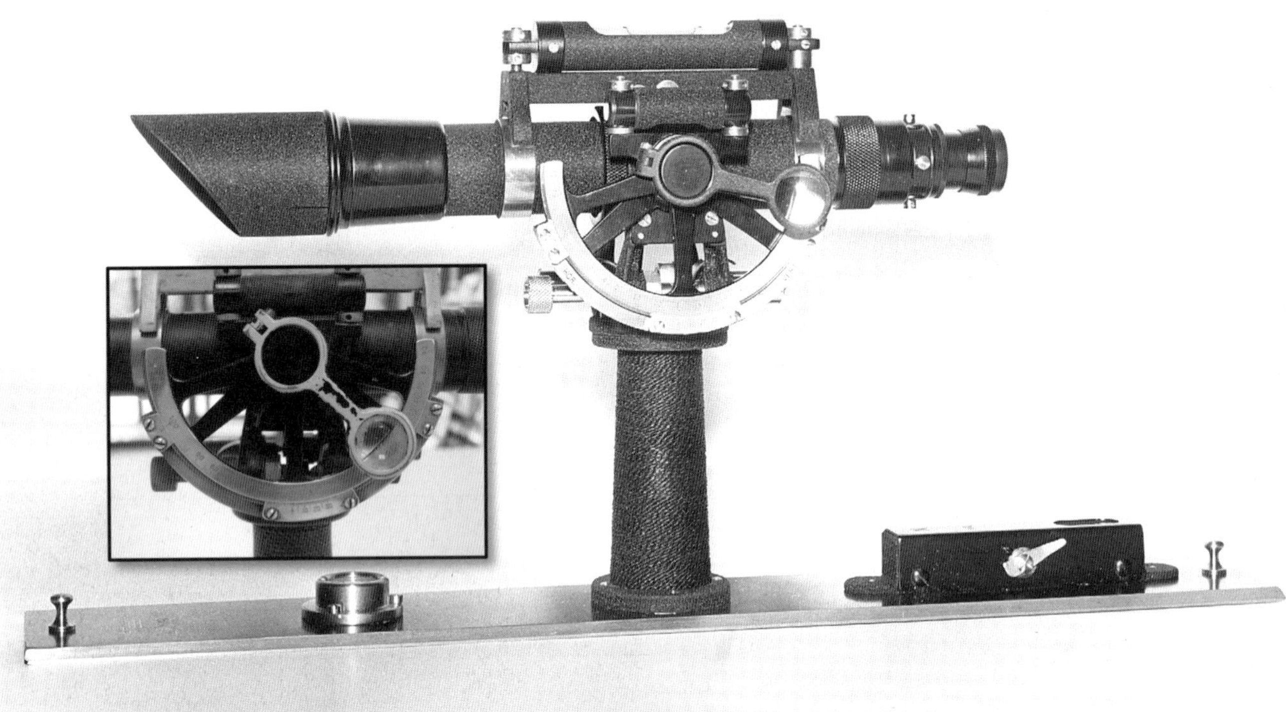

Figure 18. A Keiffel & Esser high standard alidade was used in conjunction with a plane table. The inset shows detail of W.M. Beaman's (U.S. Geological Survey topographer) improvement to the instrument, the Beaman Arc.

Figure 19. U.S. Geological Survey field crew using an automobile. The U.S. Geological Survey first used automobiles in the field in 1914 (1918).

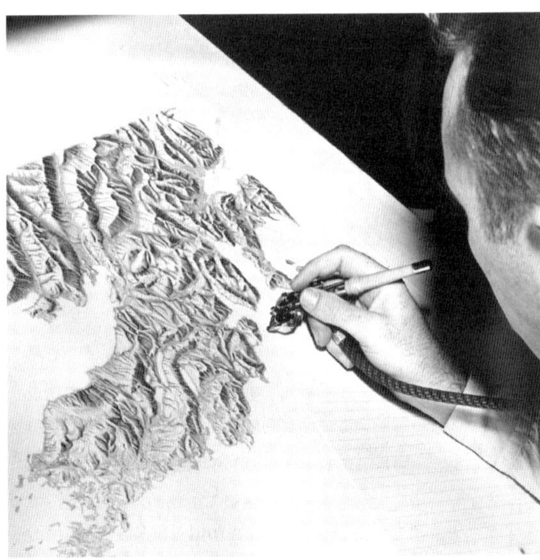

Figure 20. Shaded relief maps use the artistic illusion of depth to emphasize elevation. They were first employed by the U.S. Geological Survey in 1913.

Figure 21. U.S. Geological Survey field crew (1952).

Figure 22. A helicopter is used to transport a U.S. Geological Survey survey crew to a remote triangulation station in Utah (1953). Helicopters were first used in this capacity by the U.S. Geological Survey in 1948.

Figure 23. Pen-and-ink drafting for map production was used from the early 1940s through the mid-1950s. Inking contours by hand required a light touch to maintain consistent line weight (1952).

Figure 24. The scribing method of map production supplanted pen-and-ink drafting in the mid-1950s and produced a more legible map in a shorter time, and at less cost.

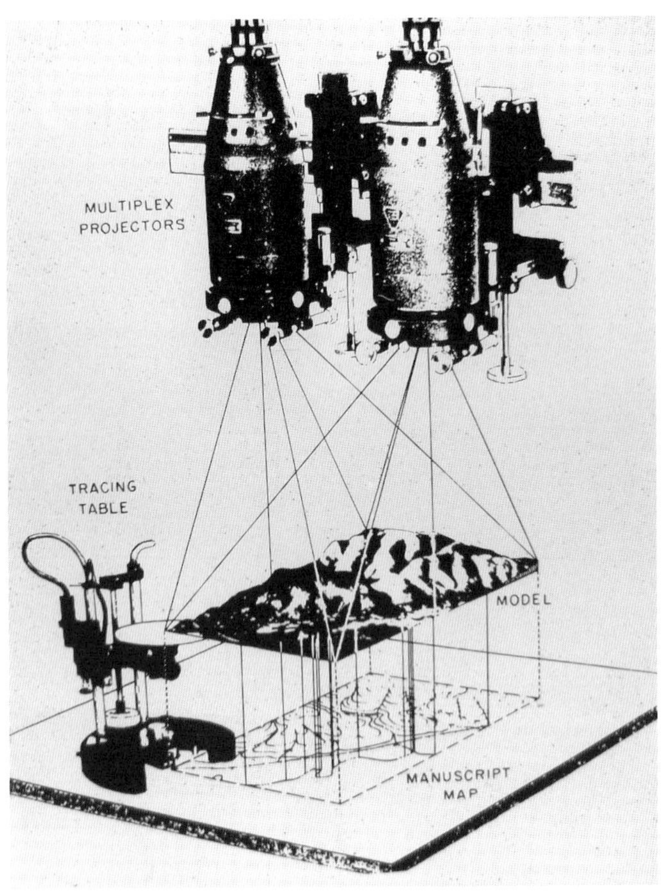

Figure 25. Convergent photography for map production employed stereoscopic models projected onto a tracing table. This allowed elevations to be determined with more accuracy, and at less cost, than former methods.

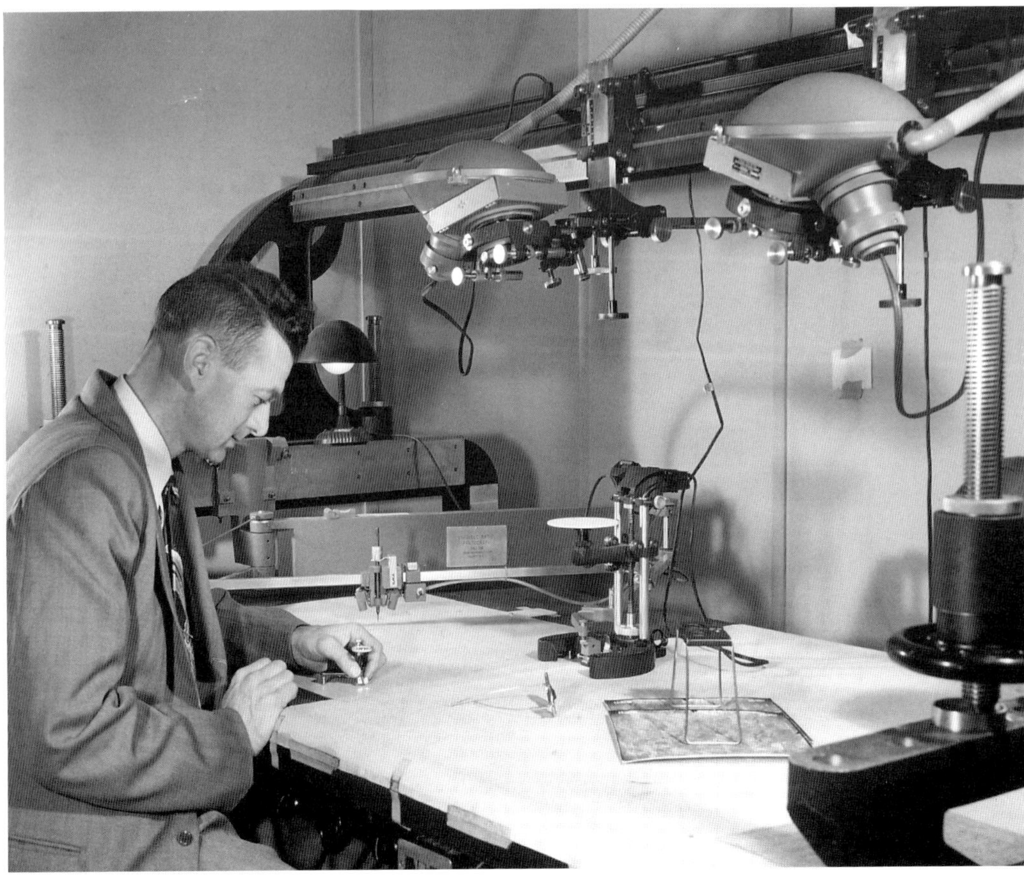

Figure 26. Russell K. Bean of the U.S. Geological Survey patented the ER–55 stereoplotting instrument, which proved superior to the multiplex plotter in the compilation phase of topographic mapping.

History of the Topographic Branch (Division)

By Richard T. Evans and Helen M. Frye

Circular 1341

U.S. Department of the Interior
U.S. Geological Survey

U.S. Department of the Interior
KEN SALAZAR, Secretary

U.S. Geological Survey
Suzette M. Kimball, Acting Director

U.S. Geological Survey, Reston, Virginia: 2009

For more information on the USGS—the Federal source for science about the Earth, its natural and living resources, natural hazards, and the environment, visit http://www.usgs.gov or call 1-888-ASK-USGS

For an overview of USGS information products, including maps, imagery, and publications, visit http://www.usgs.gov/pubprod

To order this and other USGS information products, visit http://store.usgs.gov

Suggested citation:
Evans, R.T., and Frye, H.M., 2009, History of the topographic branch (division): U.S. Geological Survey Circular 1341, 196 p.

ISBN 978-1-4113-2612-5

Foreword

This "History of the Topographic Branch" is being published to provide a view of the U.S. Geological Survey (USGS) mapping program from its inception in the various Surveys of the 19th century to approximately 1954. The manuscript has existed as a draft document within the USGS since 1955 with several unsuccessful attempts during the 1960s and 1970s to review it and make it more complete and up to date. We have chosen not to attempt to update the history for the years from 1955 to the present (2009), but rather to preserve the document as it was written in the 1950s. We have incorporated peer review comments that were developed in the 1960s and 1970s, and effort has been placed on maintaining the manuscript in the voice in which it was written with minor reorganizations to aid the reader in following the topical headings of the history.

This manuscript utilizes a series of three asterisks separated by spaces (* * *) to indicate points of ellipsis. Ellipsis is the omission of words necessary for complete grammatical construction, but unnecessary for comprehension in the context of a sentence. Ellipsis is also used in the omission of material from quotations.

Tables included in the original manuscript have been referenced in the text, as is now required under USGS publication standards. We have included a set of photographs as a separate section to illustrate some of the ideas and provide portraits of some of the people discussed. These photographs were not a part of the original document and are not the complete set that would be required to appropriately represent the manuscript; rather, they are a sample of those available from the time period and history discussed. These photographs also are referenced in the text.

We caution the reader that this is not an interpretative history of the USGS mapping program placed in the context of other events; it is more of a non-interpretative chronology including facts, dates, and events taken from USGS reports, Survey Orders, memoranda, and other documentation. The writing is somewhat uneven and not wholly consistent, some items are emphasized more than others, and all facts in the document have not been completely verified. In many cases, particularly in the latter one-half of the document, Survey Orders and memoranda are included verbatim in the manuscript. Notwithstanding these comments, we think the document provides a valuable contribution to understanding the history and methods of topographic mapping in the United States.

The USGS is indebted to Thomas C. Miller, a 37.5-year employee of the USGS. Upon his retirement, he took a copy of this manuscript and over a period of several years scanned the document, converted the scanned pages to text, performed an exhaustive edit of the manuscript incorporating peer review comments that had been marked on the paper manuscript, and brought it to the USGS in March 2009, asking if it could be published. It is somewhat fortuitous that the year he brought the manuscript forward for publication is the 125th anniversary of the USGS topographic mapping program.

About the authors: Richard Tranter Evans served more than 50 years with the USGS before his retirement in 1951 as the Survey's senior topographer. He drew the first topographic maps of Pikes Peak, the Grand Canyon, and Salt Lake City. He was an "instrumentman" in the Washington, D.C., surveys in 1910–11, and a field engineer in 1915. He also was detailed to the National

Park Service as an Acting Superintendent in Zion National Park in 1925–26, and in Hawaii in 1927. He was a member of the Cosmos Club, the Explorer's Club of New York, the American Society of Civil Engineers, and a Member Emeritus of the American Society for Photogrammetry. After his retirement, he served as secretary and executive treasurer for the American Congress on Surveying and Mapping. He received the Meritorious Service Award of the Department of Interior in 1951.

Helen M. Frye served the USGS more than 40 years. She was in the Administrative Services Section of the Topographic Branch of the USGS and contributed the Administration chapter to "A Manual of Topographic Instructions" published in 1928. She was a member of the Special Committee, which supported employee issues such as retirement notifications.

Eric W. Constance
Chief, Applied Research and Technology
National Geographical Technical Operations Center

Preface

This book is the result of many hours of research and personal recollections of Richard T. Evans and Helen M. Frye and was compiled and organized during the mid-1950s. Included are a number of documents that were considered worthy of preservation, as well as several diary excerpts from the early days. The draft had undergone one or more peer reviews but was never published; however, it is apparent that the peer review process continued for nearly 20 years. This editor, sensing the importance of this work, decided to take the old, age-blackened manuscript, and scan it. By using modern optical character recognition, and recovering text that had faded, a digital file was produced. Where practical, the peer review notes were incorporated into the final text.

This is a history of the U.S. Geological Survey, Topographic Branch (as it was then called), from its inception through the mid-1950s. No attempt has been made to extend the work to the present day (2009); this would be a monumental undertaking.

The original charter of the U.S. Geological Survey, Topographic Branch, authorized the completion of the entire country on a large scale. At the time this manuscript was compiled the task was still far from completion, and would not be realized for more than 30 years.

The editor's interest in history, especially of an organization with which he had spent nearly 40 years of a career that included most phases of topographic mapping, provided the incentive for the task of scanning and proofing the manuscript. Both content and form have been preserved from the original. In addition, an index has been compiled by the editor and appended to the work.

This history includes many warm personal glimpses of those who pioneered the organization that has contributed so much to the development of this great Nation, as well as insight into the many instrumentation and procedural inventions that were the products of a host of brilliant minds.

The reward of my work on this manuscript is simply the interest of others who might read it.

Thomas C. Miller

Contents

Figures

1–10. Photographs showing—

  1. The Aneroid barometer was used to measure differences in elevation. It was more convenient than the mercurial or cistern barometer but less reliable.

  2. The Odometer was used to measure distance traveled by counting the revolutions of a wheel (1871).

  3. The Berger theodolite was a precision instrument used for measuring horizontal and vertical angles. Manufactured by C.L. Berger & Sons, Boston (circa 1901).

  4. Clarence King, the first Director of the U.S. Geological Survey (1879–81)

  5. John Wesley Powell, the second Director of the U.S. Geological Survey (1881–94).

  6. A U.S. Geological Survey pack train carries men and equipment up a steep slope while mapping the Mount Goddard, California, Quadrangle (circa 1907).

  7. John Karl Hillers was the photographer with the Powell Surveys and later Chief Photographer of the U.S. Geological Survey (circa 1890s).

  8. Copper plate engraving of topographic maps provided a permanent record.

  9. U.S. Geological Survey printing presses for use of lithostones in printing topographic maps.

  10. Topographic information was transferred from copper plates to a lithographic stone for printing.

Tables

Conversion Factors and Datums

Inch/Pound to SI

Multiply	By	To obtain
Length		
inch (in.)	2.54	centimeter (cm)
inch (in.)	25.4	millimeter (mm)
foot (ft)	0.3048	meter (m)
mile (mi)	1.609	kilometer (km)
mile, nautical (nmi)	1.852	kilometer (km)
Area		
acre	4,047	square meter (m^2)
acre	0.4047	hectare (ha)
acre	0.4047	square hectometer (hm^2)
acre	0.004047	square kilometer (km^2)
square foot (ft^2)	929.0	square centimeter (cm^2)
square foot (ft^2)	0.09290	square meter (m^2)
square inch (in^2)	6.452	square centimeter (cm^2)
section (640 acres or 1 square mile)	259.0	square hectometer (hm^2)
square mile (mi^2)	259.0	hectare (ha)
square mile (mi^2)	2.590	square kilometer (km^2)
Mass		
ounce, avoirdupois (oz)	28.35	gram (g)
pound, avoirdupois (lb)	0.4536	kilogram (kg)
ton, short (2,000 lb)	0.9072	megagram (Mg)
ton, long (2,240 lb)	1.016	megagram (Mg)

Temperature in degrees Fahrenheit (°F) may be converted to degrees Celsius (°C) as follows:

$$°C=(°F-32)/1.8$$

Vertical coordinate information is referenced to the North American Vertical Datum of 1929 (NAVD 29).

Horizontal coordinate information is referenced to theNorth American Datum of 1927 (NAD 27).

Elevation, as used in this report, refers to distance above the vertical datum.

History of the Topographic Branch (Division)

By Richard T. Evans and Helen M. Frye

Predecessor Surveys

Introduction

From a very early period of the world's existence, man has endeavored to represent the earth's surface in a graphic form for the information of his fellow men, realizing that no oral or written description is capable of setting forth topographic facts so vividly and so clearly as a map.

Mapping of the areas of the United States began with the charting of portions of its coast line by early explorers; the need for topographic maps was first recognized during the war of the Colonies for independence from Great Britain. On July 22, 1777, Congress authorized General Washington to appoint:

> "Mr. Robert Erskine, or any other person that he may think proper, geographer and surveyor of the roads, to take sketches of the country and the seat of war."

By several acts during the Revolutionary War, Congress provided "geographers" for the armies of the United States, some of them with the pay of a colonel, amounting to $60 a month and allowances. At the end of the War, a resolution of May 27, 1785, continued in service the "geographer of the United States" for a period of 3 years. The War Department recognized the necessity of "geographical engineers" and requested Congress to authorize their appointment, but it was not until the next war that Congress authorized on March 3, 1813, the appointment of eight topographic engineers and eight assistant topographic engineers under the direction of the General Staff of the Army. These officers formed the nucleus of the first Corps of Topographic Engineers in the Army, and that Corps continued to function as an independent unit until it was absorbed by the Corps of Engineers in 1863, during the Civil War between the States.

Between the Louisiana Purchase in 1803, and the outbreak of the Civil War, more than a hundred exploring and mapping expeditions were sent into the vast territory lying west of the Mississippi River to investigate the natural resources of this newly acquired country and to find possible locations for wagon roads to the Pacific Coast. These expeditions were sent out by the War Department and were in charge of Army officers. It is interesting to note that such generals

as George G. Meade, J.C. Fremont, Joseph E. Johnston, W.F. Smith, John Pope, A.W. Whipple, J.G. Parke, G.K. Warren, and H.L. Abbott, all officers of the Corps of Topographic Engineers, had charge of expeditions and were among our earliest map makers. Unfortunately, the data obtained by these editions were not of sufficient accuracy to serve as a basis for topographic maps of value other than in illustrating their voluminous reports.

During this early period, numerous surveys were undertaken within the original Thirteen States, by the Federal government and by the States. The most important were those carried on by the U.S. Coast and Geodetic Survey, which made an accurate survey of the Atlantic Coastline and established a triangulation system that was of so high a standard as to constitute the first and only accurate data for topographic mapping obtained before the Civil War. The Coast and Geodetic Survey, while charting the coast and rivers, also mapped a strip of country extending a few miles inland, the relief being shown by means of hachures, together with contour lines, until 1846 when the first government topographic map on which the relief was shown by contours alone was made, covering an area in the vicinity of Boston Harbor. In 1835, however, the Geological and Topographical Survey of Maryland had issued a map on which the relief was shown by contours, and this is believed to be the first contoured map issued in this country.

The outbreak of the Civil War stopped all mapping activities other than those needed by the U.S. Army. During the war, topographic surveys were carried on throughout the war zone under the supervision of the Corps of Engineers, the topographers being civilian employees. After the war, the country west of the Mississippi again became the center of the mapping activities of the government, which had in view the development of the national resources of this vast area.

Although between 1867 and 1878 numerous surveys were carried on in all parts of the United States, of great value for the specific purposes for which they were planned, especially in the survey of proposed railroad, canal, and wagon routes, there were only four large, well-equipped organizations carrying on systematic topographic surveys under government supervision and support. These were the U.S. Geological Exploration of the 40th Parallel (Clarence King, 1867–79), under the War Department; the U.S. Geological and Geographical Survey of the Territories (Professor F.V. Hayden, 1867–79), under the Interior Department; the U.S. Geographical and Geological Survey of the Rocky Mountain Region

(Maj. J.W. Powell, 1869–79), under the Smithsonian Institution; and the Geographical Surveys Eest of the 100th Meridian (Capt. George E. Wheeler, C.E., 1868–79), under the War Department.

King Survey (1867–79)

The U.S. Geological Exploration of the 40th Parallel, under the War Department, was authorized by Congress on March 2, 1867, by the following act:

> "And be it further enacted, That the Secretary of War is hereby authorized to direct a geological and topographical exploration of the territory between the Rocky Mountains and the Sierra Nevada Mountains, including the route or routes of the Pacific Railroad: Provided, That the same can be done out of existing appropriations."

This Survey was because of the personal efforts of Clarence King, a 25-year old enthusiast in geology, and he was placed in charge of it.

King, a New Englander and graduate of Sheffield Scientific School of Yale University in 1862, set out the following year on a horseback trip across the continent in order to study the then practically unknown Rocky Mountains. This trip, which he and his friend James T. Gardner made on an immigrant train, began in St. Joseph, Mo., occupied several months, and was full of incidents. He traveled to California where he worked as a volunteer assistant with the Geological Survey of that State. When, in 1866, it appeared that the State would discontinue appropriations for his work, he decided to return to Washington and enlist Federal aid for his survey plan.

Gen. Andrew A. Humphreys, then Chief of Engineers, became interested in King's plan and provided for the start of the work, an unexpended balance of appropriations previously made for the survey of a military road across the continent. He instructed King[1] to examine and describe the geological structures, geographical conditions, and natural resources of a belt of country extending from the 120th meridian eastward to the 105th meridian along the 40th parallel with sufficient expanse north and south, approximately 100 miles wide, to include the line of the Central and Union Pacific railroads. He also was to collect material for detailed maps of the chief mining districts, coal fields, salt basins, etc., as well as for a topographic map of the region traversed. The latter formed the basis of the geologic work. The mapping methods accepted as standard for those days and followed by Hayden, Powell, and to a certain extent by Wheeler, will be described later.

On July 27, 1867, field work was started with a force of 11 professional men consisting of 4 geologists, including King, 4 topographers, 1 botanist, 1 zoologist, and 1 photographer. In addition to the military escort of 20 men, there were teamsters, cooks, laborers, etc., necessary for the camp equipage, transportation, subsistence, and medical attendants. Field work was continued during the open seasons of each year and was completed November 13, 1872; however, the final reports were not completed until 1879, when the King Survey was closed.

In 1873, James T. Gardner, Chief Topographer of the King Survey, was employed by Hayden. Gardner established a topographical base for the mapping by first measuring a base line between 6 and 7 miles long near Denver, Colo., chiefly along the track of the Kansas Pacific Railroad, and from its two ends extending a net of triangulation to control the mapping. To strengthen the net, he later measured baselines at Colorado Springs and in the San Luis Valley.

Hayden Survey (1867–79)

The Hayden Survey was conducted in Colorado during the field seasons of 1873 and 1876, and in Wyoming and Idaho during 1877 and 1878. A total area of 107,000 square miles was covered. The largest portion was in Colorado and a map was published on the scale of 4 miles to the inch, with a 200-foot contour interval. Hayden published 11 annual reports, each covering a season's work. When the Hayden Survey was terminated in 1879, much material remained unpublished and Hayden was appointed as a geologist in the U.S. Geological Survey, in order to prepare the data for publication.

When Nebraska was granted statehood in 1867, the legislature set aside an unexpended balance of $5,000 for a geological survey of Nebraska, said survey to be prosecuted under the direction of the Commissioner of the General Land Office; upon the recommendation of the Secretary of the Smithsonian Institution, Professor Ferdinand V. Hayden was given charge of the work.

Hayden was born in Westfield, Mass.; graduated from Oberlin College, Ohio; studied at Albany Medical College; employed as a geologist and physician with the Warren (1857) and Reynolds (1859) Explorations of the War Department; served as a surgeon in the Union Army during the Civil War; and as a professor of mineralogy and geology at the University of Pennsylvania.

In 1868, Congress supported the work with an appropriation and directed that his geological explorations be extended to Wyoming. In 1869, Hayden was instructed by the Secretary of the Interior to pay special attention to the geological, mineralogical, and agricultural resources of Colorado and New Mexico, and in 1870 were shifted to Wyoming. In 1871, Hayden presented a plan for the geological and geographical exploration of the territories of the United States. Congress approved and increased the appropriation, and changed the name to the U.S. Geological and Geographical Survey of the Territories. The scene of activities became Montana and one important result of that season's field work was the creation of Yellowstone National Park. The following year, an ascent of the Grand Teton peak in Wyoming was made.

---

[1] See: "A History of the Water Resources Branch," by Robert Follansbee, p. 3, etc.

Powell Survey (1869–79)

Maj. John Wesley Powell lost his right arm at the Battle of Shiloh, but not his zest for adventure and search for scientific knowledge. As Professor of Geology and Natural History in the State Normal University of Illinois, he made a trip with students to Colorado in the summer of 1867 for the purpose of increasing the geological and zoological collections. The trip was repeated the following summer but, instead of returning with the students, he pushed farther west with several hunters and trappers and spent the winter on the White River near a camp of Ute Indians. There he developed an interest in the history and customs of the Indians and in the geography and geology of the Colorado River Basin.

Combining his own resources with contributions from his scientific friends, Powell organized his famous exploration of the Colorado River. On May 24, 1869, Powell's party of 10 courageous men in 4 specially constructed boats, began the first voyage through the canyons of the Green and Colorado Rivers, starting from Green River, Wyo. to the mouth of the Virgin River, a distance of 1,000 miles. The party safely navigated this most hazardous and daring river journey and solved the mysteries of the mile-deep canyons.

On May 22, 1871, a second voyage was made under a Congressional appropriation obtained through the aid of Powell's associates of war days, then in Congress. Appropriations were continued each year for the explorations of the Colorado River and adjacent territory, under the supervision of the Smithsonian Institution until 1874, when the supervision was assigned to the Interior Department. That year Powell's plan to connect his surveys with those of King on the north and those of Hayden on the east was approved, and his survey acquired the official title of "Geographical and Geological Survey of the Rocky Mountain Region." Powell's territory embraced the spectacular temples and chasms of the Grand Canyon and the painted cliffs of southern Utah. His parties included such artists as John H. Renshawe and William H. Holmes, whose colored illustrations enhanced his reports. The value that Powell placed upon topographic mapping is shown by the fact that for the work of one or two geologists he employed four topographers. The topographic base maps were published as 1-degree sheets with contour interval of 250 feet. The total area covered from 1867 to 1878 was 67,000 square miles.

Wheeler Survey (1868–79)

Early in 1871, the War Department resumed explorations of the West, which had been interrupted by the war between the States, and inaugurated the "U.S. Geographical Survey of the 100th Meridian." Lt. George E. Wheeler, an engineer officer, was placed in charge of the Survey. The assigned territory was south of the Central Pacific Railroad in eastern Nevada and Arizona, and in 1872 it was extended to include the entire region west of the 100th meridian.

Lieutenant Wheeler realized the importance of topographic maps. He knew that foreign governments were having their domains topographically mapped as a military necessity and that U.S. Army officers had charge of such work. He considered "The topographic as the indispensable, all-important survey * * * which underlies every other, including also the graphic basis for the economic and scientific examinations of the country * * *. This has been the main or principal general survey in all civilized countries."

The Wheeler Survey in eight field seasons covered 333,000 square miles in Southern California, New Mexico, Arizona, Utah, and Nevada. Mapping was done on the scale of 2 miles to 1 inch. Topographic atlas sheets to the number of 75, chiefly hachure maps on the scale of 8 miles to 1 inch were issued, and the final results of the Survey were published by the War Department in 7 volumes.

Early Mapping Methods

When the King Survey began in 1867, King and several of his assistants (J.T. Gardner, A.D. Wilson, and F.A. Clark) already had some years of mapping experience in California. Soon after the Geological Survey of California was started in 1860, Director Whitney secured the services of a young German engineer, Charles F. Hoffman, who possessed artistic talents, and who devised the methods of topographic mapping and taught them to others. He was the first to carry triangulation from peak to peak without special signal stations, and the first to use a planetable in the mountains for recording sight lines graphically, instead of reading horizontal angles with a transit.

A heavy planetable, not suitable for mountain work, was used by the Army Engineers in surveying sites for the construction of fortifications[2] and by the U.S. Coast and Geodetic Survey for small open areas near the coast.[3] The one used by Hoffman was a folding table.[4] An 18-inch open sight alidade completed the planetable outfit.

By far the greater amount of the field work was accomplished with the gradienter, a small transit, having a small telescope with striding level, 3-inch vertical and 3-inch horizontal circles both reading to minutes, mounted on a planetable tripod. With it the topographer, from primary and secondary triangulation stations that usually were on the highest peaks, read many horizontal and vertical angles to peaks, summits, ends of ridges, forks of creeks, and other salient features. The angles and descriptions of points sighted were duly recorded in a notebook by the topographer's assistant, who also made a drainage sketch and a profile sketch of the surrounding country. The area covered was a complete sweep of 360 degrees,

---

[2] "Elements of Surveying (1830)," by Charles Davies, Professor of Mathematics at West Point Military Academy.

[3] "The Planetable," U.S. Coast and Geodetic Survey Annual Report for 1880.

[4] Description given under "The Planetable and Alidade," 1886.

and the stations were so selected that the limits of work from one would reach the limits from those around it. The assistant usually was an artist and his panoramic sketches were beautiful and accurate. Another duty of the assistant was to care for the barometer, and record its readings.

The four surveys based the elevations for their mapping upon a railroad elevation near the starting point for a season's mapping, and checked the elevation, if practicable, before accepting it. From there on, dependence was placed upon the mercurial, or cistern barometer, for the elevations of multiple points. Two mercurial barometers were used, one remaining stationary at a known elevation, such as at camp, while the other was carried to various points, the heights of which were to be determined. They were read synchronously, as were two thermometers, at hourly, or other intervals by prearrangement. The difference between the two barometric readings gave the data for computing the differences in height. Besides the barometric readings, the notes required the entry of the date, time of day, and temperature, and then from tables the difference in height was computed.

The Guyot's (Arnold Guyot, 1807–84) and Williamson's (Lt. Col. R.S. Williamson, U.S. Engineers) Tables, from which the computations were made, were based on several factors involved in atmospheric pressure. Mr. Grove K. Gilbert,[5] geologist, devised a new method of computing differences of elevation barometrically based on an "atmospheric gradient," involving fewer factors, but requiring two base barometer stations, one as high as the highest of the points the elevations of which are to be determined, and the other as low as the lowest point. The moving barometer is corrected by reduction, not to one base barometer, but to two.

The mercurial barometer, about 30 inches long and a delicate instrument, required careful handling. The safest mode of transportation was in a sling on a man's shoulder, whether the man was walking or riding a horse. It frequently was broken, but extra glass tubes and a flask of mercury were carried for repairs. More convenient was the aneroid barometer (fig. 1), a small watchlike instrument, and it gained in favor, notwithstanding its results were less reliable.

In exploring across country, linear distances between camps, or astronomic stations or meander stations, were determined by elapsed time taken in travel, or by counting the paces of men or horses. On the Wheeler Survey, measurements along the roads, trails, and water courses were obtained from "the number of revolutions, indicated by an odometer (fig. 2), of a single wheel attached by shafts and other rigging to an animal, usually a mule, ridden by the observer. The actual number of revolutions to the mile made by the wheel was found, from practical tests, on different classes of traversed routes, tables were prepared, and from them the number of revolutions could be quickly converted into miles."

From the combined data recorded in the field notebooks, the topographer in the office, during the winter season, plotted and drafted the topographic map.

Topographic mapping methods were described in detail when Hayden, Powell, Wheeler, Gardner, and others testified in hearings before the Committee on Public Lands in Washington on May 13, 1874. The following quotations are taken from the testimony of J.T. Gardner, who then was Hayden's chief topographer.

"As a topographer on the State Geological Survey of California under Professor Josiah D. Whitney, I made, in connection with Mr. Clarence King, Assistant Geologist, a series of experiments on the methods of surveying mountain regions, to produce maps suited to geological purposes. These experiments were continued from 1863 to 1867. We tried the meandering method—the method of a compass-triangulation resting on a meandered base; of anunsystematic triangulation, with a theodolite, the whole resting on an astronomical base; then the rectangular method of sectionizing as used by the General Land Office, and lastly a regularly developed system of well-conditioned and carefully observed triangles resting on an astronomical base. The last was the best of all the methods tried, and the meandering was the poorest.

"We found that topography taken by the meandering method was utterly unreliable, except the minor details along the routes traveled.

"Inaugurating the trigonometric method of mapping for Professor Hayden in 1873, a base line between 6 and 7 miles long was carefully measured in the vicinity of Denver, chiefly along the tracks of the Kansas Pacific Railroad. It was twice measured with a steel tape, under 20 pounds strain, and the temperature taken every 5 minutes. The base was leveled and duly corrected for temperature and level. The latitude and longitude of one end of the base had been determined by astronomic observations, as also the azimuth of the line. From the two ends of the base line a system of primary triangles was carefully expanded by observing upon signals erected for the purpose. All the angles of the triangles were repeatedly observed with a 15-inch theodolite (fig. 3) with an 8-inch circle graduated to 10 inches and reading easily to 5 inches The quality of the work was determined by the test whether the three observed angles of the triangles when corrected for spherical excess sum up to 180 degrees. That by which they vary from 180 degrees is error.

"A secondary triangulation, resting upon the primary, was carried by the topographers over the same area. The primary triangles ranged from 30 to 60 miles in the length of their sides, while the secondary averaged 8 miles. The triangulation network extended over 17,000 square miles and was con-

---

[5] U.S. Geological Survey Second Annual Report, 1880–81, p. 405–566.

nected with check bases at Colorado Springs and in San Luis Valley * * *.

"Experience has clearly demonstrated that the topography must be taken from the peaks looking down on the country, not from the valleys looking up * * *. Also, the geologist or topographer finds no difficulty in doing his large share of personal scientific labor besides guiding the general movements of the party. The energizing effect of a leader who is doing a large part of the practical scientific work is found most beneficial upon the others engaged. All are stimulated by his example."

In a report by Gardner, a few years later, he stated that in an area of about 16,800 square miles in Colorado, topographers found it necessary for their work to locate and measure about 7,000 points. For a similar area, the Wheeler Survey, doing its work from below and along its route of travel, published on its map only 40 elevations.

Inception and Early Years of the Geological Survey

Events Leading to the Creation of the Geological Survey

In the spring of 1874, the mounting rivalry of the three Surveys for territory and for appropriations became a serious concern of the Congress. The King Survey had completed its field work in 1872. The three remaining Surveys were engaged in broadly similar work in the same general region, carried on by two separate Departments, War and Interior. None had been limited to districts. Authorizations indicated vaguely that Hayden's area included the Territories, Powell's the Rocky Mountain Region, and Wheeler's simply "West of the 100th Meridian." They were duplicating one another's work in many places and were presenting claims for appropriations in a manner to provoke the Congress into discontinuing the Surveys.

King's influence with leading scientific men and his tactful handling of the situation before the Congress averted the threatened cessation of the western surveys.[6] As a result, the House of Representatives on April 15, 1874, passed the following resolution:

> *Resolved* "That the President of the United States be requested to inform the House what geographical and geological surveys, under different departments and branches of the Government, are operating in the same and contiguous areas of territory west of the Mississippi River, and whether it be not

practicable to consolidate them under one department, or to define the geographical limits to be embraced by each."

In response to this inquiry, the Secretary of War and the Secretary of the Interior each claimed that the consolidation of the work in his department would make for efficiency and economy. The Congress deferred a decision until the continued unabated rivalry reached a crisis. King again came forward and his advice was accepted by the Congress and the National Academy of Sciences was requested to act as referee and make definite recommendations.

The following rider was attached to the Sundry Civil Act of June 20, 1878.

> "And the National Academy of Sciences is hereby required at their next meeting to take into consideration the methods and expenses of conducting all surveys of a scientific character under the War or Interior Department, and the surveys of the Land Office, and to report to Congress as soon thereafter as may be practicable, a plan for surveying and mapping the Territories of the United States, on such general system as will in their judgment secure the best results at the least possible cost."

In order to lessen the rivalry between the Powell and Hayden Surveys, pending the report of the National Academy, the appropriations in the same bill limited each survey to separate areas as follows:

> "Under Professor Ferdinand V. Hayden- - - -$75,000 Provided: That the money hereby appropriated shall be expended only in prosecuting said surveys north of the 42d parallel and west of the 100th meridian.

> Under Professor John W. Powell - - - - - - - -$50,000 Provided: That the money hereby appropriated shall be expended only in prosecuting said surveys south of the 42d parallel and west of the 100th meridian."

Before formulating a plan as directed by the Congress, the National Academy of Sciences called upon the Secretaries of War and Interior for reports of the survey activities under the direction of each. Major Powell's report was lengthy and impressive.

A committee of the National Academy, consisting of: Othniel Charles Marsh, paleontologist; vice president, James D. Dana, mineralogist, naturalist; William B. Rogers, geologist, physicist; John S. Newberry, geologist; William P. Trowbridge, engineer; Simon Newcomb, astronomer; and Alex Agassiz, biologist, zoologist, an illustrious group of scientists, considered the problem. On November 6, 1878, the committee made the following recommendations, which were adopted by the Academy and transmitted to the Congress.

1. That the Coast and Geodetic Survey be transferred to the Interior Department and, in addition to its former work, be charged with the preparation of a geodetic survey of the whole public domain, a topographic

---

[6] Biographical sketch of Clarence King: U.S. Geological Survey 23d Annual Report, 1902, p. 203.

survey comprising detailed topographic mapping and rapid reconnaissance, and land parceling surveys.

2. That the Congress establish, under the Interior Department, an independent organization to be known as the U.S. Geological Survey to be charged with the study of geological structures and economic resources of the public domain.

After considerable debate, especially in the Senate, the Congress rejected the first recommendation and accepted the second. The law approved March 3, 1879, was carried in the Sundry Civil Bill, as follows:

"For the salary of the Director of the Geological Survey, which office is hereby established under Interior Department, who shall be appointed by the President, by and with the advice and consent of the Senate - $6,000. Provided, that this officer shall have the direction of the Geological Survey and the classification of the public lands, and examination of the geological structures, mineral resources, and products of the national domain.

"And that the Director and members of the Geological Survey shall have no personal or private interest in the lands or mineral wealth of the region under survey, and shall execute no private surveys or examinations for private parties or corporations.

"And the Geological and Geographical Survey of the Territories, and the Geographical and Geological Survey of the Rocky Mountain Regions, under the Department of the Interior, and the Geographical Surveys west of the 100th meridian, under the War Department, are hereby discontinued to take effect on the 30th day of June, 1879.

"And all collections of rocks, minerals, soils, fossils and objects of natural history, archaeology, and ethnology, made by the Coast and Interior Survey, the Geological Survey, or by any other parties for the Government of the United States, when no longer needed for investigations in progress shall be deposited in the National Museum.

"For the expenses of the Geological Survey, and the classification of the public lands, and the examination of the geological structures, mineral resources, and products of the National domain, to be expended under the direction of the Secretary of the Interior - - - - - - - - - - - - - - - - - - - $100,000."

Organization of the Geological Survey

Hayden, King, and Powell, the three civilian heads of the surveys abolished by the Act of March 3, 1879, were able

scientists and masterful organizers, and on their records, each was eligible for appointment as Director of the newly created Geological Survey. King was nominated by President Hayes on March 21, 1879, was confirmed by the Senate on April 3, and entered on the duties of the office on May 24, 1879. It was understood in some quarters that King accepted the appointment with the understanding that he would remain only long enough to appoint the staff, organize the work, and guide the forces into full activity. On July 19, 1879, Hayden was appointed as a geologist, and Powell was already organizing the Bureau of Ethnology, of which he became the Director in 1879.

Clarence King, Director

Director King (fig. 4) organized the new Bureau with the greatest care, gathering about him prominent members of the four predecessor surveys as the nucleus of a technical staff upon which the success of the new organization would so largely depend. By the sanction of the Secretary of the Interior,[7] appointments to the Geological Survey were divided into two classes: first, these of the regular or permanent corps, who were nominated by the Director and appointed by the Secretary; and second, those that were temporary, which the Director was authorized to make and terminate (table 1).

Region of the Public Lands in the West

The organic act creating the Geological Survey contained two ambiguities in the wording and Director King resolved their interpretation. The term "The Classification of the Public Lands" was interpreted to mean that such classification was not to infringe upon the work of the General Land Office, which was sectionizing, classifying, and selling lands of the public domain to homesteaders. The term "national domain" was interpreted to mean the region of the public lands in the West. The House of Representatives disagreed with this interpretation and passed a joint resolution on June 28, 1879, extending the operations of the Survey over the country as a whole. The resolution was not passed by the Senate, and Director King, in view of the small appropriation ($100,000 when he had requested $500,000) and the possibility of legal difficulties, considered it best to confine operations to the region of the public lands.

The region west of the 101st meridian was divided into four districts, each in charge of an eminent geologist, with headquarters established at a nearby convenient city where winter office work would be carried on instead of in Washington, D.C. This arrangement, based on Director King's past experience and observation, would save the valuable time otherwise consumed in breaking up the western camps and bringing the corps to Washington, in returning them again to the field the following year, would avoid the cost of the transportation involved and would minimize the uncertainties

---

[7] U.S. Geological Survey First Annual Report, 1879–80, p. 13.

Table 1. First personnel roster of U.S. Gological Survey.

[*, Secretary's appointment; K, King; an., annual; P, Powell; W, Wheeler; H, Hayden: NA, not applicable; p.d., per day; mo., month]

Name	Position	Predecessor survey	Date of appointment	Salary
King, Clarence*	Director	K	May 24, 1879	6,000 an.
Emmons, Samuel F.*	Geologist	K	Aug.15, 1879	4,000 an.
Hague, Arnold*	Geologist	K	April 9, 1880	4,000 an.
Gilbert, G.K.*	Geologist	P, W	July 10, 1879	4,000 an.
Hayden, F.V.*	Geologist	H	July 19, 1879	4,000 an.
Pumpelly, Raphael*	Geologist	NA	July 16, 1879	4,000 an.
Becker, Geo. F.*	Geologist	NA	July 22, 1879	2,500 an.
Wilson, A.D.*	Chief Topographer.	H, K	July 10, 1879	3,000 an.
Clark, Fred A.*	Topographer	K	July 9, 1879	2,500 an.
Lord, Eliot*	Clerk	NA	July 9, 1879	1,800 an.
Bangs, Jas. E.*	Clerk disburs.	NA	July 9, 1879	1,800 an.
Bodfish, Sumner H.*	Topographer	P	July 10, 1879	1,800 an.
Foote, Arthur D.*	Asst. geol.	NA	July 25, 1879	1,800 an.
Renshawe, John H.*	Topographer	P	July 10, 1879	1,700 an.
Goode, Richard U.*	Topographer	NA	July 12, 1879	1,200 an.
Wright, Philo B.*	Topographer	NA	July 10, 1879	800 an.
Smith, Chas. E.*	Clerk	NA	July 10, 1879	840 an.
Wilson, C.C.*	Clerk	NA	July 9, 1879	600 an.
Wallace, H.S.*	Clerk	NA	July 10, 1879	600 an.
Walcott, Chas. D.*	Asst. Geol.	NA	July 21, 1879	600 an.
Meador, Leake S.*	Messenger	NA	July 9, 1879	480 an.
Mason, Anthony*	Messenger	NA	July 9, 1879	2.25 p.d.
Blair, Andrew A.*	Geologist	NA	Sept. 10, 1879	3,000 an.
Kimball, J.P.*	Geologist	NA	Sept. 10, 1879	3,000 an.
McChesney, John D.*	Chief Disb'g Clerk	W	Jan. 1, 1880	2,400 an.
Price, Gran*	Watchman	NA	Feb.1, 1880	600 an.
Wheeler, Patrick*	Watchman	NA	Feb.1, 1880	600 an.
Baldwin, Jeremiah	Messenger	NA	Jan. 16, 1880	600 an.
Hanford, Chas. B.*	Watchman	NA	Feb. 1, 1880	600 an.
Manners, Edw'd C.*	Clerk	NA	April 1,1880	1,100 an.
Reade, Francis R.	Asst. Geol.	NA	April 7, 1880	100 mo.
Thompson, Gilbert*	Topographer	W	May 1, 1880	2,500 an.
Leffingwell, Wm. H.*	Asst. Geol.	NA	May 1, 1880	75 mo.
Jacob, Ernst*	Asst. Geo.	NA	July 1, 1880	1,200 an.
Hilebrand, W.F.*	Chemist	NA	July 1, 1880	2,000 an.
Lakes, Arthur	Asst. Geol.	NA	July 1, 1880	100 mo.
Wilson, Geo. H.	Asst. Topogr.	NA	June 23, 1880	100 mo.
Iddings, Jos. P.	Asst. Geol.	NA	July 1, 1880	75 mo.
O'Sullivan, T.H.	Photographer	NA	July 16, 1880	100 mo.

caused some years by the delay in the appropriations, sometimes weeks later than June 30.

The four field divisions, the geologist-in-charge, their headquarters, and topographic assistants are described in the following sections.

Division of the Rocky Mountains

Samuel F. Emmons, Geologist-in-Charge, Denver, Colo.—Allen D. Wilson, chief topographer, under orders of July 17, reported for duty in charge of the topographic survey of Leadville, Colo.— a detailed map of the Leadville mining district, comprising an area of 25 or more square miles, on a scale of 800 feet to the inch, with contour lines at vertical distances of 25 feet. When completed, work was to be undertaken on three other mining districts in Colorado.

Mr. Emmons was instructed to prepare to extend, in 1880, precisely similar studies over all the mining districts of New Mexico, Colorado, Wyoming, Dakota, and Montana. The importance of topographic maps to geologic studies was emphasized in his report dated October 20, 1881:

> "Experience in the case of the Leadville, Colo., report has shown that the commencing of such work, without accurate maps already prepared, results in great loss of time, and renders one liable to annoying misconceptions and delays."

Field work was completed April 1, 1881, and the whole corps returned to the Denver office and worked upon the final preparation of numerous maps and illustrations for the engraver. Mr. Wilson assisted with the Leadville report. A.D. Wilson had been appointed chief topographer on July 10, 1879, worked on the Leadville mining district from 1879 to 1881, and resigned on October 31, 1881, as Professor Raphael Pumpelly offered him a position with the Northern Transcontinental Survey. Mr. Wilson had been a topographer with both the King and Hayden Surveys and was considered outstanding in his chosen work.

Division of the Colorado

Capt. Clarence E. Dutton, Geologist-in-Charge, Salt Lake, Utah.—In order to complete the surveys that were inherited from the Geographical and Geological Surveys of the Rocky Mountain Region in southern Utah and northern Arizona, about 70,000 square miles in extent, two detailed maps of special districts in northern Arizona were required, on a scale of 1 inch equals 1 mile, with curves representing levels of 50 feet.

Sumner H. Bodfish, topographer, began the survey of 1,900 square mile sections of the Grand Canyon, a labyrinth of deep gorges and towering cliffs, and of the Kaibab Plateau, a lofty flat mass without a single peak or pronounced elevation, and densely forest clad with gigantic pines and spruces. It is loftier by 1,500 to 4,000 feet than the surrounding regions and it is extremely difficult to circumvent, for it is nearly 100 miles long, and from 15 to 40 miles wide. To the eye and mind of the topographer it is merely an obstruction which cannot be crossed by his lines of triangulation.

John H. Renshawe, topographer, began the survey of 1,475 square miles in the Uinkaret Mountains, a collection of volcanic cones and lava flows.

Richard U. Goode, topographer, was sent from the town of Kanab, Utah, on August 13, 1879, upon the "dismal" journey across the Colorado River to the San Francisco Mountains, for the purpose of accomplishing the work of second, and some primary triangulation, in that district. A year later, to the day, he set out on the identical trip, this time for the purpose of making a topographic map of the district.

Division of the Great Basin

Grove K. Gilbert, Geologist-in-Charge, Salt Lake City, Utah.—Willard D. Johnson, topographer, for the double purpose of study and illustration, made a number of local maps, and made preparations for a new, and probably final, map of ancient Lake Bonneville. His instrument was the planetable and his method that of intersection. Mr. Gilbert, in his report, took occasion to advocate the use of the planetable for work of this class.

The following year, on July 16, 1880, the corps left Salt Lake City in two parties. The topographic party comprised Gilbert Thompson, topographer-in-charge; Albert L. Webster, assistant topographer; and R.I. Gill, barometer observer. The work of the topographic party was partly geographic and partly geologic. Triangulation was extended westward into the dessert from the peaks of the Wasatch Mountains; their other duty consisted of the tracing of the ancient shorelines of Lake Bonneville.

Division of the Pacific

Arnold Hague, Geologist-in-Charge, San Francisco, Calif. Charles D. Walcott and Jos. P. Iddings, geologic assistants.—Frederick A. Clark, topographer, arrived at Eureka, Nev., on August 5, 1879, to begin mapping an area of perhaps 400 square miles. In Spring Valley, some 4 miles northwest of town, a base of 9,000 feet in length, of over 183 feet range of levels, was twice measured with a Stackpole compensated steel tape. Triangulation, with an average length of side of 1 mile, was developed, primarily upon the pentagonal plan. Owing to lateness of season and consequent small party, both triangulation and topography were carried on together—an undesired coincidence. A gradienter was used in triangulation; in topography, a planetable. The survey was made on a scale of 1:10,000 and 50 foot interval between contours. A furnished house was rented in Eureka and, during the winter a survey was made of the town, and office work of computation and compilation was performed.

In the following spring, on May 10, 1880, under Mr. Clark's direction, camp was formed and two extra parties took the field; a topographic party under George H. Wilson, and a triangulation party under G. Olivio Newman. By June 30, 1880, the total work accomplished was: Triangulation stations—188; square miles of topography—61. Mr. Hague in his annual report dated October 1, 1881, stated that the topographical party, although working under great difficulties, the mountains being covered with snow, were obliged to remain in camp till December 12 to complete their field surveys. Upon the completion of the work, Mr. Clark, with his assistants, proceeded immediately to San Francisco, and began the compilation of the map from the 64 planetable sheets prepared in the field.

The instructions issued to Mr. Clark were to prepare a grade-curve survey of 20 miles square, adopting a scale of 1:10,000 with 50 feet vertical interval between contours, this scale being deemed ample to furnish sufficient detail to lay down all desired geological formations with precision, and to express geologic structure in its relation to the varied and characteristic topography of mountain and valley. Upon this map may be accurately located all the mines of the district. By close application to office work, the map, covering a broad, handsome sheet nearly 11 feet square, was compiled by the first of March 1881, and immediately forwarded by express to the Washington office. Clark resigned on June 30, 1881.

The following sentences also were included in Mr. Hague's annual report:

> "As rapidly, however, as the planetable sheets of the Eureka, Nev., mining district were completed, copies were made on tracing linen, and photographic duplicates prepared by the "blue process" were placed in the hands of the geological worker.

> "For purposes of accurate geological work, it seems to me indispensable for the geologist, on taking the field, that he should be furnished with completed maps of the district to be surveyed. It not only gives him clearer ideas of the main structural features of the country, and consequently the problems presented to him, but saves much time in his own instrumental work. Geological structure is so intimately connected with, and dependent upon,

topographical features that the better the geologist sees the mutual relations of every hill, ridge and ravine, the better equipped is he for the work placed before him!"[8]

John Wesley Powell, Director

On March 11, 1881, Mr. Clarence King addressed the following letter to President Garfield:

"Finding that the administration of my office leaves me no time for personal geological labors, and believing that I can render more important service to science as an investigator than as the head of an executive Bureau, I have the honor herewith to offer my resignation as Director of the Geological Survey."

The resignation was accepted. On March 14, the President nominated John W. Powell (fig. 5) as Director. Congress confirmed the nomination on March 18, and on the next day he took the oath of office. Major Powell was in complete accord with the plan of operations and methods of investigations as followed by Mr. King and his wisely selected corps of geologists and specialists, and proposed to continue the work begun and in progress.

In providing for the work of the Survey during the fiscal year July 1, 1882, to June 30, 1883, Congress adopted the recommendation of the Director and extended the operations of the Survey to encompass the entire country. The Sundry Civil Appropriation Act of 1882 provided that the Geological Survey "continue the preparation of a geologic map of the United States."[9] It was not until August 7, 1882, that the act was passed and approved, but the Director, mindful of the enlarged duties in prospect, on July 1, 1882, appointed Henry Gannett Chief Geographer to take charge of the topographic mapping. As yet, there was no authority for the publication of topographic maps as such, but their production as bases for geologic and economic maps illustrating the resources and classification of the land was of such primary importance that they were given first place in the annual reports.

The Wingate Division

In the spring of 1881, the construction of the Atlantic and Pacific Railroad (later a part of the Santa Fe System) had progressed to Fort Wingate, N. Mex., so plans were made to fill in the unmapped area in the northeast corner of Arizona and the southeast corner of Utah. Professor J. Howard Gore was directed to proceed to Fort Wingate, and there measure a baseline for trigonometric expansion and determine its geographical position. He borrowed a pair of 4-meter bars, with

all accessories, from the U.S. Coast and Geodetic Survey, and started on July 25, accompanied by Richard U. Goode, topographer, and E.V. McElhone, recorder, arriving at Fort Wingate on August 2. The instruments did not arrive until August 11 and McElhone immediately prepared his barometers for observation, of which he made eight each day. Extensive reconnaissance was necessary before selection of a base site. Progress had been seriously affected by the prolonged rainy season when Gilbert Thompson and party arrived on August 28. As all details had been carefully arranged beforehand, work on the line was commenced with the assistance of some members of Thompson's party. The line was cleared of all weeds and bushes, making an unobstructed roadway about 5 feet wide, and first measured with a steel tape, then with the base apparatus. A level line was extended by Goode from a railroad bench mark to the eastern extremity, and thence over the entire course. Latitude observations were made at each extremity, and azimuth marks erected. Before each observation, local time was determined so as to know the chronometer correction. The work was finished on September 30 and Professor Gore returned to Washington.

At Salt Lake City, on July 5, Gilbert Thompson received orders to organize a party and proceed to Fort Wingate, N. Mex., and take charge of topographic mapping in that vicinity. John K. Hillers was sent as assistant to take charge of the party until Thompson could join it, as he had first to complete all drawings in hand illustrating the monograph on Lake Bonneville by Grove K. Gilbert, geologist, in whose division Thompson was at the time. Hillers arrived July 17 and reached Kanab on August 1 with 5 men, 2 wagons, and 19 animals, having followed the route along the Sevier River, a distance of 320 miles. He immediately started collecting Survey property and making necessary preparations. With A.L. Webster, assistant topographer, Gilbert Thompson joined Hillers at Kanab on August 7, making the journey from Salt Lake City in 4 days, traveling by rail and special conveyance. The following morning, they left Kanab, a party of 11 men, 4 wagons, and 34 animals, several of which had been hired. On August 20, they crossed the Colorado River at Lee's Ferry. In the distance of 1.5 miles, a descent and an ascent of 800 feet was encountered. They followed the wagon tracks of a party, which had recently passed on the north side of the river, to the Mormon settlement at Sunset Crossing. After a day's rest, they put the whole wagon train across, including that of Lieutenant Burke's party, in 3 hours, through the quicksand and swift current, without being obliged to unload the wagons. They left the Little Colorado at the Puerco, which was followed to Fort Wingate, arriving after 37 days of actual travel and a distance of 710 miles.

Mr. Thompson, and the members of his party who were not assisting Professor Gore, began the triangulation work. He was somewhat familiar with the country, having been through it in 1873 as a topographer with the Wheeler Survey, and knew that artificial stations would be required, such as rock cairns and wooden towers or tripods. Mount Taylor and Chusca Knoll, about 80 miles apart, were the only two sharp

---

[8] U.S. Geological Survey Second Annual Report, 1880–81, p. 21–22.

[9] This was undoubtedly interpreted by many of the Congressmen as meaning "topographic map." This authority was there, but King failed to recognize it originally and it died by default.

prominent peaks that were readily identifiable. The ultimate objective was the connection of the Wingate base with the Kanab base. Mr. Thompson made topographic sketches at each occupied station, reading the desired angles with a gradienter. Soon Messrs. Goode and Webster were sketching topography, using a planetable. At the request of James Stevenson, of the Bureau of Ethnology, and formerly with the Hayden Survey, Mr. Webster was sent with him to map the Moki towns, which he mapped between September 13 and October 3, on a scale of 1.5 miles to the inch. Parties were disbanded, the property boxed and stored at the quartermaster's storehouse at Fort Wingate, and Messrs. Thompson and Webster left for Washington on December 24. There, in the office, field observations were computed and topographic sketches were compiled into a map. The work executed was then summarized:

> Square miles covered by triangulation...............1,500
> Square miles covered by topography....................775
> Triangulation stations occupied 15
> Number of observations.....................................3,500
> Barometric observations 1,380
> Points determined by barometric observations.....320
>   Theodolite used, No. 172, Kubel, maker.

In July 1882, Professor A.H. Thompson was appointed by Henry Gannett, Chief Geographer, to take charge of the Wingate Division. On August 16, Professor Thompson and his assistants were ordered to the field and, upon arrival at Fort Wingate, a division was organized as follows: the triangulation party, in charge of Professor Thompson, and two topographic parties in charge, respectively, of H.M. Wilson and E.M. Douglas. A pair of barometric stations was established, one at Fort Wingate, and the other upon Mount Bradley, upon the Zuni Mountains, and the difference of elevations was obtained by means of the spirit level. Work was begun early in September and progressed favorably, though somewhat slowly, owing to bad weather and the unfavorable nature of the country for topographic and geodetic work. The two topographic parties were both assigned to work within the square degree included between the 35th and 36th parallels and 108th and 109th meridians, Wilson being assigned to the northeastern part of this area and Douglas to the southern portion. The area covered by Wilson was about 700 square miles and that by Douglas about 500. Professor Thompson occupied three primary stations for geodetic work, from which good results were obtained, and three points occupied for topographic purposes and for secondary triangulation. On account of heavy snows, field work was closed late in November.

Early in May 1883, preparations were commenced for placing parties in the field, headquarters being, as during previous years, at Fort Wingate, N. Mex. These parties were organized, two for topographic work under charge of E.M. Douglas and H.M. Wilson, and one for triangulation, under the immediate charge of Professor Thompson. A.P. Davis, Ensign C.C. Marsh, and John D. Atkins acted as assistants, respectively, in these parties. A base barometric station was established at the Fort Wingate camp with S.A. Garlick as observer.

These parties continued work until July 15, at which time W.M. Reed and Edmund Shaw joined the division. Reed was assigned to Douglas's party, as assistant, in place of Davis, and Shaw to Mr. Wilson's party in place of Marsh. Davis took charge of a third topographic party, with Samuel Schmelzkopf as assistant. Ensign Marsh became assistant in the triangulation party as Atkins had been ordered to the Washington office. The permanent camp and base barometric station were moved to Fort Defiance, Ariz., that point being more central and accessible to the work thereafter than Fort Wingate.

This organization was continued until the close of the field season with but one change, the detachment of Shaw from Wilson's party and his assignment as base barometric observer at the rendezvous camp, thus relieving Garlick, who was given charge of a small party and directed to meander, by courses and distances, all of the wagon roads in the area surveyed by the topographic parties. Douglas's party was assigned the completion of field work in the southern part of the atlas sheet 35–108 degrees, and later the survey of atlas sheets 35–109 degrees and 35–110 degrees. Wilson's party completed the work in the northern part of atlas sheet 35–108 degrees and later the whole of atlas sheets 36–109 degrees and 36–110 degrees, besides small outlying areas.

Davis's party was assigned the completion of field work in atlas sheets 37–110 degrees and 37–109 degrees. However, on account of threatened difficulties with Navajo Indians, he was recalled and assigned atlas sheet 35–107 degrees, where he commenced work about September 10.

In the early part of the season, the triangulation party extended its work over the area included in atlas sheets 35–107 degrees and 35–108 degrees. This was finished July 1, but, owing to the press of administrative duties, Professor Thompson was unable to take the field with his party again until the middle of August, and on account of Indian trouble, it was September 15 before the parties were able to proceed in earnest. From this time until December 1, work was continued by Professor Thompson and Messrs. Marsh, Wilson, and Davis. Douglas finished work about November 1 and his party was disbanded.

The party under Mr. Garlick, detailed for the purpose of traversing the roads, was continuously employed in that work from September 15, to November 20, covering some 700 square miles of route.

During the season, the following areas were covered: Douglas' party, 8,800 square miles; Wilson's party 10,400 square miles; and by Davis' party 3,800 square miles; total 23,000. Professor Thompson had been instructed, upon the close of field work, to remain at Fort Wingate and determine its astronomical coordinates, connecting for longitude with the Washington Observatory at St. Louis, Mo.; he was assisted in this work by A.P. Davis.

It was effected during the month of December, telegraphic signals being exchanged upon 5 nights, with the requisite observations for time. Latitude was determined by zenith distances from observations upon 87 pairs of stars, obtained during 12 nights. Upon the completion of this work, Professor

Thompson and Davis returned to Washington. The reductions of the observations for latitude and longitude were made by S.S. Gannett with satisfactory results.

Previous Federal, State, etc. Surveys

In order to accomplish a topographic survey in the shortest time and with the greatest economy, advantage had to be taken of all work previously done by the Federal government, by the several states, counties, townships, etc., and by industrial corporations and individuals. The early military surveys and two-thirds of the work of the Wheeler Survey were known to be unsuitable, due to the use of scales as small as 8 miles to the inch, and to the use of hachures instead of contours. But the remaining one-third of the Wheeler Survey, about 115,000 square miles, and the work done by the King, Hayden, and Powell Surveys was suitable. Other sources of material are listed herewith.

The U.S. Coast and Geodetic Survey was organized on February 10, 1807, during the administration of President Thomas Jefferson, as the U.S. Coast Survey, and began operation on June 18, 1816. Except for periods 1818–32 and 1834–36, this Bureau of the Treasury Department carried out its functions without interruptions. In 1878, the Bureau was assigned the continuation of control surveys across the continent and its name was changed to "U.S. Coast and Geodetic Survey".

The U.S. Geological Survey has always looked to the U.S. Coast and Geodetic Survey for primary horizontal and vertical control and, during many years, has maintained cordial and effective cooperation.

The Lake Survey had mapped the shores of the Great Lakes, carrying triangulation around them, and connecting the head of Lake Michigan with the foot of Lake Erie. Hydrographic surveys were conducted at the same time as the small amount of shoreline topography, done generally at scales of 2 inches and 4 inches to the mile, and extending inland about 0.5 miles. The work was begun in 1841, as authorized by an act of Congress, and was in charge of the Corps of Topographical Engineers until that Corps was consolidated with the Corps of Engineers in 1863.

The Engineer Corps had completed a number of small pieces of topographic work in different parts of the country and had also made numerous surveys in connection with river improvements in many other parts of the country. In 1879, an act of Congress provided for the creation of the "Mississippi River Commission" whose function was to improve the Mississippi River from its mouth to its headwater near Lake Itasca in Minnesota. Of first importance was an accurate survey of the entire length of the river upon a foundation of primary geodetic control and precise levels. Topographic maps were being prepared showing the shoreline of the river and a narrow strip of contiguous territory on a scale of 1:10,000 with contour interval of 5 feet.

The General Land Office already had accumulated township plats covering about 1,500,000 square miles, plats that were nominally topographic maps but lacked longitudes, latitudes, and altitudes except at rare intervals. The plats and final field notes constitute the official records of the cadastral surveys made for the parceling of public lands to homesteaders and others.

After the Revolutionary War, the original Thirteen States ceded to the Federal Government lands extending westward from their western boundaries to the Mississippi River, and the Continental Congress, by the ordinance of May 20, 1785, inaugurated the rectangular system of surveys, providing for townships 6 miles square, containing 36 sections of 1 mile square.[10] As the primary purpose of the surveys was the disposal of the public lands by sale, these activities eventually were placed under the supervision of the Secretary of the Treasury. By the Act of March 3, 1849, the General Land Office was transferred to the Department of the Interior.

The township plats, generally accurate along the surveyed lines, the section lines, and the monuments on the ground marking the corners of sections and the 0.5-mile points, were to become indispensable aids to the topographers of the Survey.

The principal regions surveyed by state organizations, usually the State Geological Survey, were as follows:

New Hampshire.—Entirely mapped under Professor C.H. Hitchcock, and the results were published on a scale of 2.5 miles to an inch with contour interval of 100 feet.

Massachusetts.—Entirely mapped under Henry F. Walling, who, with much additional original work, assembled by compilation, various scattered maps, some produced between 1830 and 1842 by the Borden Survey.

New York.—A geodetic survey in progress under James T. Gardner, presumably for the purpose of fitting local maps of counties and townships to a scheme of triangulation and producing a map of the State.

New Jersey.—One sheet of the state map, comprising 1,250 square miles, already published on a scale of 1 mile to 1 inch and 20-foot contour interval. The field work was prosecuted with the planetable and level.

Pennsylvania.—Topography was being mapped of coal and iron regions, on diverse scales, by traverse lines, by odometer, and by stadia.

North Carolina.—A state map was published in 1882 on a scale of 10 miles to the inch, the product of compilation and additional original surveys under the supervision of Professor W.C. Kerr and Professor Arnold Guyot.

Georgia.—The abrupt termination of topographic work had left the accumulated material unpublished.

Michigan.—Maps of certain small areas in the iron and copper regions of the Upper Peninsula were published with contours.

---

[10] Although this may be true, a number of the earlier townships were 5 miles square and contained 25 sections, each 1 mile square.

Missouri.—Maps were published of the iron regions of Iron Mountain and Pilot Knob.

Most of the state maps enumerated above possessed good geographical position as determined by the U.S. Coast and Geodetic Survey, but small scales and general lack of vertical relief rendered them of restricted value to the U.S. Geological Survey. A map of the United States showing areas surveyed on a scale suitable for the purposes of the U.S. Geological Survey accompanied the Fourth Annual Report (1882–83) of the Director.

Also, many counties and towns of other states had been mapped and much map work had been done by private enterprise, but the maps and atlases, although on an amply large scale, were entirely wanting in two elements, namely, general geographic position and vertical relief.

Throughout the United States there were many maps of railroad, canal, and turnpike surveys, which could be utilized to great advantage. The railroad maps, especially, furnished important material. A vast network of trial and located lines had been run, which, by judicious adjustment, could be used to advantage. They were control traverse and level lines combined. All the railroad companies, of whom the maps were asked, responded promptly and favorably, although in many cases they were put to much expense thereby.

In addition to the work of compilation and adjustment of existing source material, it was found practicable to conduct field operations to a limited extent during the season of 1882. For convenience of administration, but controlled by geologic considerations, the area of the United States was divided into seven districts as follows:

District of the North Atlantic.—Comprising Maine, New Hampshire, Vermont, Massachusetts, Rhode Island, Connecticut, New York, New Jersey, Pennsylvania, Delaware, Maryland, and the District of Columbia.

District of the South Atlantic.—Comprising Virginia, North Carolina, South Carolina, Georgia, Florida, Alabama, Tennessee, Kentucky, and West Virginia.

District of the North Mississippi.—Comprising Ohio, Indiana, Illinois, Michigan, Wisconsin, Minnesota, Dakota, Nebraska, Kansas, Iowa, and Missouri.

District of the South Mississippi.—Comprising Indian Territory, Arkansas, Mississippi, Louisiana, and Texas.

District of the Rocky Mountains.—Comprising Montana, Wyoming, Colorado, part of Utah, New Mexico, and part of Arizona.

District of the Great Basin.—Comprising parts of Washington Territory, Oregon, California, Utah, Arizona, Nevada, and Idaho.

District of the Pacific.—Comprising part of Washington Territory, part of Oregon, and the greater portion of California.

In later years these Districts were modified, or combined, to accommodate varying administrative requirements, and they became known as Divisions. In 1882, work was started in five Districts but not in No. 1 (District of the North Atlantic) and No. 3 (District of the North Mississippi).

District of the North Atlantic

It was decided to commence a topographic map of New England to be published upon a scale of 1:125,000. The first step was to compile and collate the large body of valuable material already in existence. This compilation, though not complete, was carried so far that certain areas could be selected for field work; upon one of these, in the Berkshires in western Massachusetts, a single party, comprising R.D. Shepard, S.R. Duval, H.S. Selden, and William T. Griswold under the direction of Professor Henry F. Walling, began work. The State was covered with the triangulation of the old Borden Survey, which required slight adjustment to fit it to the geodetic control of the U.S. Coast and Geodetic Survey. Furthermore, the traverse work that had been completed in the State, mainly under the direction of Professor Walling, promised to relieve the Survey of the necessity of surveying much of the level country, leaving as the principal work to be accomplished the proper connection of the old traverse work with the geodetic points, the filling in of the drainage in the hills, and the addition of the element of relief contours throughout the country.

District of the South Atlantic

Topographic mapping in the East was started in the southern Appalachian Mountains through the personal appeal of Professor Washington C. Kerr, State Geologist of North Carolina. Professor Kerr had charge of triangulation and was assisted by John W. Hays and Sam S. Gannett. Charles M. Yeates, in charge of one topographic party, with Robert C. McKinney as assistant, was assigned that portion of western North Carolina and eastern Tennessee lying south of the Smoky Mountains, east of the south fork of the Holston, and west of the Blue Ridge, with instructions to work as far to the northeast as the season would permit. Morris Bien's area lay around the heads of the Watauga and New Rivers, and Frank M. Pearson's area was in Virginia and eastern Kentucky.

For this work Henry Gannett, Chief Geographer, put into use Dr. Grove K. Gilbert's new system of barometric hypsometry. Dr. Gilbert, as a geologist, in his early investigations, often was compelled to make his necessary preliminary maps, and in using thermometers for the determination of altitudes, evolved a new method that eliminated from the intricate formulas the two annoying factors for temperature and humidity, elements of the intervening air column between two barometric stations that caused oscillations in the air pressure. Barometric observations in a area being mapped were referred to synchronous observations made at some distant base station maintained by the Signal Service, U.S. Army, at the end of the field season, and corrections then made to the altitudes.

Six sources of errors[11] are recognized in barometric observations, and Dr. Gilbert listed them in table 2, in which are presented the probable errors in feet arising from each of the indicated sources, and in the second column the possible error.

---

[11] U.S. Geological Survey Second Annual Report, 1880–81, p. 435.

Table 2. Error sources in barometric observations.

Source	Probable (feet)	Possible error (feet)
Annual gradient	6	20
Diurnal gradient	8	30
Non-periodic gradient	20	50
Temperature (by thermometer)	100	300
Moisture (by psychrometer)	10	20
Imperfection of observation	10	No limit
Total	154	420

The new system[12] proposed a new method of observation and a new method of computation. Two base stations are established, one high, the other low. Their difference in altitude is made as great as practicable, and their horizontal distance is made as small as practicable. Each is furnished with a barometer, and a barometer only, and observations are made at frequent intervals through each day, as in the ordinary system. At each new station a barometer is observed, and no other instrument; the psychrometer and all thermometers, except that attached to the barometer, being discarded. The difference in altitude of the two base stations is determined by spirit level and constitutes a vertical base line by which all altitudes are gaged.

The field notes thus consist of three series, of barometric readings, pertaining respectively to the upper base station, the lower base station, and the new mapping stations.

To inaugurate the new system of barometric hypsometry, a group of four barometric base stations was established upon and around Roan Mountain, N.C. These stations were under the general charge of Arthur P. Davis, with directions to see that they were properly maintained, to obtain, by leveling, the difference in height between them, and by connecting them with some station upon the nearest railroad (Roan Mountain Station on the East Tennessee and Western North Carolina Railroad) to establish their height above sea level. Further, he was directed to establish the heights of neighboring mountain summits by means of levels, in order that comparison might be made between them and heights determined by barometer. Among the seven men detailed as barometric observers at these stations were Robert H. Chapman and Abner F. Dunnington.[13]

Mr. Gannett, who personally superintended the outfitting of this division (Appalachian) and spent several days with each party in the field in order to get them well started in their work, noted "that the dense forest cover extending over the highest mountain summits made it necessary for the triangulation party to erect artificial signals upon all stations; that it rendered unprofitable the use of planetables; that the parties were considerably handicapped by not being outfitted with pack trains, which would enable them to go anywhere over the country, but with wagons, which necessarily confined them to the roads and therefore necessitated long trips away from the wagon in order to reach mountain stations. The topographic work was done very largely by means of traverse lines and by sketching from any and all points from which an outlook could be obtained."

Mr. Yeates, a native of North Carolina, continued topographic mapping toward the Great Smokies in succeeding field seasons and was eager to determine the elevations of the high peaks. In his office in the Survey building in Washington in the winter of 1884–85, he was visited by a fellow North Carolinian, General Clingman, a graduate of University of North Carolina, a member of Congress and the U.S. Senate with Clay, Webster, and Calhoun, who suggested that a line of levels be run to the top of Clingman's Dome. It might be shown to be higher than Mitchell's Peak.

The elevation of Mitchell's High Peak (as it was then called) in the Black Mountains had been established as 6,711 feet by a line of levels extended from a railroad bench mark at Swannanoa; that was about 1853. Professor Guyot, of Princeton University, using a single barometer, had reached the figure 6,707; Mr. Yeates, 30 years later, using Dr. Gilbert's double barometric method, made the elevation 6,717 feet.

Though not sharing General Clingman's views, Yeates was glad to receive authority to resolve the question of the relative heights of the two peaks, something like 75 miles apart. He employed a competent civil engineer, named Ramseur, who ran a line of levels from a railroad bench mark on the Tuckasegee River to Clingman's Dome and back to the bench mark, making the height of the peak 6,615 feet above sea level. General Clingman was very much disappointed at the result.

It was almost 50 years later that revised elevations were determined for the two peaks. A Survey Press Bulletin of January 22, 1930, stated:

> "Accurate level lines have now been run by the Geological Survey to six of the high peaks in the eastern part of the United States. Mount Mitchell, in North Carolina, which is probably the highest point in the United States east of the Mississippi River, is 6,684 feet above mean sea level.
>
> Three peaks in the proposed Great Smoky Mountains National Park are Clingman's Dome, elevation 6,642 feet; Mount Guyot, 6,621 feet; and Mount Le Conte (Myrtle Top), 6,593 feet.
>
> Mount Washington, in New Hampshire, is 6,288 feet.
>
> Mount Katahdin, in Maine is 5,267 feet.
>
> Mount Marcy, in the Adirondacks, N.Y., is 5,344 feet."

---

[12] The first publication of the new system was made in a communication by Dr. Gilbert to the Philosophical Society of Washington, D.C., in 1877.

[13] Mr. Yeates' letter to Director Mendenhall, dated November 30, 1931, published in Topographic Division Bulletin, June 1951.

South Mississippi District

Preparatory to commencing geodetic and topographic work in Arkansas, Professor J. Howard Gore, astronomer, with Sam S. Gannett as assistant, was detailed to measure a base line in the Ozark Hills. After a careful examination of the country about the Hot Springs, they decided that the most practical location for this base was along the tracks of the St. Louis, Iron Mountain, and Southern Railroad, commencing at the town of Malvern and running southwestward for approximately 4 miles. The line was measured August 16–31, 6,406 meters in length, by means of a pair of 4-meter secondary basebars belonging to the U.S. Coast and Geodetic Survey, kindly loaned to the Geological Survey for the purpose. The base was not favorably located for expansion, as the country to the west was of low relief and timbered, necessitating the erection of high signals.

District of the Rocky Mountains

Work in this district was started in May 1882, by John H. Renshawe, who recovered the terminal points of the base that was measured near Bozeman, Mont., in 1877 by the Wheeler Survey, and effected a most admirable system of expansion. About September 1, he organized two topographic parties, under the supervision of Paul Holman and H.S. Chase, who succeeded in covering a small area in the Gallatin Valley. He continued triangulation up the Yellowstone to the south, but, like the others, was held to slow progress by severe snow storms. In the course of his work, he climbed Electric Rock and also Mount Washburn three times, but each time was prevented by stormy weather from doing any work.

In Colorado the survey of the mining region, Silver Cliff, Rosita, and Querida, was assigned to Anton Karl. The area was about 50 square miles, to be mapped on the scale of 1:20,000 with contour interval of 25 feet. The base line measured for the inception of the triangulation was 4,051 feet in length. The instruments employed were for the triangulation, a gradienter; for the topography, the planetable, stadia, and level. All the roads, trails, beds of gulches, and main water courses were meandered.

Upon completion of this work, Karl proceeded to Golden for the purpose of completing the survey of a small area comprising the two Table Mountains and the coal outcrops west of Golden. The scale of this map was 1:10,000 with contours 50 feet apart.

Work on the "Denver map", projected to cover about 1,000 square miles, with Denver slightly east of the center, was commenced during the winter by compilation of a large amount of topographic material from various sources.

Work was resumed early in September, 1882, at Fort Wingate, N. Mex., under the supervision of Professor Almon H. Thompson, who continued the triangulation begun the previous year. Two topographic parties were organized to work in the "square degree' included between the 35th and 36th parallels and 108th and 109th meridians, in charge, respectively, of

Edward M. Douglas and Herbert M. Wilson. Before heavy snow stopped the work about November 20, the former had mapped about 500 square miles, the latter about 700 square miles.

District of the Great Basin

Geographic work was prosecuted in Nevada under the general supervision of Dr. Grove K. Gilbert, geologist. Topographers Albert L. Webster and Willard D. Johnson mapped about 8,000 square miles around lakes Pyramid, Winnemucce, and Walker.

District of the Pacific

Gilbert Thompson was placed in charge of the work in California, with headquarters at Red Bluff. He left Washington on June 15, 1882, in order to get the work started from the point established by the Wheeler Survey and the U.S. Coast and Geodetic Survey on the summit of Lassen's Butte. He made three ascents in July, but was able to obtain but few observations on account of the haze and smoke from forest fires. A fourth ascent, after several days of waiting for storms to pass, was made on October 18, a beautiful day overhead with 3 feet of snow on the ground. The required observations were made.

About September 1, Thompson organized a second topographic party under the supervision of John D. Hoffman, with Albert Noerr as assistant, and Eugene V. McElhone as an aid, for work in the mountains to the north and northeast. Thompson, with Mark B. Kerr (formerly with the Wheeler Survey) as assistant, and Charles C. Bassett as an aid, extended triangulation and sketched topography to the northeast and east. Both parties had a cook, teamster, and packer, with wagon and pack train (fig. 6).

A barometric base station was established at Red Bluff for the field season and observations were made hourly from 6 a.m. to 7 p.m., and in addition, one at 9 p.m. One of the two observers was relieved from this monotonous duty later in the season by Robert F. Cummins.

During the working field season of about 2 months, which was broken by frequent storms, some 2,000 square miles were mapped, two primary triangulation stations occupied, and the altitude of 125 points determined barometrically.

Mapping of the four quicksilver districts of California was started in July, 1882, by Sumner H. Bodfish, who completed areas about the Sulphur Bank and the New Almaden by October 16, when bad health compelled him to give up further field work. The Knoxville district was mapped in the winter, and work on the New Idria was begun in the following June by John D. Hoffman, assisted by Mark B. Kerr, Charles C. Bassett, and W.H. Robertson.

The map of each district comprised about 12 square miles, on a scale of 800 feet to 1 inch, with a suitable contour interval, and was based an a line of 2,000–3,000 feet measured with a steel tape. The details of the topography were worked in by the use of the planetable and telemeter.

The gathering of pertinent map information, such as elevations and profiles of railroads, occupied the time of some topographers during 1882, in the spring months and even further.

During the summer of 1883, fieldwork was actively prosecuted in the districts of the previous season, and also was extended into District No. 1.

First State Cooperation—Massachusetts

In order to expedite the work, the Massachusetts State Legislature, in the spring of 1884, passed an act appropriating $40,000, which amount was estimated to be one-half of that required for making a map of the State in 3 years in sufficient detail to be published on a scale of 1:62,500. This was arranged on the condition that the Geological Survey should pay the other one-half. The interests of the State were placed in the hands of a commission, which agreed to the proposition of the Director of the Survey that 2 months work be done as a trial, which upon examination would be accepted or rejected. The commission's acceptance followed in October and the surveying was continued as the first project of State cooperative mapping.

The work, begun in the northwestern part of Berkshire County, was extended southward over nearly the whole of the county. During the winter E.W.F. Natter and J.H. Jennings were left in the field, the former to begin the survey of the Boston 15-minute quadrangle, the latter to make traverse surveys and run level lines in the valley of the Housatonic. The following summer Natter and Jennings completed the Boston map. To complete the mapping of Plymouth and Bristol Counties three members of the field parties were left during the winter to survey, while frozen over, the many areas of swamp, marsh, ponds etc.

The areas mapped, for publication at a scale of 1:62,500 with 20-foot contour intervals, were: in the fiscal year 1885, 1,250 square miles; in fiscal year 1886, 2,500 square miles.

Anton Karl, Sumner H. Bodfish, and John D. Hoffman spent the winters in Washington finishing their maps, making transcripts of them for deposit with the state of Massachusetts, and compiling the surveys of the U.S. Coast and Geodetic Survey, on and near the coast, for future use.

The mapping of the State proceeded: 2,215 square miles in 1887, and 2,350 square miles in the fiscal year of 1888, completing its total area of 8,315 square miles. In addition to the regular atlas sheets, a small map of Massachusetts was compiled on a scale of 1:300,000, with 100-foot contours, reduced from the atlas sheets.

Base Map of the United States

The general plan adopted by Director Powell for the great basic map of the United States was a very simple one. The country was divided into quadrangles bounded by parallels of latitude and meridians of longitude. As experience had shown that the intelligent presentation of the principal facts of structural geology required a scale of 1:250,000, or about 4 miles to 1 inch, it was decided to continue the square degree type of map.

Mining districts of intense development would be mapped on large special scales, such as 1:10,000, as had already been the practice. However, it was soon realized that densely populated and culturally developed areas in the eastern states could not be mapped adequately on the 1:250,000 scale, so two larger scales were adopted, the 1:125,000 and 1:62,500, making a set of three different scales. These were to be the publication scales and were to be printed on the same size paper, 16.5 by 20 inches.

The maps published on the smallest of these scales, 1:250,000, each cover, on an average, about 3,600 square miles of territory (the areas naturally decrease gradually northward, as the meridians converge). The maps on the scale of 1:125,000, or approximately 2 miles to the inch, each comprise a quadrangle measuring 30 minutes of latitude by 30 minutes of longitude, and covering on an average about 900 square miles. The maps on the scale of 1:62,500, or approximately 1 mile to the inch, each cover, on an average, about 225 square miles. Each map is usually designated by the name of a city, town, or prominent natural feature within it.

A map, as representing on a flat sheet of paper a section of the earth's spheroidal surface, needs to be made on a projection that best preserves relations of features on the earth's surface regarding area, distance, and direction, in their true values. For the quadrangle maps, the projection adopted was the polyconic, which is based on the development of a large number of cones, one of which is conceived to be tangent to the earth's surface at each parallel of latitude to be represented on the map. Meridional distances and abscissas and ordinates to the arc of the developed cones were computed for every minute of latitude from zero to 90 degrees. Published tabular values make the construction of the polyconic projection easy.

Experience also had shown that vertical relief can best be expressed by contours, with varying vertical intervals grading from 200 feet in rugged mountain areas down to 25 feet in broad valleys and prairie areas of gentle slopes. The objections to the use of hachures and brush shading are manifold: hachures, particularly, obscure the map and conceal the conventions employed for the representation of other facts and features of the terrain; they lack definiteness and easily degenerate into generalized conventions for imperfectly ascertaining facts of relief; they lead to the development of special artistic styles by the several draftsmen employed upon the work, and thus do not have a uniform meaning from sheet to sheet; the cost of drawing and engraving hachures is excessive, almost equalling the expense of the field work by which the facts are collected. The contours need not be run out on the ground but can be placed with sufficient accuracy by skillful topographers from salient and controlling points whose position and elevation have been previously determined. A contour line on the map represents an imaginary line on the ground, all of which are of the same elevation above some adopted reference plane, usually mean sea level. For instance, the 25-foot contour would be the coast line were the sea to rise 25 feet. Or, if a plane 25 feet above the parallel to sea level were passed through a section of the earth, its intersection with the ground surface would trace the 25-foot contour. A series of such parallel planes at successive vertical intervals of 25 feet is what is assumed, or imagined, when a map is drawn with 25-foot contours. Besides showing the form and inequalities of the ground at their levels, contours permit one to read from the map the approximate elevation of any point on the ground and also the slope of the ground by using the map scale.

The organization set up to execute the work consisted of: first, an astronomic and computing division, the officers of which were engaged in determining the geographical coordinates of certain primary points; second, a triangulation corps, engaged in extending a system of triangulation over various portions of the country from measured base lines; and third, a topographic corps, much the largest, engaged in collecting topographic data in field surveys and drawing the quadrangle maps in the office the following winter.

The work of the first two divisions consists largely of mathematics and trigonometry for the reduction of the limited and accurate observations in the field to the geographic coordinates of two or more selected points on each quadrangle to be mapped. Starting with these two or more control points, the work of the topographer consists of two parts, clearly distinguished, one from the other, in character. These are, first, the geometric control, which furnishes the skeleton of the map, and, second, the sketching, which furnishes the map itself. The geometric control, or the location of many secondary points in their proper relation to the first two or more primary points, is affected by one or the other of two methods, either by angular measurement alone or by the measurement of direction and distance, the former being generally known as that of intersection, the latter as the traverse, or meander method.

The former is the preferable one, being accurate and rapid, but in flat, timbered regions, it cannot be economically employed, and for the purpose of reaching the details of roads, streams, and other obscure features, the latter method is found to be a useful adjunct to it. In practice, both methods are used as supplying the desired number of instrumentally located points, distributed in each square mile or in each square inch of map surface, so as to reduce the errors of sketching to a minimum. The building of the control skeleton is mathematical and comparatively easy to master, but the sketching is artistic and to a large degree is dependent upon the topographer's natural talent in that direction.

The methods of topographic surveying, as finally developed, and described by James T. Gardner, in 1874, and thereafter followed, were considered the best so far devised, and were adopted. It was expected that the methods would be varied and modified to meet peculiar conditions in different portions of the country, and that refinements would constantly be introduced. The purpose for which the map was to be constructed was the representation of the areal geology of the United States but it was realized that in showing the

geographic distribution of phenomena many other important purposes would be served.

Once constructed and engraved, the plates could serve for new editions from time to time, to be used for such purposes as the study of drainage systems, the regimen of rivers, the great subject of irrigation, the distribution of forests, the distribution of artesian water, catchment areas for the supply of water to cities, drainage of swamps and overflowed lands, soils and the classification of lands for agricultural purposes, the laying out of highways, railroads, and canals, for strategic and administrative purposes in the event of war, and as authentic and fundamental data for subsidiary maps and atlases for commercial, statistical, and instructional uses. All these purposes would be subserved by maps made to meet the more exacting demands of the geologists.

The magnitude of the undertaking made it of prime importance that the survey should be conducted with the utmost regard for economy, and at the same time to be so well executed that frequent resurveys would be avoided. This dual requirement called for the daily exercise of the topographer's ingenuity and judgment. Satisfaction was his only if he felt reasonably sure that his work was done sufficiently well to please the critical geologist and to stand for all time without need for a resurvey unless on a larger scale.

The economical and time-saving procedures for mapping the United States would have been to establish one astronomic station, then measure a base line, then from it expand a triangulation net, and from this control center have topographic parties begin their surveys and work outwardly in all directions, completing quadrangles and leaving none unmapped behind them. In this way a state could be completely mapped. However, some quadrangles of little or no importance at the time would be mapped. Moreover, in several states and in widely separated sectors in a state, there arose requests for mapping. The needs of each locality had to be given careful consideration in order to extend the mapping over the United States. The quadrangles selected for mapping were listed in the Survey's plan for the year's work, which the Director would submit to the Secretary of the Interior for examination and approval at the commencement of each fiscal year. The Secretary then submitted them to Congress.

Mapping in Texas, Missouri, and Kansas

The enlarged appropriation for 1884–85 ($489,040) with the consequent increased allotment for geographic work ($83,950 to $144,665) rendered it possible to commence mapping in Texas, Missouri, and Kansas. In Texas the work consisted of the measurement of a base near Austin, 6.2 miles in length, and an expansion made for the immediate purpose of furnishing topographers with located points. As soon as the control party under Edward M. Douglas had determined the needed locations, the topographic parties in charge of Abner F. Dunnington and Charles H. Fitch, thus supplied with locations, began their work and, during the season, covered about 4,000 square miles.

The new work in Missouri and Kansas was based on the plats of the General Land Office. These plats had been found, by tests in various parts of the country, to furnish at least the major part of the drainage in sufficient detail for maps upon a scale of 1:125,000. Geographic position was usually lacking. To supply this control for the area selected to be mapped, Professor Robert S. Woodward, formerly assistant on the U.S. Lake Survey, was appointed and detailed to make astronomic stations (five) near the limits of the area and connect them by triangulation with township corners.

The mapping was done by townships, the assistant topographers following section lines and adding all missing drainage and determining the placing of contours by recording slope angles and other vertical angles with a gradienter. The initial elevations were established by mercurial barometers set at the nearest railroad elevation. The chief of the party, besides following selected section lines, went into the sections to obtain topographic details, carrying a barometer for the determination of elevations as a check upon the gradienter.

The topographic work, which consisted almost entirely of hypsometry, went on with great rapidity. Topographers Richard U. Goode and William J. Peters surveyed 13,600 square miles. The cost of this work in both the field and the office was less than $1.00 per square mile.

The total area surveyed during the fiscal year ending June 30, 1885, was 57,508 square miles, distributed among the several areas of work projects shown in table 3.

Table 3.   Areas surveyed during fiscal year 1885.

[The average cost of all the work done was about $3 per square mile]

Region	Scale of publication	Area (square miles)
Massachusetts	1:62,500	1,250
New Jersey	1:62,500	1,286
Appalachian	1:125,000	17,640
Missouri–Kansas	1:125,000	13,600
Texas	1:125,000	4,000
Plateau	1:250,000	15,000
Yellowstone Park	1:125,000	1,000
Northern California	1:250,000	3,750
Total		57,750

Astronomic and Computing Section

The importance of geodetic control and the computation of geographic positions, and the necessity for the adoption of uniform standards for the work, led to the organization in 1885 of an Astronomic and Computing Section, and Professor Robert S. Woodward was placed in charge.

The preceding year, the increased allotment for geographic work made it possible to commence work in Kansas, Missouri, and Texas. Much of Kansas and Missouri had been surveyed by the public land system and it seemed probable that the mapping could be expedited by a proper use of the plats of the General Land Office. It appeared that by applying a proper system of correction for these surveys, by supplying the missing drainage wherever it was necessary, and by adding the vertical element, the resulting maps would be fully up to the proposed scale of 1:125,000 in point of accuracy and detail.

For the control and correction of the surveys of the General Land Office it was decided tentatively to use astronomic locations, and during the field season Professor Woodward made five determinations of positions: Elk Falls, Oswego, and Fort Scott, Kans.; and Springfield and Bolivar, Mo., using St. Louis Observatory as the base station for longitude determinations. The stations selected were near the limits of the area to be surveyed and at the same time as near as possible to correction lines and guide meridians of the land surveys. These stations were connected by triangulation with township corners.

During the winter and spring months, Professor Woodward was engaged in Washington computing his astronomic work. In June 1885, he was detailed to establish an astronomic station at Albuquerque, N. Mex. Again, for the determination of longitude, telegraphic connection was made with the Washington Observatory, St. Louis, Mo., where the observations were taken by Professor H.S. Pritchett, Director. This arrangement and cooperation of Professor Pritchett were depended upon for many such longitude determinations in subsequent years. For latitude at Albuquerque, observations were made on 10 nights upon 32 independent pairs of stars.

The scope of the work of the section may be indicated by excerpts from two annual reports of Professor Woodward, the first one for the fiscal year ending July 1, 1887. Of the various computations that have been made or completed within the year were:

1. A set of tables giving the coordinates for the polyconic projection of maps for each of the scales 1:30,000; 1:62,500; 1:63,360; 1:125,000; 1:126,720; and 1:250,000.

2. A table giving the areas in square miles of quadrilaterals of the earth's surface bounded by meridians and parallels.

3. A table for facilitating the computation of differences of height from angles of elevation or depression.

4. A table giving differences in elevation from telemeter measures.

5. A table similar to 4, but adapted to a special scale for planetable work.

6. Two sets of tables for facilitating the determination of azimuth from observations of Polaris.

Tables 3, 4, and 5 were reprints of tables already in use by topographers in the field.

Another activity that was directly contributory to the division of geography was the collection, classification, and arrangement for ready reference of geographical positions of points throughout the country. The chief sources from which these were drawn were the reports of the U.S. Coast and Geodetic Survey, the U.S. Lake Survey, the various U.S. geographical and geological surveys, the Mississippi River Commission, and the state surveys of Massachusetts and New York. The list of such positions embraced upwards of 7,000 entries, of which about 200 were added during the year. Constant attention was needed to keep the list up to date.

In August 1886, Professor Woodward made a survey of the Falls of Niagara,[14] the object of which was to determine their rates of recession by comparing the positions of the crests with their positions as determined by the New York State Geological Survey in 1842 and by the U.S. Lake Survey in 1875. The work embraced the fixation by intersection of 30 points on the crest of the Horse Shoe Falls and 25 points on the crest of the American Falls. The heights of both falls were also measured.

During the winter, the coordinates of the points fixed on the crests of the falls were computed and prepared for publication. The survey indicated that the average rate of recession along the whole crest line of the Horse Shoe Falls had been about 2.4 feet a year since 1842; in the central portion of the channel, the rate had been about twice as great. The recession of the American Falls during the same period had been slight.

During the two following years the work of Professor Woodward's Mathematical Division continued to fall under the following two classes: First, that which was directly auxiliary to the operations of the Division of Geography; and second, that miscellaneous work, of a mathematical character chiefly, which arose in the various branches of the geologic and irrigation work of the Survey. Professor Woodward prepared for publication:

- The U.S. Geological Survey Bulletin No. 48 (1888), a treatise on the form and position of sea level;

- Bulletin No. 49 (1889), the determination of latitudes and longitudes of certain points in Missouri, Kansas, and New Mexico; and

- Bulletin No. 50 (1889), 124 pages of tables required in the work of geographers and draftsmen.

During the fiscal year ending June 30, 1890, the work of the Mathematical Division was more varied. The usual attention was given to current computations to aid topographic work, and much time was given to consultations with topographers and geographers of the Survey concerning the technical details of their operations, and the following activities were noteworthy.

---

[14] Assisted by P.C. Warman, loaned from Mr. Gilbert's party.

Professor Woodward examined and tested six new theodolites made by Fauth and Co. in accordance with specification drawn by him, and prepared specifications for two new astronomical transits of the best modern type. As a member of the Survey board of special examiners for the U.S. Civil Service Commission, he drew up several sets of examination papers for use in the topographic aid examinations conducted by the Commission, examined the papers submitted by candidates, and rated them.

In October 1889, he went into the field and established astronomic stations at: Speareville, Kans., where he was assisted by Harry L. Baldwin, Jr.; Boise City, Idaho, where William T. Griswold assisted him and then continued with the first topographic mapping in Idaho; and Cisco, Tex., where Mr. Richard U. Goode was the assistant. In May 1890, Professor Woodward was again in the field, in El Paso County, Tex., where with Arthur P. Davis' assistance, he established Sierra Blanca astronomic station and from it extended triangulation for the purpose of locating a point on the 105th degree meridian. The longitudes of the four stations were determined with reference to St. Louis, Mo., where the observations were made by Professor Pritchett.

Upon the resignation of Professor Woodward, to accept an appointment with the U.S. Coast and Geodetic Survey, Sam S. Gannett was given charge of the Astronomic and Computing Section. As a preliminary to topographic work in South Dakota he established an astronomic station at Rapid City, October 23 to November 20, 1890, being assisted with his observations by Abner F. Dunnington. The determination of longitude was made by exchanging time signals with Professor Pritchett at St. Louis.

Instrument Repairs

The supply and repair of instruments had been, from the inception of topographic work, unsatisfactory and the necessity for having a mechanic connected with the Survey had been long apparent. Finally, on March 1, 1885, Edward Kubel, a well known instrument maker who, from long association with the Survey and its predecessors, was well acquainted with the instruments and methods in use, was appointed mechanic. The work connected with the adjustment and repair of instruments warranted the employment of two assistants, and during the spring months, a third assistant. Even so, though it was necessary to purchase nearly all new instruments in the next 3 years, Mr. Kubel found time to make 6 first-class telescopic alidades and 24 traverse planetables with ruler alidades. In the fiscal year ending June 30, 1890, he made 6 alidades and an 8-inch theodolite. This arrangement was discontinued on July 1, 1891, and repairs to instruments were made again by contract, on an hourly basis. Mr. Douglas, Geographer, was now in charge, assisted by Stephen A. Aplin, Jr., who, in 1896, was made custodian of instruments. In December of that year, Mr. Aplin, under the general direction of Mr. Wilson, Geographer, began the indexing and cataloguing of topographic records. These records were divided into six separate classes, namely:

astronomic and baseline measurement notebooks, primary triangulation notebooks, spirit level notebooks, field sheets, and miscellaneous map material. Each book or map was numbered and filed and a cross index card catalogue was prepared for all the separate entries. There were about 10,000 entries; annual accumulations amounted to about 1,000.

In 1898, Mr. Aplin was placed in charge of the purchase and repair of instruments, relieving Mr. Douglas of that work. A more rigid system of accountability had been established and instruments were returned from the field in better condition. Repair work, during the year ending June 30, 1899, was given to private instrument makers, at hourly labor. G.W. Saegmuller of Washington, D.C., and W. Gurley and L. Gurley of Troy, N.Y., being the principal contractors.

During the fiscal year 1899–1900, the plant of Ernest Kubel was acquired for the use of the Survey, and he was employed as an expert instrument mechanic on May 2, 1900. Kubel was in immediate charge of the repairs not only of the surveying instruments, but also of the presses and machinery belonging to the Section of Engraving and Printing. The chief engraver, who was the brother of Ernest Kubel, had administrative charge of the broadened enterprise.

In July 1903, the Section of Instruments and Topographic Records was organized and placed in charge of Mr. Aplin, with Powell P. Withers as assistant. Mr. Aplin was transferred to the Accounts Division on June 1, 1907, and Mr. Withers became custodian of instruments and topographic records. W.H.S. Morey, topographic engineer, was transferred to the Section of Field Equipment as custodian on July 1, 1920. He returned to the Topographic Branch on July 1, 1928, and was assigned to the Computing Section.

William H. Griffin's Diary

The broad open prairies of Kansas offered a natural training ground for young topographers. For the season of 1889 there came a youth from a southeastern state filled with enthusiasm for topographic surveying, based on a few months of instruction, and for new and strange places. Travel and new experiences were such a delight that he undertook to keep a diary, which he was to continue faithfully throughout a long and honorable career. As hundreds of young men starting out with the Geological Survey had similar experiences, but kept no diary, excerpts from his diary are considered pertinent to this history.

William H. Griffin, born and raised in Lake City, Fla., and educated in the public schools and Florida State College, was 21 years of age when, in early September 1888, he boarded a train for Washington, D.C. to seek a job with the U.S. Geological Survey. The journey, in old time day coaches, consumed 3 days and 2 nights, with frequent stops while health inspectors examined all passengers to prevent the spread of yellow fever, which that summer was worse than in previous summers.

In the late afternoon he arrived in Washington, at the old Pennsylvania, or Union Station, Sixth and B streets, found lodging in the nearby Howard House, and that evening called on his Senatorial sponsor. The Senator was surprised to see him but the Sundry Civil Bill, which carried the appropriations for the Geological Survey, had been passed and Griffin's appointment need not be delayed.

The next forenoon, the Senator introduced him to Major Powell, who turned him over to Henry Gannett. After reporting to the Survey office every morning for 10 days and seeing the sights of the Capital by day, and attending the theaters, the first real theaters he had ever seen, at night, he received instructions to report to Louis Nell at Columbiana, Ala. Railroad transportation was furnished.

Louis Nell had been a member of the Wheeler Survey and is mentioned in the reports for 1871 and 1875. With Lieutenant Birnie, and F. Brockdorf, he climbed Mount Whitney in California on October 13, 1875.

Nell established a single-handed record, for seasonal production, which has never been equalled. His remarkable achievement was cited by Major Powell in his Annual Report No. 10 (1888–89).

> "On the 14th of August (1888) Mr. Louis Nell was ordered to the field in Alabama, and to his party were assigned Messrs. R.L. Longstreet and Charles W. Goodlove as assistants, and subsequently Messrs. Wm. H. Griffin, Duncan Hannagan, F.P. Metzger, and B. Peton Legare; he was instructed to survey the Bessemer and Talladega sheets with certain unfinished parts of sheets lying immediately north of them, an area of about 2,300 square miles. Shortly after he had commenced work Mr. Longstreet, his principal assistant, met with a serious accident, breaking his wrist and rendering him incapable of affording further assistance during the season. In spite of his serious loss Mr. Nell succeeded in finishing the area assigned him and disbanded his party in November."

Nell cordially greeted the new arrival and found out that Griffin knew how to hitch, unhitch, saddle, ride, and drive a horse or mule, and Griffin soon became Nell's recorder and orderly. What he needed to know about the camp curriculum was imparted by members of the camp, especially by Goodlove, an 18-year old veteran of two seasons. Life in the Survey camp was a cherished hope realized, in daytime, a ride over mountains with stops at commanding lookouts, a healthy appetite for a well-cooked supper, an hour or two around a big blazing fire in the long, fall evenings when each would exploit his daring escapades and extol the virtues of his home town, and finally a sound sleep in a real Army tent.

The season ended and after a 2 days' trip, Talladega was reached. The camp equipage was stored then and the animals pastured until the next field season. All technical assistants returned to Washington, but Griffin accompanied Nell to Cincinnati, where profiles and elevations were to be obtained in the railroad offices.

In Washington, for the winter, the hours at the office were from 9 a.m. to 4 p.m., and were devoted to inking drainage on the manuscript sheets, then contours, and finally margin lettering and names of features. Six newcomers, including Griffin, Goodlove, Hannagan and Hersey Munroe, definitely in pursuit of careers as topographers, requested Nell to give them instructions. He had them at his home, 3 nights a week, for lessons in mathematics and contour construction. For the latter, he made a helpful plaster relief model.

Other activities of Griffin's during the early months of 1889 are indicated by the following excerpts from his diary.

Diary—1889

January 14 (Monday)—Went around to Corcoran Cadet Armory with Hersey Munroe.

January 20 (Sunday)—Saw my first snow today—a fall of 2 inches.

February 16 (Saturday)—Left the office at 2 and went to the White House to Mrs. Cleveland's last public reception. Near the head of the line which extended way out on the Avenue, I had just gotten under the porte cochere when a heavy rain fell * * * . In the East Rooms people were wedged together in a tight mass * * * Mrs. Cleveland was sweet and pretty and had a smile for every handshake. She looked strange, though, without a bustle.

March 4 (Monday)—Inauguration of Benjamin Harrison and Levi P. Morton, President and Vice-President. Goodlove and I were together all day and we did not miss a trick between the White House and Capitol, and Capitol and White House. Snow in morning, turning to sleet and then rain * * * Pennsylvania National Guard paraded 10,000 strong * * * Saw Mary Anderson in Pygmalion and Galatea at Albaugh's Theater.

March 7 (Thursday)—saw Mary Anderson again, this time as Hermione and Perdita in "A Winter's Tale." America's greatest and most beloved actress, she did not again appear on a Washington stage and shortly afterward retired.

May 31 (Friday)—Received official orders transferring me from Appalachian (Maj. Gilbert Thompson) to the North-central division (Mr. John H. Renshawe); also travel orders to Wichita, Kans., where I report to W.H. Herron whom I have met and like. Too, I like the prospect of going West to new scenes * * * Mr. Nell was mad as the devil at the transfers, losing thereby one he had spent so much time training.

June 1 (Saturday)—Met a traveling companion at B. and 0. depot for the 3 o'clock train, but no train was going out on account of great flood * * * Went to Monument to look at river and had to ride across C Street on 15th on account of water * * * Extras told of the devastating Johnstown, Pa., flood which took more than 10,000 lives.

June 2 (Sunday)—Joined crowds on Pennsylvania Avenue to see the high water. From Peace Monument to 9th Street the wheels of the horse-cars were under water. Around Center Market and south of the avenue, in deeper water, sight-seeing gondolas did a rushing business.

June 4 (Tuesday)—Left Washington on B. and O. at 3:30

June 6 (Thursday)—Kansas City, where I spent 3 days with a chum who left home a year before I did. I had built up a yearning to come here to see the metropolis of the cowboys and Robin Hoods of the prairie land—the hideout particularly of the Dime-novel hero, Jesse James.

June 10 (Monday)—Wichita, Carey House. Three Parties here and all at train when I arrived this evening. Lamar and self with Herron, Legare, and Seely with Towson. All leave here tomorrow and live on the country.

June 13 (Thursday)—Have been out with Mr. Herron watching him sketch. Sketched 34 square miles today and got here (Mount Hope) at 7. Saw my first binder at work and also a woman and three girls with crocus bags in pasture picking up "cow-chips". For fuel, Herron says. Later I was to see this fuel stacked in pyramids beside sod houses as meticulously as a Yankee's woodpike. Saw, too, for the first time, straight roads; and a few days later a prairie-dog town, a windmill; and learned that, though I was from the South, friendly greetings and cordial hospitality were invariably extended one.

June 15 (Saturday)—Went out alone for first time. Missed Herron. Saw first real cowgirl—big hat, boots and chaps—tough looking mug.

June 21 (Friday)—Am with a Hoosier family tonight * * * Old man was in the Oklahoma run on April 22 and says he got away at sound of bugle but when he reached his claim a fellow had a couple of furrows plowed round it.

June 24 (Monday)—At Norwich. Worked on sheets till midnight.

July 4 (Thursday)—In Kingman. Big parade, heard for first time "Marching through Georgia" and saw fireworks on 4th of July. Dance on main street with caller who sings the figures, and in "Opry House" in evening.

July 10 (Wednesday)—Got news of fight held in Richberg, Miss. on Monday. Sullivan won over Kilrian in 75th round. Possible last big fight with bare knuckles * * * Worked 32 square miles and never saw so many sod houses!

July 26 (Friday)—Wichita—Baldwin's, Herron's and Towson's parties here getting outfit and stock for camp. H.L. Baldwin, Jr. was straw boss for Kansas and only did triangulation; other two parties, topography. The shipment of buckboards was 4 days late, but we got the tents up and moved into camp on second day. We assembled the buckboards on depot platform and pushed them over to a blacksmith to have odometers put on * * * got our two wagons packed and away at 10:30, we following at 5. This our first real camp and they made good time to get this far, 15 miles, over much sticky mud * * * Our Negro cook, Charlie, was a good cook, but tough. Our teamster whom we called "Pa" wore a big bunch of whiskers and was quite a character.

July 31 (Wednesday)—Runnymede. English neighborhood and large store stocked with fancy pipes, riding habits, jams, marmalade, etc. Alluring ads in English papers attracted young people to come and enjoy life on ranches hereabouts.

August 4 (Sunday)—Tents blown down at night.

August 6 (Tuesday)—Emry, Mr. Herron came in afoot. His buggy horse had played out and died.

August 12 (Monday)—A tough day. Creeks up all morning. Swam one where bridge was out and went 200 feet downstream before I found place for horse to get out. Less than an hour later crossed bridge covered with water and horse got front foot caught in floor and fell. Thought I'd never get him up * * * Creeks down in afternoon.

August 13—Found a 5-inch centipede when we got up. On my telescope between heads of our cots * * * . We're camped at Caldwell, not far from Cherokee Strip line. Some tough looking hombres on street.

August 20 (Tuesday)—A horseback from 8 a.m. to 7 p.m. Awful sore. Always thought kindly of General McClelland, (sic) but not of the McClelland (sic) saddle. Many times today had to let down barbed wires and fence posts, some rotten.

September 15 (Sunday)—at that farmer's where I stayed last night owing to a storm. Husband and wife, 2 kids, another man and I slept in one room.

September 17—Left home 1 year ago today. I've sure seen a lot since.

September 23 (Monday)—Had prairie chickens (Charlie killed) for supper. First I ever ate. Fine. Charlie had served prairie dogs before. Delicious.

September 25–27—Camps at Calista, Ninnescah and Cairo.

October 4 (Friday)—Pratt. Went to baseball game after supper. Pratt defeated Cairo 31 to 17.

October 10 (Thursday)—Haviland. Three 50-pound watermelons for 5 cents. For some days now we've been working in the sandhills which extend from the buffalo-grass plateau to the Arkansas River, 10 to 20 miles. These glorified anthills vary in elevation from 10 to 60 feet. Many of them were perfectly bare with deep blowouts on top, but the greater percent were sparsely covered with bunch-grass. Through this grass we could detect the section line furrow going over some high hill ahead and thus keep oriented. We carried neither compass nor tripod and sketched drainage and 20-foot contours on a lap board. Elevations were determined on an aneroid which was set at the morning starting point and checked on the river. The wind blew hard from sunup to sunset which added 50 to 100 additional feet to be adjusted in elevations each night, and incidently caused sore eyes.

The steepness of the hills and deep sand made it impossible to keep a straight course, and to re-orient oneself after detouring the base of dunes was not an easy job. The traverseman (Griffin or Lamar) started from a located section corner on the flat, followed line from that corner to the river, and jogged up or down a mile to the next section line. To follow that angling river and find the line and get properly oriented for the return trip, without compass or companion, was as exasperating a job as ever puzzled a green beginner. And he had better get out of the sand dunes before dark!

Mr. Herron, chief of party, laid out the work each week. His part of the mapping consisted of running vertical angle lines with a gradienter (and seldom with a rodman) about every 4 miles, establishing elevations at section corners with which to check the two traversmen's aneroids, and sketching the interior of sections.

October 21 (Monday)—Cold rain with wind. Worked on sheets all day by stove in dining tent. Fuel: cobs, corn and cowchips, the latter gathered yesterday. Corn, bringing only 6–10 cents a bushel, was burned instead of coal all through the country.

October 26 (Saturday)—East on Section 7, R. 16, T. 26. Invited to lunch. The hospitable people don't like to see a man eating lunch alone outside!

October 29 (Tuesday)—Cold rain. Got turned around in sandhills. Old Bo, the horse, given his head, couldn't help.

November 4 (Monday) Two wagon loads of our scattered neighbors in the Quaker settlement came by this evening and took George Lamar and me to their little sod "meeting-house," a mile north of camp.

November 5—Cold. Worked, and slept in nearby store instead of in tent. (Later moved into boxcar with stove.)

November 17 (Sunday)—Camp at end of W. and W. Railroad (later the Chicago, Rock Island, and Pacific) on the Kiowa-Ford County Line.

November 25 (Monday)—Heaviest snow yet, still falling. George and I drove into Dodge City to hear Gilmore's Band at the Old Skating Rink. The Band was America's best then and made the 1-night stop as a break in a long trip.

November 28—Thanksgiving. Our last day's work, and the coldest, windiest. Camp was moved into Dodge City.

November 29—Charlie, the cook, celebrated the end of the season with a bang-up Thanksgiving— Farewell dinner of eight big, brown stuffed prairie chickens, ham, cranberry sauce, canned vegetables, big blazing aromatic plum pudding, etc., etc.

During the season, June 15 to November 29, Mr. Herron with two assistants, or traversemen, mapped seven 30-minute quadrangles or about 6,500 square miles: Spearville, Kinsley, Pratt, Kingman, Cheney, Anthony, Caldwell.

Before breaking camp Mr. Herron, took some pictures with one of the new Eastman Kodaks. Members of the party were shown in their 4-months old beards and camp with Old Glory pushed straight out by a Kansas patriotic but devilish wind and displaying on its blue shield but 32 stars.

December 1—Left Dodge City at 5:25 a.m. on the Santa Fe, after a night in the Delmonico Hotel which was still the best but nothing to brag of. Stopped in Wellington to get county data, plats of towns, villages, etc.

The return trip to Washington was circuitous for the purpose of more sightseeing. On Mr. Herron's invitation I spent December 6 and 7 at his home near

Homer, Ill. Spent the 5th in Springfield visiting the Lincoln Monument, and museum.

December 9–10—In Chicago. Met Dabney Harrison and visited Marshall Fields, the Palmer House, the stockyards, etc.

December 11 (Wednesday)—Enroute to Niagara Falls in Wagner Sleeper (compartments).

December 12—New York. The Battery, Central Park, Brooklyn Bridge, Delmonico's. Streets crowded. Horses of all weights. Saw Denman Thompson in the "Old Homestead."

December 13—With Charley (Slick) Moncrief at his home in Brooklyn.

December 15 (Sunday)—Back to Howard House in Washington.

December 16—Back to office to meet field friends and settle down to drafting.

December 22 (Sunday)—To Unitarian Church with Herron.

December 23—Got promotion to $840.

December 31 (Tuesday)—Clear and cold. Holiday in afternoon. Rented a bicycle for first ride on one. Herron too * * * to Albaugh's to see Crane in "The Senator." New Year's Eve on the Avenue.

Section of Editing

A manuscript map was inked, usually by the topographer who had made the original field surveys, was then checked by another topographer for gross errors, and was submitted as completed to the geographer of his division who examined it and certified that it was ready for engraving. This practice prevailed until 1894, when it was realized that the maps should be subjected to closer examination. They were produced by many unequal in character and quality. It was decided to have every topographic manuscript, before publication, referred to an editor for his close scrutiny. Marcus Baker was assigned to this position.

He was responsible for the original manuscript topographic maps as they were given to him upon completion. The maps were to be carefully examined and prepared for engraving, the examination consisting of comparison with other maps and published material relating to the region. The preparation consisted of the correction of such errors or defects as were discovered during the examination, and in proper uniformity, especially in geographic nomenclature, which included research into the origin and authenticity of place names and geographic features. The editorial work covered not only the new maps but all the old ones when a revision was made for the purpose of printing a new edition, or when any sheet was prepared for use as a topographic base for maps of one of the geologic folios. After the sheets had successfully undergone examination and criticism, they were approved for engraving and sent to the Division of Engraving and Printing, through the office of the Director. After engraving, proofs were submitted and carefully read, before the sheets were approved for printing.

Mr. Baker found the accumulated work so great that he secured the assistance of a proof-reader (Herbert W. Elmore) from the Division of Engraving and Printing. Mr. Elmore was transferred on December 1, 1894. Other transfers were made in the next year or two so that the group constituted a section—the Section of Editing, composed of Mr. Baker, Mr. Elmore, Wm. Stranahan, John W. Brashears, and James McCormick who, in May 1896, was permanently assigned and eventually became chief of the editorial work.

During the fiscal year ending June 30, 1895, the manuscripts of 125 new maps were critically examined and approved for engraving. Proofs, usually two, but occasionally three, were read of 41 new maps. Old maps, to the number of 91 were revised, corrected, and the correction proof read. All known correction material was collected and systematically arranged.

Mr. Baker was able to exercise only a general supervision over the map editing at times, for he continued to serve as Secretary of the U.S. Board on Geographic Names, the work of which was so closely related to that of the Geological Survey and so useful to it. He also prepared a map for the Venezuelan Boundary Commission and on May 11, 1896, at the request of the chairman of that body, he was detailed to the Commission and, since that date, devoted most of his time to its work. On July 29, 1899, he was relieved of the duties of editor of topographic maps by Mr. Kubel, chief engraver, who continued in charge of the editing through 1901.

A part of the work of the Section was preparation of 11 maps by July 1, 1899, for publication on a scale of 1:125,000 by combining maps that had originally been published on a scale of 1:62,500.

During one of Mr. Baker's longer absences, December 1896 to June 30, 1897, Bailey Willis served as acting editor.

Photography

Photography early became an important factor in the work of the Survey. Many negatives were taken in the field by geologists for the purpose of illustrating geologic phenomena. A photographic laboratory was organized, with J.K. Hillers (fig. 7) (who was the photographer with the Powell Survey on and after 1871) in charge, in which photography was used in preparing various illustrations for the reports, for changing the scale of charts, for copying unfinished topographic maps and charts for use in the field during succeeding seasons, and for the guidance of editors and engravers. During the fiscal year ending June 30, 1884, there were made in the laboratory 536 negatives, 6,231 prints and 94 transparencies. Mr. Hillers

made occasional trips to scenic areas and secured some fine photographs.

The laboratory was a unit in the Division of Illustrations, which at first was in charge of Wm. H. Holmes, then DeLancey W. Gill, from 1889 to his resignation July 16, 1898, and then John L. Ridgeway. In 1899, the laboratory had the following personnel: J.K. Hillers, chief photographer, C.C. Jones, John Erbach, Charles A. Ross, Nelson H. Kent, Edgar M. Bane, and Ernest A. Shuster, Jr.

The Photographic Laboratory, after study and recommendation by a committee, on November 23, 1899, was placed in the Division of Engraving and Printing, S.J. Kubel, was chief engraver and on May 1, 1900, Norman W. Carkhuff was appointed chief photographer. J.K. Hillers retired, but remained on a per diem basis until about 1919.

Section of Topographic Drafting

Although the topographers employed by the Survey were necessarily competent draftsmen, and usually would pencil and ink their own manuscripts, it was found desirable to organize a section of topographic drawing, in order (1) to handle tasks of a miscellaneous nature, (2) to obviate the necessity of employing topographers in the office during months in which they might more profitably be employed in the field, and (3) to secure uniformity in style and character of drawing.

During the winter of 1883–84 the compilation of a map of the United States upon a scale of 16 miles to an inch (1:1,000,000) was commenced, almost all of the material for its construction being at hand, or easily accessible. Also, index maps showing the progress of the field surveys were prepared. Other duties related mainly to the preparation of map illustrations for the geologic divisions, to the revision of maps of the Hayden Survey, with a view to engraving them on copper, and to the transcripts of the original maps of the Massachusetts Survey for deposit with the State authorities in accordance with the terms of the contract. The work was keeping from five to seven draftsmen busy and the section was placed in charge of Harry King.

Map Scales and Symbols

By the spring of 1885, a plan had been established for map scales and for cartographic representation. Whereas, in 1879, the mapping was accomplished for publication on a scale of 1:250,000, or 4 miles to 1 inch, with special large scale maps for small mining areas, it was recognized that the small scale of 4 miles to 1 inch was suitable only for deserts and uninhabited mountains. Various sections of the country required the larger scales of 1:125,000 (2 miles equals 1 inch), and 1:62,500 (1 mile equals 1 inch), as determined by considerations given to (1) present or prospective density of settlement; (2) economic importance; (3) complexity of geologic phenomena; and (4) degree of detail in topographic features. A contour interval, whether 10, 20, 50, 100 or 200 feet, was selected for each map after consideration for the scale of the map and the magnitude of relief features.

A system of colors, conventional signs, and lettering was established for topographic maps. Culture (works of man) such as towns, cities, roads, railroads, and boundaries were printed in black; water features (hydrography), such as streams and canals, were represented in blue; relief (hypsography) was shown by contour lines printed in brown. Lettering was in black, of various styles and sizes, which were set forth in a chart of conventional signs and symbols. Such cartographic conventions were reduced to the greatest possible simplicity, in order that the maps might be easily understood and be of value to all classes of people.

Engraving of Maps

Under the statutes relating to the Geological Survey, there was no provision for the general publication of topographic maps. They could be published only as a basis for "geologic and economic maps illustrating the resources and classification of the lands." A small edition (250 copies) was necessary for the use of the employees and collaborators of the Survey in the prosecution of the field and the office work. After careful study, plans were formulated for the printing of these maps.

The cartographic conventions—the symbols to be used on the maps for the representation of the cultural, hydrographic, and hypsographic facts—were reduced to the greatest possible simplicity, in order that the maps be easily understood and be of value to not only scientists and engineers, but to all classes of people.

The maps were to be engraved on copper (fig. 8), three plates being required for each. On one was to be engraved the hydrography, for printing in blue; on the second, the hypsography, as represented by contours, for printing in brown; and on the third, the projection lines, lettering, and public culture, to be printed in black. Plans were matured by the Public Printer and, in the spring of 1885, a contract was made with Messrs. Bien & Co., of New York, for the engraving of 100 maps. Of these, 17 were transmitted to Bien & Co. before June 30, 1885, while the manuscript maps for the rest were then completed and were ready for transmission. It was believed that, with the current appropriation for topographic work, areas comprising 100 maps could be surveyed each year and the manuscripts prepared for the engraver.

The first maps issued by the Geological Survey as topographic maps only and on which no geology was shown were photolithographs of areas in northeastern Arizona and northwestern New Mexico. Small editions of these maps were printed in 1883 and each map included an area of 2 square degrees of latitude and longitude.

The engraved maps for these areas were published as regular 1 degree quadrangles. The first was the Canyon De Chelly, Ariz., Quadrangle, dated March 1886; the second was the Marsh Pass, Ariz., Quadrangle, dated April 1886. The maps of the remainder of the area—Fort Defiance and

Tusayan, Ariz., and Wingate and Mount Taylor, N. Mex., quadrangles—were dated May or June 1886.

By June 30, 1886, the engraving firm been furnished with the manuscripts of 76 maps, and had delivered to the Survey completed copper plates of 55 maps, also small printed editions (250 copies) of 49 of them. Of the 55 maps, 35 were degree quadrangles in the far western states: Arizona—11, California—1, Montana—3, Nevada—2, New Mexico—2, and Utah—16. The other 20 maps were 30-minute quadrangles: Alabama—2, Kansas—6, Missouri—6, Tennessee—4, and Texas—2. The locations of the quadrangles mapped were indicated on a Progress Map of the United States, printed by T. Sinclair & Son, Philadelphia, and attached to the Survey's Seventh Annual Report.

The engraving of maps continued under the contract system. The Public Printer entered into contracts: in October 1888 with Messrs. Sinclair & Son of Philadelphia, for engraving 100 maps of the general atlas; in April 1889, with H.C. Evans of Washington, D.C., for engraving 30 maps; in June 1889, with Harris & Co. of Philadelphia, for engraving 21 maps. Competition between engraving firms for contracts was keen, price levels had been greatly lowered, and with them the quality of the product, uniformity of execution was out of the question; vexatious delays in the delivery of the work were frequent.

As the topographic maps were engraved, proof copies were placed in the hands of geologists, to use in connection with their work; they also were given to engineers and other persons whose work demanded local topographic data. Errors and omissions were discovered and corrected. Moreover, the cultural elements of topography—such as names, towns, roads, and railroads—undergo changes, and so far as these changes supervene between the engraving of maps and their publication, the plates should be revised so as to conform thereto. It was estimated then that the general atlas of the United States would comprise about 7,000 maps. In 1887, consideration for the publication of the first maps of the geologic atlas developed the fact that the old classification and color symbolism must be abandoned and that, in order to devise and establish a satisfactory and enduring scheme, an extensive study of, and experimentation with, colors and processes would be necessary. These operations, in addition to the modification of the topographic plates by way of correction and revision, constituted work which could not advantageously be performed under the contract system. While the topographic features of a quadrangle receive three impressions, each with a different color, for the delineation of the geology, each sheet will receive from 2 to 20 different impressions.

A Division of Engraving and Printing (fig. 9) within the Survey was deemed necessary. The first legislative step in this direction was taken when Congress, by Act of March 2, 1889, made an appropriation of $45,000 for engraving and printing the maps of the Geological Survey and placed the same under the control of the Director. Stephen J. Kubel, of the Hydrographic office, U.S. Navy, was appointed chief engineer on February 15, 1890, and he organized this important Division.

By June 30, 1890, the plant of the Division consisted of two Parke lithographic presses, a full supply of lithographic stones, (fig. 10) inks, colors, papers, engraving desks, and instruments, and arrangements were made for the purchase of a stone-grinding machine, a routing machine, a plate press, and a motor. As soon as the plant was installed, the work of engraving and printing was begun and by the above date, revision and correction of different kinds had been made on 60 different maps, including the entire set of Rhode Island maps. Four new maps had been taken up for original engraving, viz.; Tipton and Durant, Iowa, the Larned and Cheney, Kans. The first two were finished and a proof edition of the Tipton was printed. Preliminary studies of methods, experiments with materials, a system of cost accounting, the care and handling of copper plates, etc., were all given close attention. When appointed, the chief engraver was made custodian of the engraved copper plates for 240 topographic maps, the plates of a wall map of the United States, and those of a wall map of Massachusetts, which had been engraved under the contract system.

During the following year, new contacts were made with three private firms for the engraving of 146 manuscripts, and the map of the state of Connecticut in four sheets. The Survey's new engraving Division engraved 28 quadrangle maps. The force had grown to seven engravers, two printers and two assistants, and considerable additional equipment had been acquired. The Survey expected to build the Division to a size capable of doing all its engraving and printing, and after July 1, 1894, no new contracts were made.

Sale of Topographic Maps

On February 18, 1897, Congress authorized the publication and distribution, by sale of the topographic maps, at a price which would cover the cost of printing and materials. The price fixed for single copies was 5 cents, and for large orders 3 cents per copy. Editions of 1,000 copies were printed. In a few years, under a stimulating demand, editions were increased to 3,000.

Topographic maps engraved by the Division in its 10th year, 1899–1900, numbered 72; in 1902–03, 101. The total number engraved by contractors and office force by 1904 was 1,450. The total number of topographic maps printed during the first 10 years was 533,655; by 1904 the total was 3,434,424.

Investigation by Congress

The rapidly growing appropriations for the Geological Survey, as well as for two other large scientific bureaus of the government—the Signal Service (which then performed the functions vested in the Weather Bureau), and the Hydrographic Office, became, beginning in 1883, the subject of discussion in Congress and in 1884, provision was made in the sundry civil appropriation act for a joint congressional commission to "consider the present organization" of the bureaus mentioned, "with the view to secure greater efficiency and economy in the administration of the public service in said Bureaus." The commission appointed held numerous hearings in 1885 and 1886, Director Powell appearing for the Survey.[15]

The main questions regarding the Survey's work raised in hearings related to the value of topographic maps executed on so small a scale as 1:250,000, the rate of progress toward completion and publication of the Survey's geological maps, and the propriety of the wide scope of the Survey's scientific investigations and publications.

While the minority of the commission, consisting of two of the six members, submitted a report criticizing the Survey, the majority expressed the opinion that the Survey as a whole was "well conducted, and with economy and care, and discloses excellent administrative and business ability on the part of its chief."[16]

The investigation demonstrated that the Survey's work was built on sound principles and, further, that Major Powell, well known as an excellent artillery officer and fearless explorer, was also a fighter and great leader. Every year he was called upon to justify before the appropriations committee, and sometimes on the floor of the House and Senate, the continuation of the Survey's work.

Some members of the Congress disapproved of the Survey, as a new scientific Bureau, and were ready to reduce its appropriations or discontinue them. When these reverses happened, Major Powell furloughed topographers, as well as others, assuring them that he was sorry to have to do so and would recall them as soon as he obtained sufficient funds to carry on the work. He always succeeded. Topographers, more than once, worked late into the night to prepare charts and diagrams showing the progress of topographic mapping for Major Powell's use before Congress the following day.

Appropriations depended upon a good showing of square miles mapped the previous season. This put a premium on speed and undue emphasis on the small scale of 1:250,000 as yielding the most mileage at less cost.

Excerpts from "Testimony before the Joint Commission on the Signal Service, Geological Survey, Coast and Geodetic Survey and the Hydrographic Office—U.S. 49th Congress, 1st Session, Senate Miscellaneous Document 82." (1886) Hearings began Thursday, December 4, 1884.

From Major Powell's testimony:

Q. How many years do you think that it will take to make the geological map of the compass and scope that you contemplate now?

A. About 24 years.

---

[15] Service Monographs of the U.S. Government No. 1 (Geological Survey) by Institute for Government Research, 1919, p. 15.

[16] S. Misc. Doc. 82, 49th Cong., 1st sess. (v. 4 of misc. doc.) S. Rept. 1285, 49th Cong., 1st sess.

Q. That will be all-embracing; it will take in the entire area of the United States?

A. Yes sir.

Q. Including States and Territories?

A. Yes, sir.

Q. Including Alaska?

A. Not including Alaska * * *

Q. Major Powell, have you in your mind any plan of operations for the completion of this map? How do you propose to carry on the work; where first, where next, and in what order?

Q. In the first place, I am trying to connect detached portions of work in the far West. Parties are at work in that region. Then I expect to expand from these districts where work has already been done (indicating upon the map).

Q. Please describe in words, so that the stenographer can take it.

A. It is proposed, then, to start from several centers where base lines have been measured, and where base lines are yet to be measured, in the eastern portion of the United States, and to extend from these several places until the work shall come together. The number of places already selected are about 15, but more are yet to be chosen.

Q. Please state where they are.

A. There is one in Massachusetts, one in New Jersey, one in Pennsylvania, one near where the five great States meet—Virginia, West Virginia, North Carolina, Tennessee, and Kentucky. Another of these starting points is in Alabama, still another in Texas, still another in southwest Missouri, northwest Arkansas, northeast corner of Indian Territory, and southeast corner of Kansas. Those mentioned are the only ones established at present.

Q. What was the purpose of establishing these different bases of operations?

A. The fundamental work of the survey cannot be properly executed from one base line with sufficient accuracy. The distance between base lines should be about 200 miles, in order that the triangulation of one may check that of the other. Should the work be extended to a great distance from one base line and an error be made in the work, it could not be detected, and would be propagated and multiplied, and thus subsequent work would be vitiated; that is, an error would propagate itself indefinitely and multiply itself.

Q. The principle is that you should begin different points and come together so as to check errors?

A. Yes, sir; to secure accuracy.

Q. On what principal did you select these bases? Why did you select these particular places for them?

A. First, I have selected districts for the beginning of operations that are of special importance by reason of their being mining districts, districts in which coal and iron or other minerals are found. Second, districts are selected at proper intervals, and positions are determined by certain topographic characteristics—places where base lines can be easily measured and the triangulation most economically expanded therefrom. Third, the particular spot for the beginning of a base line is selected so that it shall be convenient to some telegraph station, in order that telegraphic connection may be made with the Observatory in Washington, or some other observatory, the latitude and longitude of which has been accurately determined; and this is done that the longitude of the spot may be obtained by the most accurate method.

Q. That is what I wanted to know. You spoke about the appropriation made by the state of Massachusetts, but you did not explain fully how you were connected with that survey. Does the state of Massachusetts have its officers and employees by whom and through whom this money is expended?

A. The state of Massachusetts appoints a commission to take charge of this work. This commission arranges with me to supervise the same; to furnish plans for carrying on the operations; to appoint the persons to do the work; to supervise it in its progress; to audit and approve the accounts which are sent to the commission and paid by the State.

Q. And that commission pays for it?

A. Yes, sir.

Q. The state of New Jersey, also?

A. Yes, sir; but by a somewhat different plan. Instead of being managed by a commission in New Jersey it is managed by the State Geologist.

Q. Do they go on with the work?

A. I had commenced work in Massachusetts before the State determined to make a state survey. Without my solicitation and without my knowledge, certain gentlemen in Massachusetts presented this matter to the governor of the State. The governor in his message recommended that the State seize the opportunity to do work conjointly with the United States and make a more elaborate map than the United States was doing. It was for the purpose of serving certain local interests. The legislature took up the matter and made provision for a commission, as already explained, and also made an appropriation. In the meantime I was called before the legislature and explained to them the methods of operation, and gave them estimates of the cost. In the previous years, the Legislature of Massachusetts had discussed the matter of the construction of a topographic map of the State but the scientific men of the State who were called in as experts for their opinions disagreed as to the method upon which it should be made, and also as to the probable cost; so that the State was already ripe for action when the U.S. Geological Survey commenced operations in the commonwealth. I have telephoned for the sheets of a part of the work to brought here, and they are now present for examination.

Q. I understand you were proceeding to make that geological map on a scale of 2 miles to 1 inch, and the state of Massachusetts said, "If you will enlarge that to 1 mile to 1 inch we will pay for it."

A. Yes, sir; that they would pay the additional cost which I estimated at $40,000 * * *.

Q. Major Powell have you completed any section of this map which you are preparing?

A. Oh, yes, sir; a good many sheets.

Q. As it will be in the end?

A. Not of the geologic work.

Q. That is what I mean?

A. The topography goes first * * *.

Q. What is your idea as to the particular form that the geological map of the United States will take for public use?

A. I thank you for asking that question. I expected yesterday to explain some things of that nature which the general conversation prevented. I wish to explain that subject so that it may be understood just what we have in view and what we are planning. In the first place, we, too, make triangulation; in the second place, we make topographic maps as preliminary and incident to the Geological Survey. Now, to go on and answer your question, the map of the United States, as planned at the present time, contemplates one sheet for each degree of territory—each degree of latitude and longitude. The work is being prosecuted on that plan for a large part of the territory of the United States. The scale of work to be represented on such sheets is 1:250,000, that is, the distances on the map multiplied by 250,000 will represent the distances

in nature. The size of the sheets will be 20 by 16 1/2 inches, and they will represent the topography as a basis for areal geology.

Q. About the size of that (indicating)?

A. I can show you the sheets (after getting sheets) that will cover a degree of latitude and longitude. The general plan is one sheet for every degree; but there are certain districts of country where the topography is too intricate to be properly done on that scale, and it is proposed that in such districts the scale shall be enlarged to 1:125,000; that is, that a distance on the map must be multiplied by 125,000 to equal the distance in nature. There are reasons other than intricacy of topography which determine the enlarged scale; certain districts are important because of the great mining interests, and more elaborate maps are demanded; other districts are important because their watersheds must be accurately delineated in planning works for the supply of water to cities and towns. Again, in other areas where the topography is exceedingly intricate, and especially where assisted by state funds, it is proposed to enlarge the scale to 1:62,500, or about one inch to the mile. This is the scale adopted in Massachusetts and in New Jersey, and will probably be adopted in Pennsylvania, though as yet no final arrangement has been made with regard to this State. These are the three scales upon which the work is being carried on, and upon which a little more than 600,000 square miles have already been surveyed. The United States, exclusive of Alaska, has an area of about 3,000,000 square miles, and more than 600,000 square miles has been thus mapped, or surveyed for the construction of maps. It is thus that I have a fair basis for an estimate of the ultimate cost and it is upon this experience that I made the statement yesterday that I thought the work would be completed in a time not greater than 24 years.

Q. How many sheets on the plan that you have described will be required in the finally completed work of the geological maps?

A. About 2,600 sheets.

Q. Not to be a map, then spread out like this (indicating)?

A. No, sir; only a map in small sheets. The purpose is eventually to have them for sale. As soon as we get them done they will be called for in separate sheets. The people of the country will want more sheets of their own districts, but they will be in such a shape that they can be bound, if so desired, by states.

Q. I was about to ask you that. If you make these in sections comprising the latitude and longitude, taking, for instance, the state of Massachusetts, with which State you seem to be in accord, won't they also be glad to have you get into one section the whole state of Massachusetts there?

A. Yes, sir; but when that is done the, preparation of maps on smaller scales is a trivial matter in regard to expense. That is only a question of executing it in the office and having it lithographed. It is a trivial matter under present methods.

Q. Out of this any state can get a complete map of its own territory at very small cost?

A. Yes sir.

Q. When this map is completed will it be only a nature map, or will it take culture too?

A. It will take certain features of culture. It takes all the highways, roads, and railroads, and it takes county line, town lines, etc. It does not take all the particulars of culture.

Q. Does it take bridges?

A. No sir.

Q. But it takes 24 years to complete it. What do you think of the accuracy of the first part of a map which you have taken 24 years to complete?

A. The culture will always have to be kept up. So far as the natural features are concerned, it will never have to be changed. So far as changing the culture is concerned, after you have established natural points, the change of the culture part of the map from time to time is a small expense. Further, and in connection with that particular point, it may be stated that up to the organization of the present Geological Survey, that work was put on stone. The work of Wheeler, Powell, King, etc.—that work was put on stone. That was due to the vigor with which this work was progressing and the demand for immediate results. That was one of the evils of the system and one of the arguments which I made for a coordination of the systems, which was finally accomplished. One of the arguments was that we were doing ephemeral work. Since that, these sheets go on copper, and the copper belongs to the Government. In the printing on stone, one edition is printed off, and this stone is then cleaned off and smoothed down, and a new edition must have a new engraving, but the work is now permanent. On the present plan, those sheets go on copper. Now, copper is of such a nature, you can make changes at any time. It is a soft metal, malleable and mobile, so that you may erase a line or put a new line upon it, as the case may be, so that a practical mechanician can cut

that town out or put a new town in, or if a road has been changed, he can change that road on the map; the natural features will remain—the rivers, the valleys, the hills, and so on; and it is easy to change the artificial features, because you already have the natural features to which they are related.

Q. Twelve years, do you say, or 24, to complete it?

A. Twenty-four; I have given that as the longest time.

Q. Of course that time could be shortened by enlarging the appropriations?

A. Yes; it is only, a question of appropriations.

Q. That is to say, you have an appropriation of $300,000. That enabled you to have five or six parties in the field?

A. I have 27.

Q. But you have them in five different portions of the country?

A. Yes sir.

Q. If you double the appropriation, of course you would double your force in the field?

A. Yes sir; but much would depend upon the division of the appropriation. At present a little more than one-third, I think, of the appropriation is used topographically * * *.

Letter from the Director of the Geological Survey, December 6, 1884, furnishing certain information to the committee.

Sir: In response to your request of the 5th instant I have the honor to submit herewith a statement showing the appropriation and expenditure by fiscal years of the several surveys of Hayden, Powell, King, and Wheeler, and the number of square miles covered respectively by each of those organizations.

Appropriations for fiscal year 1884.

Recipient	Amount
Wheeler Survey	$599,316.72
King Survey of the 40th parallel	386,711.85
Hayden Survey of the Territories	720,000.00
Powell Survey	279,000.00
U.S. Geological Survey, 1880	106,000.00
U.S. Geological Survey, 1881	156,000.00
U.S. Geological Survey, 1882	156,000.00
U.S. Geological Survey, 1883	254,940.00
U.S. Geological Survey, 1884	339,640.00
U.S. Geological Survey, 1885	489,040.00

Area Surveyed

U.S. Geographical Surveys West of 100th Meridian (Wheeler).—Gross area surveyed by this organization, 359,000 square miles. Of this area, 227,500 square miles was of the nature of reconnaissance work, and was published on a scale of 8 miles to 1 inch. The remaining area, 131,500 square miles, was of a more exact and detailed nature, and was published on a scale of 4 miles to 1 inch. Nearly all work was published in hachures.

U.S. Geological Exploration 40th Parallel (King).— Gross area surveyed by this organization, 86,390 square miles. This was published on a scale of 4 miles to 1 inch in 300 foot contours.

U.S. Geological and Geographical Survey of the Territories (Hayden).—Area surveyed in systematic manner, 100,000 square miles. This was on a scale of 4 miles to 1 inch in 200 foot contours. Besides this there was, during the years 1869 to 1872, inclusive, much reconnaissance work done, the area comprised in which, however, it is impracticable to estimate.

U.S. Geological and Geographical Survey Rocky Mountain Region (Powell).—Area surveyed, 67,000 square miles; published on a scale of 1:250,000, or about 4 miles to 1 inch, in 250 foot contours.

U.S. Geological Survey.—Area surveyed up to date, 184,200 square miles. This is embraced in the following States and Territories: Massachusetts, Connecticut, New York, New Jersey, Maryland, Virginia, West Virginia, North Carolina, Kentucky, Tennessee, Alabama, Georgia, Missouri, Kansas, Texas, New Mexico, Arizona, Nevada, Montana, Idaho, Wyoming, California, and Washington. That portion of this area lying in the sparsely settled Western States and Territories is to be published on a scale of 1:250,000, or 4 miles to 1 inch. That in the Mississippi Valley and the Southern Appalachian Region on a scale of 1:125,000, or 2 miles to 1 inch, while that in the densely settled districts, when states assist, is to be published on a scale of 1:62,500, or about 1 mile to 1 inch. The interval between contours depends on the scale of the map, and amount of relief in the country, ranging from 200 feet in the mountains of the West to 10 feet in low coast regions.

On December 18, 1884, Maj. J.W. Powell, Director of the Geological Survey, then appeared and made a further statement as follows: (relating to transfer of topographic work of the Geological Survey to the Coast and Geodetic Survey) * * *.

"Some 4 years ago, when it was found that the wording of the organic law of the Geological Survey did not permit that survey to come into the eastern portion of the United States, and when a proposition was introduced into Congress to correct that defect, the Coast Survey antagonized that proposition, and for the first year it was defeated. During this time the Coast Survey commenced a new class of work, namely that small piece of topographic work proper to which I called the attention of the Commission the other day, the tract in Maryland and Virginia. But, as soon as the work of the Geological Survey was extended into the eastern third of the United States, from which it had been previously excluded, the Coast Geodetic Survey immediately withdrew from the class of work. It will thus be seen that that organization has, with the single exception I have mentioned, not attempted to do the class of work which the Geological Survey is doing; and, more than that, it has never had authority to do the class of work which I have defined as topographic.

"On the other hand, the geologic surveys have executed topographic work to the extent of carrying it over about one-fifth of the United States, exclusive of Alaska. The geodetic work does not require the construction of a final map; the geologic work does require the construction of a final map. The Geodetic Survey requires the determination of many points, a few of which it uses in its grand triangulation; the Geological Survey requires a vast multiplicity of points; the number of points necessary for the Geodetic Survey has to be multiplied by thousands to subserve the purposes of the Geological Survey; but both surveys need more or less of this work as preliminary to their fundamental functions. * * *.

"As military engineers were the inventors of early topography, they developed methods of constructing maps suitable for their science. In later years, topographic methods and plans of mapping have been changed; and these changes are radical, and are due to the influence of geologists, who have demanded better maps than those of the old military engineers. The military engineers have yielded to this demand by reason of the change in military science which I have previously mentioned. I will confine my statements now to the United States. The Coast Survey has constructed charts of the coast. These maps are very elaborate and represent a vast system of cultural or artificial details; they are, therefore, very expensive; they are also difficult to understand. The geologic organizations of the Government have made topographic surveys in the interior to the extent of one-fifth of the territory. In the course of these surveys, they have invented new methods of exhibiting the work, and, to a very important extent, new methods of doing the work, and to a very important extent, new methods of exhibiting the results on maps; by which the maps have become greatly simplified, and much more easily understood by geologists, engineers, and the common people.

"Topography is no longer occult, and a topographic map is no longer a system of hieroglyphics; it is a plain representation of facts that all people may easily read. The experience in the United States, up to the present time, warrants this statement, that, if the topography is kept under control of geologists, it will be executed better and at less expense than if placed under some other organization. * * *.

"For more than 15 years, topographic work has been connected with geologic surveying under the auspices of the General Government. The Hayden, Wheeler, King, and Powell Surveys were conducted in this manner. When the present organization was formed, it was not merely a nominal, but a real consolidation. All of the experts of the four antecedent corps were ultimately brought into the new work, and the men who originated the new topographic methods are still employed. I beg permission to express this opinion, that there has never been another such a corps organized in any country or at any time. The vigor and enthusiasm with which they push this work is tempered only by a scientific knowledge which comes thorough training. The topographers constituting this corps were trained in their work by geologists, and some of them have acquired much knowledge in geology, and can render valuable assistance in the simpler portions of the geologic work. * * *.

"Again, in the topographic survey, the topographers are draftsmen, for the man who does topographic work in the field must be able to put it upon the map. The geologists who perform the work of the Geological Survey are specialists of a high grade, to whom the highest salaries are paid. Such specialists are rarely draftsmen, and when they are, it is not wise to employ them in drafting work, from the very fact that they are high-salaried men. But for the construction of their sections and other illustrations necessary to set forth the geology of the country, the topographers of the topographic survey are employed as draftsmen for the Geological Survey. Again, geology demands more refined and accurate topography then any other science, and no other industrial need demands better work than economic geology * * *.

"There is yet another reason why the Geological Survey should control the topographic survey. The problems which arise in a geologic survey do not appear in some systematic geographic order, as from east to west, or from north to south; but they arise in the course of research relating to geologic formation * * *.

"I think that the facts which I have laid before the Commission warrant the following statement: First, the topography is more essential to the Geological Survey than to the Geodetic Survey; second, that the Geological Survey can execute the work more economically than the Geodetic Survey; third, that the efficiency of the Geological Survey will be greatly impaired by having the topography taken from it; fourth, that a topographic survey should precede a geodetic survey, and that in the prosecution of a geodetic survey economy will result therefrom; fifth, that if the topography be placed under the Geodetic Survey, the relations between the two organizations will be rendered more complex than at present. For these reasons it would appear that the recommendation of the committee of the National Academy is wise. That recommendation is as follows: 2. The Geological Survey (should) comprise the present Geological Survey with its organization unchanged.' * * *.

With regard to the comparative cost of map making by the two organizations.

"It is necessary to state that a proper comparison cannot be made. It is much like comparing the cost of a horse with the cost of a house, as the two organizations make maps of a different character and for different purposes With regard to the general triangulation or geodetic survey of the interior of the United States, the cost would greatly vary according to the plan adopted. If it is proposed to extend over the country a grid-iron of triangulation belts, say at a distance of about 100 miles apart, the cost of such a geodetic survey would probably not be far from $2 a square mile, including in that all the astronomic work, the hypsometry, or leveling, measuring of bases, triangulation, and gravity determinations. Such a system would perhaps be sufficient for purely geodetic purposes, but in order that this triangulation may ultimately be used for cadastral purposes, as I have explained in a former communication, it must eventually be developed over the whole country instead of in narrow belts across it, in which case it will cost probably not less than $12 per square mile.

"The cost of the topographic survey, which is precedent to the geological survey, and which makes maps subserving the purposes above enumerated, is greatly variable in different portions of the country. There are very large areas of the prairies and plains where the cost will not exceed $1.50 per square mile. In the mountainous and arid regions of the country west of the 100th meridian, the cost will vary from $1.50 per square mile to $3.00 per square mile.

"In those portions of Oregon and western Washington Territory which are densely covered with forests the cost will be about $5 per square mile. In the eastern portion of the United States where forests prevail, but where the rectangular surveys have been extended so that the roads, in the main, follow right lines, the cost will be from $2.50 to $3.50 per square mile. Still farther east, where forests prevail, and where the roads meander through the country, the cost will be about $5 per square mile.

"These estimates include all elements of cost in the production of topographic maps—leveling work, topographic work, astronomic work, triangulation, etc.—but do not include engraving and printing. These estimates are based upon experience in surveying, upon various scales, areas comprising some 300,000 square miles. The above statements are made on the theory that maps are to be constructed in the far West on the scale of 4 miles to the inch. But in some portions of the East the Geological Survey is, in conjunction with State surveys, constructing maps upon a scale of 1 mile to the inch. These maps will cost $10.00 per square mile, only one-half of the expense being borne by the United States.

"It may be asked whether the scales adopted by the Geological Survey are sufficient for practical purposes, and the reply to this should be that for all purposes except geology the scale is probably ample, but that there will be limited localities where this scale will be insufficient for geologic purposes, and to some slight extent, maps on a larger scale must be constructed. It should further be noted that should it be found advisable in the future to enlarge the scale of the maps of any district of country, by reason of the development of the country and the dense settlement of certain districts, the topographic maps constructed on the scales now adopted will not be thrown away; the new surveys can take up the work from the basis of the old surveys, and all the work of the old surveys can be utilized, and the new surveys can be made with relative diminution of expense. This has been proved by actual experience in many parts of the world. It must be understood that this statement depends on the fact that the topographic work of the Geological Survey deals chiefly with natural features; if it dealt chiefly with cadastral features the statement would not be true. But it should be stated that in the event of a new survey by reason of a change of scale, a re-engraving of the maps becomes necessary, and to this extent there is an additional expense.

"The estimates above given for the cost of a topographic map do not include the engraving and printing of the maps. At the present stage of the work it is impossible for me to give estimates of this element of expense. For 10 or 12 years I have been carefully studying this branch of the subject, and making many experiments for the purpose of discovering the best methods of representing topographic features and the most economic methods of reproducing maps from the drawings of the topographers, and have not yet reached a conclusion. The experiments made within the past 3 years have cost $5,580, but the work produced is not quite satisfactory. The following are the fundamental considerations that enter into the problem: first, the maps should represent those characteristics which will be more permanent, in order that it may have more enduring value; second, the maps should represent all the important characteristics that are needed for industrial and scientific purposes; third, the system of representation should be simple so that the maps may be intelligible to all people; fourth, the method of representation and engraving should be economic, for the area to be mapped is great, and the total expense will be considerable. I am sanguine that the problem of the reproduction of maps is practically solved, and that experiments now nearly complete will prove satisfactory and fulfill the conditions enumerated above.

"In planning the work for a topographic survey of the United States it has been necessary to consider that the area to be surveyed is very large—about 3,000,000 square miles, exclusive of Alaska, and with Alaska about 3,500,000 square miles. The survey of a region so vast must necessarily be at a great aggregate cost, and the economics relating thereto should be duly weighed. The area and cost being great, the time for the execution of such a survey must necessarily be long, that the cost may be distributed over many years; but the needs for topographic maps are so many and so great that the time for its completion should be shortened as much as possible. All of these considerations have been carefully weighed, and the result is the plan which I have presented to the Commission. I have stated to the Commission that the map can be completed on the present plan, with the present organization, within 24 years, but the demands for such a map are so urgent that the work ought properly to be completed in a shorter period. Since the organization of the Geological Survey, Congress has, by increased appropriations, expended its work from year to year, and the Director has earnestly desired and hoped that this growth would continue, so that a map of the whole country could be completed by the year 1900, and he has steadily and vigorously worked to that end. He has tried hard to develop a plan which should not be impracticable on account of excessive cost, and which should not be without substantial value by reason of imperfect methods and results. He has also endeavored to develop the plan in such a manner that no work would be lost, though the needs of the distant future should demand more elaborate work than the wants of the present time, and that all work done should have enduring value."

Q. Is your work benefited practically by the geodetic work of the Coast Survey or not?

A. If the geodetic survey was extended over the United States precedent to the topographic survey, it would reduce the cost of the topographic survey from 5 to 10 percent.

Q. But the cost of the geodetic work would be a great deal more than it would save your work?

A. Yes sir * * *.

Q. You stated that the topography of one-fifth of the United States is complete. That includes the Western Territories and some portions of the older regions of the country?

A. Yes sir.

Q. State, in a general way, what portions of the East have been completed.

A. You will remember, perhaps, Mr. Chairman, that I presented a map to the joint commission some days ago, setting forth this matter very fully, and as directed by yourself I have had that map engraved so that it will accompany my former statement.

Further statement by the Director of the Geological Survey, February 6, 1885: * * *.

The General Systems of Maps Needed

It has been mentioned that topographic maps were originally made for military purposes. As wants were developed for scientific, industrial, and cadastral maps, an attempt was made in various countries to construct a single map that should subserve all wants, and maps constructed on this theory have been made from time to time. But experience has abundantly shown that all of these purposes cannot be subserved by one map. The facts that it is necessary to represent cannot be included on one sheet, as they obscure each other and render the map unintelligible.

Experience also has shown that these maps cannot be constructed on the same scale. Three general classes of maps have been developed and pretty thoroughly distinguished from each other, namely general topographic maps, cadastral maps, and military maps. General topographic maps are usually constructed in such a manner that many subsidiary charts may be derived therefrom. Thus, from the general map, drainage maps, hill maps, road maps, etc., may be directly compiled and the general maps may be used as a basis for geological maps, agricultural maps, forestry maps, sanitary maps, statistical maps, etc., on the same, or on reduced scales. Such maps require no additional expense for topographic surveying in the field.

The Best Method of Constructing Topographic Maps

The Facts to be Represented

The first question of importance is a plan for a topographic survey (that) relates to the facts that would be represented upon the map. For all purposes for which a topographic map is made, except the functions of a military map, the natural features of the country are of prime importance. The natural features, therefore, should be represented. Of the artificial features there are two classes. Those which may be included under the term public culture, relating to the sites of towns and cities, and of highways of all classes—canals, railroads, and wagon roads; all of these should be put upon the map.

There is another class of artificial features that relate to private properties. These are houses, fields, orchards, etc., and as they relate to private property, or estates, they may be called cadastral features. These should not be placed upon general topographic maps because they are so minute that they cannot be properly represented without enormously increasing the scales, and their delineation upon the map obscures the conventional characters used to represent natural and public topography and, to the extent that they are introduced, the more important features are obscured; the map becomes more complex and less easily interpreted.

There is yet another potent reason for their exclusion. When a map has been constructed it must be maintained to be of permanent value. The natural features remain, the cultural features are changeable, and those cultural features that have been denominated "cadastral" are the most changeable. If they are represented upon the topographic map constant revision becomes necessary, and this can be accomplished only at great cost. For all these reasons it is manifest that cadastral topography should not be placed upon a general topographic map.

Method of Representing Topographic Facts

The exclusion of many facts from topographic maps greatly simplifies the system of conventional signs necessary for use. Wide experience has led to a system from which there is little departure in the various countries of the world. The more important points to be mentioned are the following:

First, relief features are no longer represented by hachures, but by contour lines, which more thoroughly convey specific facts relating to elevation, and a contouring system is much less expensive than a hachuring system.

Second, to clearly distinguish the different elements of topography, colors are introduced upon maps. There is yet some diversity in the methods of using colors, but the best systems adopt blue for the drainage and all bodies of water, black for culture and names, and brown for contours. This is the system adopted by the Geological Survey.

The Scale

The next point of importance to be considered is the scale on which topographic maps should be made. It is necessary to consider the subject from two aspects, namely, the best method of representing the facts, and the most economic method of accomplishing the purpose. With respect to the representation of the necessary facts upon a topographic map, these two fundamental principles are controlling: First, the scale should be sufficiently large to clearly represent all the necessary facts; second, the scale should be as small as possible consistent with the proper representation of facts, in order that the largest possible area may come under one view, because all the important purposes for which topographic maps are made include large areas of country * * *.

The Planetable and Alidade

Six years of topographic surveying were enough to convince the Geological Survey that the planetable and alidade constituted the best instruments for mapping at the time. These instruments were not new, but they had been considered suitable only for mapping small areas, such as sites for military fortifications. The U.S. Coast and Geodetic Survey, engaged in mapping a narrow strip of coast terrain, devoted pages of its annual report for 1879–1880 to the description and use of the planetable (fig. 11). It was a heavy and cumbersome instrument for mapping operations over a large area where much travelling over rough country was required.

Among the instruments inherited by the Survey in 1879 were folding planetables. They were made of a series of pine slats 1.5 inches wide by 2 feet long, glued to heavy canvas so as to roll up to a diameter of 6 inches. When in use heavy wooden cross pieces were attached to the under side by means of thumbscrews, and a light folding tripod was screwed into a plate on the cross pieces. The planetable sheets were 12 inches square, of single mounted paper, and were taken off the table each time it was packed for traveling. When the table was in place at a station the sheets were attached to it by special tacks, officially named B and BD (blow and be d--) tacks, which were similar to ordinary thumb tacks, but had tapered screw threads in place of sharpened stems. The heads of these screw-tacks were one-half-inch in diameter with two small holes in each, and a two-pronged steel pin was used in screwing them in or out of the planetable. An 18-inch open-sight alidade completed the planetable outfit.

Vertical angles were read with a small transit, called a gradienter, having a small telescope with striding level, 3-inch vertical and 3-inch horizontal circles, both reading to minutes. When in use, the gradienter was mounted on the planetable tripod. The gradienter in the early surveys had been used for both vertical and horizontal angles, but the folding planetable had come into use as possessing definite advantages for secondary triangulation and graphic intersection work. Topographers found it easier in the field to read horizontal angles to a recorder for writing down in a notebook than to draw fine lines with a sharp pencil on drawing paper, stations usually being on summits exposed to a wind. Experience enabled the reading of angles rapidly and speed in covering square-mileage was achieved.

It was a matter of routine, in a comfortable office during the winter season, to plot the horizontal angles on a map sheet, work out the elevations from the vertical angles or from barometric readings, and with the profile sketches before one, to compile the map. However, one often found that he wished for more locations and elevations in order to compile a satisfactory map. Such a dilemma would not exist if the map details were drawn in the field on a planetable while the topographer was on the ground with the country in clear view.

Improvements in instruments and methods were constantly sought. Solid planetable boards were substituted for the folding ones. In 1886, Almon H. Thompson, Edward M. Douglas, and Willard D. Johnson were assigned to a study of instruments. After a lot of planning, they designed a telescopic alidade that was adapted to the survey use. Mr. Douglas went to Troy, N.Y., and engaged W. Gurley and L.E. Gurley's most expert instrument man to make the final full-scale drawing.

The alidade consisted of a ruler of brass 18 inches in length, and carrying upon a column, 3 inches in height, a telescope having a focal distance of 12 to 15 inches, and a power of about 15 diameters. It had a vertical arc reading by vernier to single minutes, and a delicate level upon the telescope. To the reticule of the telescope were added two fixed horizontal wires, above and below and equidistant from the center horizontal wire, and so placed as to subtend one foot on a stadia rod at a distance of 100 feet.

The Survey proceeded, as fast as funds permitted, to acquire alidades for its field parties. The instrument, as then designed, proved satisfactory though slight changes and refinements have been made in later years.

Mr. Johnson invented the tripod that bears his name—a tripod that combined the elements of stability, lightness, and facility of operation in a remarkable degree. The movement was essentially an adaptation of the ball and socket principle, and consisted of two cups, one set inside the other, the inner surface of one, and the outer surface of the other being ground so as to fit accurately to one another. The inner cup was in two parts, or rather consisted of two rings, one inside the other, the one controlling the movement in level and the other that in azimuth. From each of these rings there projected beneath the movement a screw, and upon each screw was a nut by which it was clamped. There was no tangent screw for either the leveling or the azimuth motion, as none was required. The movement was sustained by a light hardwood tripod with split legs.

Mr. Johnson was granted a patent on the tripod dated May 3, 1887, but waived all claims to royalties. A plate, countersunk in the board, secured the board firmly to the tripod. Boards were of three sizes: 24 by 31 inches to accommodate a full atlas sheet and used for planetable control; 18 by 24 inches to accommodate a one-half size atlas sheet and used for mapping; 15 by 15 inches for traverses. Still a smaller board,

9 by 9 inches, was used for traverses in timber and brush, but the tripod (Bumstead) was lighter and the sighting instrument was an open-sight alidade. Mr. Johnson also eliminated the despised B and BD tacks for holding the drawing paper to the board and substituted instead corner holding screws in metal sockets, which continue to be standard.

The studies revealed that vertical angulation with the telescopic alidade was better for station elevations than elevations from mercurial barometer readings, so that much despised instrument was discarded. Vertical angle tables were prepared so that recorders as well as topographers became adept at computing differences in elevation. For the determination of critical elevations or distant points, readings were observed with the alidade and striding level in four different positions and the mean of the four resulting quantities was accepted as the mean elevation within a maximum error of 5 feet.

Though some types of country were more economically mapped with other instruments, the Survey, as rapidly as possible, installed the planetable with Johnson tripod and telescopic alidade as its chosen mapping outfit. The alidade was light, easily carried in the crook of one arm, extremely durable, and quickly adjusted. The planetable and Johnson tripod was not as bulky and cumbersome as one might suppose. Topographers soon became accustomed to transporting the outfit in their daily fieldwork whether they traveled in a buggy, ahorseback, or afoot. One topographer, working in western mountain, prairie and desert areas, estimated that during the period 1900 to 1914 he performed stadia traverse ahorseback in excess of 10,000 miles, covering more than 10 miles on some days.

The Bumstead Tripod

Further research resulted in the development, of a lighter, simpler tripod (Bumstead), suitable for tape and stadia traverse work. As this work concerned only the details of topography, and the traverse line, made up of short sights subject to constant checks by means of points located by intersection, a lesser degree of accuracy, was required than in intersection work. The tripod head, fitted to the board with clamp for azimuth motion, has no arrangement for leveling the board other than movement of tripod legs.

The usual board was 15 inches square, or 9 inches square for use in brush, and had a compass set in one edge of it, in a narrow box, by means of which it is oriented by reference to the magnetic meridian. The alidade for direction was a simple brass ruler with raised sights at the ends. An aneroid barometer was used for elevations and the traverseman usually carried a Locke hand level. By June, 1889, there were fully 100 traverse planetables in use in the Geographic and Irrigation Surveys, and some were being used by geologists.

Base Lines

Efforts were constantly made to achieve efficiency in mapping operations—the accomplishment of desired results in less time and cost. By 1887, it was found that the required degree of accuracy in the measurement of base lines was easy to attain by the use of long steel tapes, and it was decided to drop the use of bars. The metal bars, 4 meters in length, required many contacts to be made in each mile of measurement, each contact afforded the possibility of a trifling error, a number of men were needed, and it often took 4 to 6 weeks to measure and remeasure a base 5 miles in length.

With the 300-foot steel tape (fig. 12), carefully compared, at an observed temperature, with the standard of the U.S. Coast and Geodetic Survey, both before and after its use, a base could be measured in 1 week by a party of four. Preferably, the site for the base line should be selected along a railway tangent, as such a location is approximately level, and the railway ties afford an excellent support for the tape. If such a location cannot be obtained, it should be selected in comparatively level country, cleared of brush and undergrowth, and its sharp inequalities should be leveled. The tape should be supported by a series of low stools, whose legs are pressed into the ground at intervals of not more than 25 feet, while similar stools should sustain each end of the tape.

On the first measurement, a board about 5 feet long and 4 inches in width is nailed to two stools or railroad ties, so that the end of the tape falls near its center where is nailed a block of wood about 8 inches long and 1 inch in width, covered with a numbered zinc strip. The end of the tape is marked on the zinc strip. For insuring a uniform tension of the tape, an end spring balance is attached to the forward end of the tape, where a tension of 20 pounds is applied. Harry L. Baldwin, Jr., devised an apparatus consisting of a wheel, worked by a lever and held by ratchets in any desired position, for attachment to the spring balance to control the tension and prevent the tape from slipping.

The personnel, required in an ordinary case, was as follows:

1. The chief of the party, who exercised a general supervision over the work, marked the extremities of the tape, and provided the necessary precautions against errors in the measurement.

2. The rear chairman, who adjusted the rear end of the tape to the contact marks and who carried and read one of the thermometers.

3. The head chainman, who adjusted the forward end of the tape, exerted the requisite tension upon it, and carried and read the second thermometer.

4. A recorder.

The base should be measured at least twice, at night or during cloudy days, and the two results compared by sections of 1,200 feet, or four tape lengths. The base, whose two ends must be intervisible, could be nearer 10 miles in length to permit easier expansion. Wherever possible, the base line should form a side or diagonal of a closed quadralateral or pentagonal figure.

Computations included: (1) reduction to standard; (2) correction for inclination, the data for which are obtained by running a line of levels over the base line; (3) the correction for temperature, as steel expands for each degree of temperature 0.0000063596 of its length, and this factor multiplied by the average number of degrees of temperature above or below 62 degrees, which is taken as the normal temperature, gives the proportion in which the base line is to be diminished or extended on account of this factor; and (4) the reduction to sea-level datum.

The diminished cost attendant upon the use of the 300-foot tape made it practicable to measure much longer bases, and to measure many more bases at shorter intervals in the triangulation net.

Traversing with Wheelbarrow

Toward a more accurate alignment of roads on topographic maps, the idea was adopted for experimentation, of using a wheelbarrow, which was often used in measuring roads for county maps in the New England States. The wheelbarrow was an apparatus that only slightly resembled its household namesake. It consisted of a lighter, larger wheel whose perimeter was one half a rod, with tire and axle bearings of brass; an odometer that accurately recorded the revolutions of the wheel; a sight compass mounted on a staff that attained a vertical position when the wheelbarrow was stationary on its two legs; and a box in which was carried lunch, a canteen, notebook, etc. Bearings of foresights and backsights along the road were read on the compass; corresponding distances were noted on the odometer, both being recorded in the notebook so that the traverse could be plotted at night or when rain prevented work out of doors.

Traversing roads with the wheelbarrow was tried out on the quadrangles during the season of 1888 and 1889, by Merrill Hackett, W.T. Griswold, and Charles W. Goodlove. It was recognized as the one undeniably humorous operation in the careers of the three topographers. To the picture of stalwart, dignified Charles Goodlove pushing a wheelbarrow was added, one day, his arrest for trespass. He was traversing along the B and O Railroad near Point of Rocks, Md., when he was stopped by a railroad detective and detained until he succeeded in establishing his identity and convincing the railroad officials that he was not a scout for a rival railroad. It was the age when new railroads were being built by enterprising individuals and small companies in all parts of the United States.

A.E. Murlin bore another distinction: that of having established the only Survey camp in the District of Columbia—near Anacostia. These distinctions deserve mention because they contrasted so vividly with the many subsequent seasons that Mr. Murlin set up camps in the remote mountains of the West where there were few roads and not many trails. His last field assignment, in 1913, was on the Arlington Quadrangle, in the Cascade Mountains of Washington–Oregon; there he had a near serious experience with a bear and endured the maximum of hard physical labor. He retired October 2, 1925, after 42 years of faithful service, and lived to enjoy 20 years of leisure.

Washington Clubs and Societies

The Geological Survey is essentially a Field Service performing original work in various sections of the country. However, the investigations of geologists and ground surveys of topographers required considerable time for completion and preparation for publication. The Survey tried to have the results of its fieldwork in the hands of the public within 2 years. The finishing touches to the field work were performed during the winter season in Washington (figs. 13*A* and 13*B*). Then the field men, who perhaps worked alone in a wilderness area, had opportunities to become acquainted with their fellow workers, to compare notes and practices. Geologists, scientists, and topographers were individual authors with professional reputations for quality and accuracy to be attained and sustained.

There was much to be learned about their work and how it fitted into the work of others. So, meetings of professional clubs and societies were combined with social parties and entertainment.

The Cosmos Club

In Washington, D.C., in 1878, a growing desire among scientific men for the establishment of a social club composed of men interested in science, professionally or otherwise, led to the formation of the Cosmos Club. Henry Gannett was one of a dozen men who assembled at the home of Maj. John W. Powell, on the evening of November 16, 1878, and after a mutual exchange of views, effected a preliminary organization, with Major Powell as temporary president, which, the following January, took on a permanent character.

Membership in the Club was restricted by high qualification requirements and candidates were admitted only if they (1) had performed meritorious original work in science, literature, or the fine arts; (2) though not occupied in science, literature, or the fine arts, were well known to be cultivated in a special department thereof; and (3) were recognized as distinguished in a learned profession or in public service.

Topographers, as authors of particular maps, which they had executed in the field, were eligible for membership under the first classification.

The membership of the Club increased steadily. When it celebrated its 25th Anniversary on November 16, 1903, the membership numbered 567 (408 resident, 159 non-resident), the Geological Survey being represented by 32 geologists and 22 topographic members.

The National Geographic Society

Called by a group of 6 men, among whom were Henry Gannett and Professor Almon H. Thompson, 33 scientists met at the Cosmos Club on January 13, 1888, and organized the National Geographic Society. Its purpose was to establish a bond for the geographers in Washington, and to circulate a technical magazine among professional geographers. After 10 years of struggle and obscurity, the magazine was vitalized by the adoption of Alexander Graham Bell's idea of utilizing it for popularizing the science of geography. It grew into a powerful, popular agency for "the increase and diffusion of geographic knowledge."

In 1914, the Society numbered 350,000 members in all countries of the globe. For years, its income was sufficient to enable the Society to financially aid geographical explorations, as chosen by a Research Committee of which Henry Gannett was the chairman. He was the first recording secretary, and continued to serve the society as manager, treasurer, vice president, and as the fifth president, from 1910 to his death in 1914.

The Triangle Club

The three inspectors lost no time in imparting their mature knowledge of topographic mapping to the corps of topographers. During the field season of 1907, every party was visited by one of the inspectors who spent a day or more with each instrument man, and during the following winter, the inking of the field material was closely watched. There seemed to be a need for supplemental instruction also, for a plan was devised, with the approval of Mr. Marshall, for the formation of clubs where members could exchange ideas, discuss them, and become better acquainted socially.

The first of the clubs was the Triangle Club, composed of assistant topographers and topographic aids. A constitution was drawn up that provided for a president, elected for a term of 1 year. Regular meetings were held on two Monday evenings in the month, in the Director's conference room, or elsewhere on special occasions. The first regular meeting was held on January 27, 1908, attended by about 30, among whom were the inspectors and several geographers. They all approved of the club and believed that unlimited benefits could be derived from the meetings. Until the time arrived for departures on field assignments, club enthusiasm ran high. Many interesting papers were presented and discussions were lively.

The Quadrangles

The ferment that bubbled over to produce the Triangle Club spread to a small exclusive group of five upper classmates who thought it wise to form a higher association to which eventually the youthful members of the junior group might graduate. So Robert H. Chapman, William M. Beaman, Albert H. Sylvester, B. Harvey Sargent, and Francois E. Matthes organized themselves and on January 28, 1908, adopted a constitution, and called themselves "The Quadrangles." The society would devote itself to the study and advancement of the art and science of topographic mapping, and only members of the Geological Survey who received a salary of $1,800 or more would be eligible for membership. The meetings would be presided over by the Chief of Party. The host of one meeting was to be the presiding officer at the next meeting. A recorder would be elected annually, and later the Chief of Party was similarly elected. Wm. M. Beaman was elected the first recorder. Regular meetings were held fortnightly on Tuesday evenings, and the order of proceedings were:

- Reading of notes of previous meetings

- Selection of next camp site

- Addition to membership

- Reports from side camps

- New observations

- Technical discussions

- At 10 p.m., adjournment to the dining tent

Twenty-Year Club

In the winter of 1910–11, the 20-Year Club was formed in the Topographic Branch. Any topographer who had served 20 years or more in the U.S. Geological Survey, the greater part of this time having been spent in topographic work (the beginning of service dating from his first regular employment in office or field, and service in any 1 calendar year, whether continuous or not, being counted as 1-year's service) and who was still with the Topographic Branch, was eligible for active membership; there were 35.

Any topographer with a similar length of service but who had been detached from the organization or Topographic Branch was eligible for associate membership only and was not entitled to vote or hold office; there were 18.

The sole object of the Club was the promotion and perpetuation of good fellowship, which was celebrated in an annual banquet, held at the old Shoreham Hotel, 15th and H streets, the last, or next to last, Saturday night in February.

The gentlemen, attired in tuxedos, were seated about the banquet table in the order of their years of service and dined to the strains of selected music by the U.S. Marine Band. Afterwards, there were toasts, original poems, oratory, and tales of glaciers, torrents, swamps, deserts, and forests encountered and conquered during the past field season in the remotest parts of the United States.

John E. Renshawe, as having the longest service, was the first president, and the annual banquets were continued through 1917. Members included in the 1917 Banquet Programs are listed in table 4.

Our beloved John H. Renshawe paid the following tribute to:

"The 20-Year Topographers"

"A little band of men brought together for the promotion and perpetuation of good fellowship, and for the renewal of happy associations incident to years of companionship in field and office.

"As each new member joins the band, having qualified by his allotted time of service, he brings with him twenty years of progress, twenty years nearer perfection. And so the band marches on, the thinning ranks always renewed by vigorous manhood.

"And where can there be found a finer body of men in all that goes to make up brave and sturdy manhood! Year after year they go out into the world, often into untrodden wilds, and hold themselves ready to meet every danger and to overcome every obstacle that may lie in the way of the accomplishment of their allotted tasks.

"So it must be that all can look backward with self satisfaction for duties well done; and as the end must come for each one, he can, as he goes over the last divide, look back across the dark canyons of hardship and toil and see only the sunlit peaks of good fellowship."

Powell Irrigation Survey

The act approved October 2, 1888, creating the Irrigation Survey, at Major Powell's request, made the money immediately available and he was prepared to begin the special surveys of the western one-half of the United States; an undertaking of unprecedented magnitude, novelty, and urgency. He organized the work under two divisions. The Topographic Survey was to prepare the necessary topographic maps; the Hydraulic Survey was headed by Capt. Clarence E. Dutton, Chief Engineer. The Hydraulic Survey was divided into the Hydrographic Survey to measure stream flow and ascertain the water supply, and the Engineering Branch to locate and design the necessary dams and canals.

The first division (Topographic) of the Irrigation Survey was set up through the reorganization of the Geographic Branch. To the first division was transferred all that part of the personnel of the Western Division, including its chief, Professor A.H. Thompson, which was working within the limits of the arid region, and its force

Table 4. Individuals listed in the 1917 banquet programs.

[--, unkown]

Active Members			
A.E. Murlin, President			

Past Presidents			
Member	**Year**	**Member**	**Year**
John H. Renshawe	1911	Henry Gannett	1914
Hersey Munroe	1913	A.B. Searle	1915
Robert H. Chapman	1912	R.C. McKinney	1916

Glenn S. Smith, Chairman			
1 John H. Renshawe	1879	40 C.F. Urquhart	1887
3 Robert H. Chapman	1879	41 Robert Muldrow	1887
4 Hersey Munroe	1880	42 W.H. Griffin	1888
7 A.B. Searle	1882	43 Glenn S. Smith	1888
10 S.S. Gannett	1882	44 Duncan Hannegan	1888
13 E.M. Douglas	1882	47 R.B. Marshall	1889
14 R.C. McKinney	1882	48 Jas. McCormick	1889
16 Merrill Hackett	1883	49 W.M. Beaman	1889
21 A.E. Murlin	1883	50 Albert Pike	1890
23 Chas. E. Cooke	1884	51 A.M. Walker	1890
25 Geo. L. Hawkins	1884	52 W.J. Lloyd	1890
26 Frank Tweedy	1884	53 J.M. Whitman	1892
27 J.H. Jennings	1884	55 T.G. Gerdine	1893
34 W.H. Herron	1885	56 Oscar Jones	1895
35 W.L. Miller	1885	58 J.E. Blackburn	1895
36 J.H. Wheat	1885	59 E.L. McNair	1896
37 Frank Sutton	1886	61 T. Foster Slaughter	--
38 Basil Duke	1886	62 D.C. Whitherspoon	--
39 C.W. Goodlove	1886	63 J.F. McBeth	1897

Associate Members			
5 H.S. Wallace	1880	24 C.H. Fitch	1884
8 S.A. Aplin	1882	28 E.T. Perkins	1884
9 C.C. Bassett	1882	29 H.L. Baldwin, Jr.	1884
11 H.M. Wilson	1882	30 E.C. Barnard	1884
12 A.P. Davis	1882	31 W.J. Peters	1884
15 A.F. Dunnington	1882	33 Van H. Manning	1885
18 W.T. Griswold	1883	45 R.A. Farmer	1888
19 R.D. Cummin	1883	54 J.R. Ellis	1893
20 Jeremiah Ahern	1883	60 F.E. Matthes	1896
22 D.C. Harrison	1884		

Deceased Members			
17 L.C. Fletcher	1883–1911	6 Henry Gannett	1882–1914
32 W.H. Lovell	1885–1913	57 Sledge Tatum	1895–1916
46 T.M. Bannon	1888–1917	2 W.D. Johnson	1879–1917

was increased by numerous transfers of personnel from eastern divisions.

The Geographic Branch of the Survey became limited to that part of the country lying east of the Rocky Mountains, more specifically east of the 100th meridian, and consisted of the following divisions and chiefs of divisions (table 5):

Table 5. Geographic Branch, Rocky Mountain Divisions and Chiefs.

[Henry Gannett, chief geographer, in charge]

Division	Chief
Northeastern	Marcus Baker
Southeastern	Gilbert Thompson
Central, of four sections	John H. Renshawe
Southern Central, of two sections after May, 1889	Richard U. Goode
Drafting	Harry King

The Topographic Division of the Irrigation Survey consisted of four sections (table 6).

Table 6. Topographic Division Sections.

[Professor Almon H. Thompson, in charge]

Area	Personnel	
1. California–Nevada	Herbert M. Wilson	Sundry Civil Bill August 30, 1890 7 Sections
2. Colorado	Willard D. Johnson	(1) = E.M. Douglas
3. Montana	Edward M. Douglas	(3) = Frank Tweedy
4. New Mexico	Arthur P. Davis	(6) = Kansas–Texas = R.U. Goode
5. Idaho, added in 1889	Wm. T. Griswold	(7) = North Dakota Morris Bien

The topographic mapping of quadrangles that previously has been executed in the arid regions with a view to the problems of irrigation was now done with particular attention to the areas of drainage basins, the position of possible reservoir sites, the location of dams and canal lines, and the altitude, position, and general character of irrigable lands. The mapping was prepared on planetables at the scales of 1 inch equals 1 mile, or 1 inch equals 2 miles.

Though the topographic surveys were to be conducted generally as during preceding years, Major Powell had the following instructions distributed to party chiefs:

Maps.—All field work for the general map to be sketched on planetables on a scale of 1 inch to 1 mile, and in contours having a vertical interval of 100 feet in high mountains, 50 feet in the lower or less rough country, and 20 feet in all areas of possible irrigable lands or sites of possible reservoirs. A larger scale and smaller contour interval might be used for special maps.

The work being in a sparsely settled or entirely uninhabited region, it was necessary for the parties to subsist in camps. The organization for this purpose was nearly the same in all localities. The means of transportation was usually one large four-mule wagon for camp equipage and supplies, and buckboards or saddle animals for the persons engaged in the map work.

Usually each party employed, in addition to the regularly appointed assistants, one or two persons as traversemen or rodmen to assist in the field work. One cook, one teamster, and one laborer usually furnished sufficient force for camp duties.

Horizontal Control.—To be secured by primary triangulation, secondary triangulation, planetable intersections and sketches, and planetable traverses between located points. Primary triangulation to be executed with accurate instruments, and at least two stations should be located on each quadrangle. All roads, streams, and important topographic features must be traversed and sketched by planetable methods.

Vertical Control.—One or more points on each quadrangle to be accurately determined in altitude and used as primary reference, or bench marks, for the area, the "Y" spirit level to be used in determining the gradients of streams, reservoirs sites, etc. Gradienters and altitudes might be used in angular leveling, aneroid and mercurial barometers in traversing.

The Hydraulic Branch of Powell's Irrigation Survey

The Hydraulic Branch of Powell's Irrigation Survey prosecuted its phases of the work with equal vigor. Within a little more than a year it segregated 127 reservoir sites, having an area of over 2,500 square miles, and, in addition, about 46,875 square miles of irrigable land located in five distinct basins—the Snake River, Bear River, Upper Missouri and Yellowstone, Owens River Valley, and the Rio Grande Valley. Vested as it was, for the first time, with powers of direct administration with respect to the public domain, the Survey's sweeping action provoked a widespread protest from persons in the arid region who were adversely affected. As a result, a special committee of the Senate was appointed to investigate the entire subject, and its report severely criticized the policy adopted by the Survey.[17] By the Act of August 30, 1890, the whole of the Act of 1888 was accordingly repealed except as to reservoir sites, the segregation and reservation of which was expressly continued.

The short-lived Powell's Irrigation Survey terminated through congressional disapproval, achieved results of enduring value—topographic maps of the arid regions and stream measurements, the basis upon which the subsequent work of the Survey, in connection with reclamation, was largely based.

---

[17] The report of the committee and hearings, was printed as S. Rept. 928, 51st Cong., lst sess.

A month before the act of Congress that terminated the Irrigation Survey, Capt. (now Maj.) Clarence E. Dutton received orders from the War Department to report to the Ordnance Office on July 23, 1890, terminating his detail of 15 years with the Geological Survey and one of its predecessors. The training camp at Embudo, N. Mex., was closed, and the equipment, mules, and horses in use by hydrographers were transferred to the nearest topographic field parties. Of the nine hydrographers, Thomas M. Bannon, W.A. Farish, A.C. Lane, and Wm. P. Trowbridge, Jr., were transferred to the Topographic Branch and continued in that Branch for a year or more. No fieldwork was carried on during the remainder of 1890 except on the Rio Grande at El Paso, where measurements were made by the local hydrographer.

Stream gaging was too important to abandon, and Frederick H. Newell contrived early in 1891 to secure a small fund from the Topographic Branch. Newell's title was changed at the end of September 1890 from assistant engineer to topographer, and a little later, in order to relieve the strain on topographic funds, he was transferred temporarily (6 months or more) to the Census Office to direct a census of irrigation in the United States. While so employed, he was still able to supervise the small amount of possible stream gaging.

Newell felt that if he was to succeed in obtaining future funds for hydrography it would be necessary to enlist support by showing the value of stream-flow records. The first of his proposed gaging stations in the East was attempted at Chain Bridge on the Potomac above Washington, D.C., April 13, 1891, while an unusual flood was raging, and was successfully established a month later, on May 1.

Through Newell's persistence, untiring zeal, and resourcefulness, a specific appropriation of $12,500 was secured by the Survey in the Act of August 18, 1894, "for gaging streams and determining the water supply of the U.S., including the investigation of underground currents and artesian wells in arid and semi-arid sections."

The item in the bill was originally $25,000, but as appropriations generally were reduced throughout the Government, the cut was neither surprising, nor too discouraging. As small as this amount was, Mr. Gannett deemed it necessary to have it pay for the services of one topographer, and Arthur P. Davis, who was Major Powell's nephew, was selected for the transfer.

The Division of Hydrography remained in the Topographic Branch for the first year and was transferred to the Geologic Branch in 1895. Interest in the value of stream gaging and the demands for records grew apace and appropriations increased to the figure of $50,000 in 1897, and to $100,000 in 1900. Not only did the Reclamation Act of June 17, 1902, greatly increase the Survey work, but in the Sundry Civil Bill, enacted June 28, 1902, the Congress increased the appropriation for stream gaging from $100,000 to $200,000. To meet this situation, the Division of Hydrography was expanded into the Hydrographic Branch, ranking equally with the Geologic and Topographic Branches, "and surpassing them then and at other times in the amount of available funds."

The Hydrographic Branch was divided into (1) the Division of Hydrography, which carried on the stream gaging; (2) the Division of Hydrology, which studied the occurrence of ground water; (3) the Division of Hydro-Economics, which examined the quality of water in its relation to agriculture and industry; and (4) the Reclamation Service, which selected and constructed irrigation projects in accordance with the Reclamation Act, the administration of which was entrusted to the Secretary of the Interior and by him delegated to the Director of the Geological Survey.

Mr. Newell was the logical man to direct the greatly enlarged work of the Hydrographic Branch and was placed in charge of it by the Director, with the title of chief hydrographer. He was also designated chief engineer of the Reclamation Service. In less than 2 years it became apparent to Newell that he could not give the required supervision to the enlarged work of the water resources investigation, and July 1, 1904, Nathan C. Grover was transferred from the field to Washington with the title of assistant chief hydrographer. On July 1, 1906, Marshall O. Leighton, who was specialized in chemistry and biology at Massachusetts Institute of Technology and, at the time was health officer at Montclair, N.J., was appointed chief hydrographer. One of his fist acts was to obtain the approval of Director Walcott to have the name of the Branch changed from "Hydrographic" to "Water Resources," which more correctly defined the character of work performed and which could be more easily understood by the public.

The Water Resources Branch maintained cooperation with the Reclamation Service, the Forest Service, and, to a limited extent, with the Indian Service, all Federal agencies. Beginning in 1894, cooperation had been carried on informally with State engineers and other State officials. In 1907, with an appropriation of $20,000, State cooperation became formal and continuous. The work of the Branch is well told by Mr. Robert Pollansby in his "History of the Water Resources Branch." (He died on July 22, 1952, aged 73, at Denver, Colo.) The First Irrigation Congress was held in Salt Lake City, September 15–17, 1891; the second in Los Angeles in 1893; and the third in Denver in 1894.

Tenth Anniversary

The Geological Survey approached its 10th year with the approval of its plans and accomplishments indicated by enlarged functions and increased appropriations. The appropriation for the fiscal year ending June 30, 1889, reached a total of $737,240 ($605,240 Survey Service Monies, Topography $220,066), and, instead of being granted in one lump sum, was for the first time, itemized under eight heads. The date of the approval of the Sundry Civil Bill (October 8, 1888) is considered the beginning of three branches of the Geological Survey; the Geologic, the Topographic, and the Water Resources[18] (known as the Hydrographic Branch

---

[18] The Reclamation Service became a Bureau March 9, 1907. The Branch moved into the Interior Building, second floor, May 1917.

until July 1906). The name Topographic was first applied to the first division of the Irrigation Survey, while the Geographic Branch, with name unchanged, still conducted the mapping operations east of the 100th meridian.

During the year ending June 30, 1889, the work of the Geographic Branch was carried on in 23 States and Territories. The entire area surveyed was 43,222 square miles; 6,472 was surveyed for the scale of 1:62,500, 26,850 for the scale of 1:125,000, and 9,900 for that of 1:250,000. In addition to the area specified above, the Topographic Division of the Irrigation Survey surveyed 21,000 square miles, which, as a contribution to the general map of the United States, may be added, making a total of 64,222 square miles, and thus completing 99 atlas sheets. The entire number of atlas sheets completed by survey and compilation was 406. Of these, 138 are at the scale of 1:62,500; 207 at 1:125,000; and 61 at 1:250,000.

The area surveyed by the Geographic Branch was distributed as listed in table 7.

Table 7.   Area surveyed by Geographic Branch.

State	Field scale	Publication scale	Contour interval (feet)	Area surveyed (square-miles)
Alabama	1:63,360	1:125,000	100	2,300
Arkansas	1:62,500	1:125,000	50	3,500
California	1:63,360	1:125,000	100	2,200
Connecticut	1:45,000	1:62,500	20	550
Georgia	1:63,360	1:125,000	50	1,500
Iowa	1:31,680	1:62,500	20	1,500
Kansas	1:62,500	1:125,000	50	4,200
Kentucky	1:63,360	1:125,000	100	800
Maine	1:45,000	1:62,500	20	320
Maryland	1:63,360	1:125,000	50	400
Montana	1:126,720	1:250,000	200	4,400
New Hampshire	1:45,000	1:62,500	20	575
New Mexico	1:126,720	1:250,000	200	2,500
New York	1:45,000	1:62,500	20	645
Oregon	1:250,000	1:250,000	200	3,000
Pennsylvania	1:45,000	1:62,500	20	842
Rhode Island	1:45,000	1:62,500	20	1,000
Texas	1:62,500	1:125,000	50	6,250
Virginia	1:63,360	1:125,000	50–100	5,550
West Virginia	1:63,360	1:125,000	100	600
Wisconsin	1:31,680	1:62,500	20	590

State Cooperation

The idea of cooperation in public surveys between the Federal and State governments originated in 1884 (fiscal year 1885), in connection with a plan to make a topographic map of the state of Massachusetts. The work was conducted under the direction of three topographic survey commissioners appointed by the governor. The total cost of mapping the state of Massachusetts was $107,845. This sum, however, does not include the cost of much of the primary triangulation, which was executed by the Coast and Geodetic Survey. The average cost of the completed work was $12.90 per square mile.

At the time the cooperative survey was begun in Massachusetts, the state of New Jersey was engaged in making a topographic map of its area, under the direction of the State Geologist. On July 15, 1884, (fiscal year 1885), the state of New Jersey asked the U.S. Geological Survey to cooperate in completing the map of the State. The Federal Survey took up the work but proceeded on lines different from those on which cooperation had been conducted elsewhere. About half of the State having been accurately mapped, the results were turned over to the Federal Survey, which took charge of the organization and of the personnel and carried the work to completion in 1887.

The State Geologist reported that the total expense of making the topographic survey was $54,744.58. Of this sum, the U.S. Geological Survey expended $35,073.98. The average cost of mapping the State was $6.93 per square mile, exclusive, however, of three-fourths of the primary triangulation, which had been previously executed by the Coast and Geodetic Survey.

The topographic survey of Rhode Island was provided for by an act of the general assembly passed March 22, 1888, and immediately thereafter the governor appointed three commissioners to represent him in the work. Field work was commenced in June 1888, and the survey of the State was completed in the fall of the same year. The total cost of this work was $9,732.51, or about $8.97 per square mile.

Mapping in Idaho

As no topographic work had ever been attempted in Idaho, Wm.T. Griswold was directed to measure a baseline near Boise City. This was accomplished in July 1889. The location chosen for the line was on the mesa near the railroad station south of the city. The ground was first cleared of sage brush and measurements made with a steel tape for a distance of 25,066 feet. Together with E.T. Perkins, Jr., the mapping of the Mountain Home and Camas Prairie Quadrangles, scale of 1:25,500 with 50- and 100-foot contour intervals, was completed by November 1.

Mr. Griswold remained until the end of November to assist Professor R.S. Woodward, Chief Geographer, in determining the geographic coordinates and azimuth of the baseline. In determining the longitude, time signals were exchanged through the courtesy of the Western Union Telegraph Company with the Washington University of St. Louis, through an arrangement with Professor H.S. Pritchett. The latitude was determined by the method of zenith distances, observations being made on 26 pairs of stars on 3 different nights.

Mapping Accomplished (1889–90)

Mapping the western states during the year that mapping began in Idaho are shown in table 8.

Table 8. Mapping completed from October 8, 1888, to June 30, 1890.

[--, no data]

State	To June 30, 1889 (square miles)	To June 30, 1890 (square miles)	Totals (square miles)
California	4,848	3,250	8,098
Colorado	23,200	9,900	33,100
Idaho	--	1,900	1,900
Montana	5,270	1,600	6,870
Nevada	1,892	1,600	3,492
New Mexico	6,370	2,600	8,970
Total	41,580	20,850	62,430

Mapping in North and South Dakota

Field work in North Dakota was accomplished between August 15 and December 1, 1890. Morris Bien organized two level and two transit parties at Minot, and lines were extended primarily to ascertain the position and height of the lowest passes in the divide between the Missouri River and the Mouse and James Rivers. About 730 miles of levels were executed and many bench marks were established for use in the topographic surveys of that region.

As a preliminary to future topographic work in South Dakota, an astronomic station was established at Rapid City. Observations were made by Sam S. Gannett, assisted by Abner F. Dunnington, between October 23 and November 20, 1890. For the purpose of determining longitude, signals were exchanged with Professor H.S. Pritchett, of the Washington University, St. Louis.[19]

Appropriations and Reorganization

The name "Topographic Branch" became established in place of "Geographic Branch" in the reorganization that followed the Act of August 30, 1890. By that act, a portion of the work of the Irrigation Survey was discontinued and the appropriation for topographic mapping was divided equally between the country lying east and west of the 100th meridian. In conformity with this division of the appropriation, the Topographic Branch was organized into the Eastern Division with Mr. Gannett in charge and the Western Division with Professor Thompson in charge. One half of $325,000, a reduction in the amount heretofore appropriated for work on each side of the 100th meridian required an immediate reduction in force, changes in plans, and consolidation of Northern Central and Southern Central Sections into Central Section with Mr. Renshawe in charge. On November 1, 1890, Marcus Baker was made the General Assistant of the Director and H.M. Wilson was placed in charge of the Northeastern Section.

Appropriations continued to be made for specific purposes. For the fiscal years 1892 and 1893, two other meridians besides the 100th were made dividing lines. For the latter years:

1. $90,000 for mapping east of the 97th meridian;

2. $90,000 for mapping west of the 103d meridian; and

3. $60,000 for mapping west of the 97th meridian in the States of North Dakota, South Dakota, Nebraska, Kansas, and the Territory of Oklahoma. The $60,000 was divided equally between the Eastern and Western Divisions, and each Division had to keep separate accounts of the two funds.

In addition, the seven statutory salaries (formerly aggregating $16,200) were reduced to four (aggregating $9,200): a chief geographer, a geographer, and two topographers.

Three of these persons to be in the Eastern Division—$7,200, and one of these persons to be in the Western Division—$2,000.

The Eastern Division remained unchanged with three sections, but the Western Division consisted of seven sections: Southern California, Texas, central California, Washington, Idaho–Oregon, Wyoming, Colorado–South Dakota.

The Cuvier Prize

The Geological Survey was accorded a high honor when the Academy of Sciences of the Institute of France, the foremost scientific body of that nation, awarded to it the Cuvier Prize for the year 1891. Commonly, the prize was awarded triennially for noteworthy work in science to investigators in France or neighboring countries, but in 1891 the Academy chose to honor collectively the scientific corps of the Geological Survey for its contribution to geological, botanical, ethnological, and geographical knowledge of large areas of the United States. The citation included the mapping operations of the numerous personnel of topographers and engineers under the direction of Mr. Henry Gannett, and mentioned that more than 600 maps had been surveyed and drawn, of which about 400 already had appeared.

---

[19] U.S. Geological Survey 12th Annual Report, 1890–91, p. 49.

The value of the Cuvier Prize was 1,500 francs. Since the acceptance of pecuniary gifts by Federal employees is prohibited by law, the Academy substituted a gold medal, which was accepted, and the splendid souvenir is preserved in the Survey library.[20]

Primary Traverse

The determination of the geological positions of the two, or more, points required for the minimum horizontal control for a quadrangle was effected by triangulation. However, triangulation was not practicable in certain types of country—hilly country of low relief whose summits were covered with timber, and broad flat prairie lands. Extensive clearing for sight lines and the erection of tall towers for instrument stations, if not prohibitive, were expensive and time consuming.

Such types of country were encountered by Mr. George L. Hawkins in western Arkansas. During 1889, 1890, and 1891, he extended triangulation from Little Rock northward to the Missouri state line, thence westward to near the Indian Territory line, thence southward, and tied to triangulation near Fort Smith, Ark. It was intended that this belt of triangulation be tied to the Indian Territory line, but about 50 miles before this line was reached, timbered country intervened. Mr. Hawkins, having experience with railroad survey methods, devised a plan of running an accurate transit traverse line from the most westerly triangulation station along the St. Louis and San Francisco Railroad to the Indian Territory line. This line, known as a "Primary Traverse Line", is believed to be the first horizontal geodetic control of this character to be used by the Geological Survey, or by any other geodetic organization, and from this beginning the Survey developed a system of transit traverse that produces the entire horizontal control for many quadrangle maps. Such control in the prairie states consists of a line around the perimeter and one across the center of a quadrangle.

The instruments used were the theodolite, the same as for triangulation, having a circle 6 to 8 inches in diameter upon which angles could be read by vernier to ten seconds, and the transit which shortly superseded the theodolite; one 300-foot steel tape graduated to feet at either end; one spring balance for assuring a 20-pound tension on the tape; one 100-foot steel tape for the measurement of short lengths on slopes; two thermometers, as the temperature of the tape was required to be read and recorded at least once per hour; and four hand recorders, two flag poles; and one good watch, which had to be compared with standard time often enough to determine its error within ten seconds.

A line of transit traverse consists of a continuous series of straight courses, the length of each course being measured with a steel tape, and the angles between them being measured with a transit. The measured angles, and observations upon Polaris, preferably at elongation, at intervals of 10 to 20 miles,

depending upon the number of courses to the mile, provide the means of determining the azimuths, or bearings, of all courses. With the lengths and bearings of the courses, the increase or decrease in latitude and in longitude may be computed for successive parts of the line. Then, with the line originating from a triangulation station, or possibly an astronomic station, and closing upon a similar station, the positions of all points established by the line can be computed.

The lines followed railroads, wherever practicable, in order to make use of the tangents and also obtain the easiest grades and thus avoid errors incident to slope. The record of each day's work contained the location of each railroad station, mile post, and switch passed; every wagon road, stream, land or county line crossed; and connection with corners of the public land surveys in order to establish as close a relation as possible between the traverse line, which served as ultimate control, and the township system of surveys, which served as secondary control. At first it was required to locate by intersection, prominent houses, and natural features of any kind in sight from the line, as additional useful locations for the topographers who were to make the map.

Primary traverse parties consisted of: one chief, as transit man; one recorder; two tape men, either of whom could act as front or rear flagmen, and one flagman; also a cook and teamster when camping was necessary. In the early 1890's primary traverse parties fairly raced along the railroads and main wagon roads, bent on supplying control for topographic maps of small scales, without much thought of tape tension, temperature, or permanent marks. A speed of 15 to 18 miles a day was usual, and there were occasional days when as much as 22 miles of traverse were completed. During the first 6 fiscal years, the linear miles of primary traverse run were as follows:

1890—758 1892—480 1894—633
1891—362 1893—680 1895—500

The figures are estimated from the description of topographic mapping done in those years.

In later years when greater care was observed, and permanent marks established, transit men aspired to accomplish records for speed. The following records are selected from the notebooks:

- E.L. McNair ran 22 miles in 1 day in 1917 along the Atlantic Coastline Railroad between Savannah and Jacksonville (tangent all the way). He was first to discover for the Survey that observations on Polaris could be made in daylight.

- H.S. Senseney ran 12 miles, with two observations on Polaris, over a dirt road in Illinois in 1 day in 1922. S.E. Clement was recorder. He made 100 setups in about 5.5 miles in Wisconsin in 1 day in 1919; a record. R.H. Runyan was recorder. He was the first to introduce the standard stadia rod into a traverse party in 1916, to check against gross errors in chaining. Before then, distances were checked by readings on the range rod or by pacing.

---

[20] U.S. Geological Survey Annual Report No. 13, p. 61–66.

First Map Revision

It will be remembered that Captain Wheeler declared that many areas would require remapping every 25 years in order to keep pace with development. The Survey found that in the much shorter space of 10 years, portions of the Appalachian Mountain Region had developed so rapidly as to render it advisable to undertake revisions, more or less extensive, of parts of the following quadrangles, which were mapped on the scale of 1:125,000 (table 9).

Table 9.   Appalachian Mountain Region mapped at 1:125,000.

Quadrangle	States	Year mapping begun	Topographers
Pocahantas	Va.–W. Va.	1883	Morris Bien and W.A. Shumway
Tazewell	Va.–W. Va.	1883	Morris Bien and W.A. Shumway
Abingdon	Va.–N.C.	1882	Morris Bien and W.A. Shumway
Wytheville	Va.–N.C.	1883	Morris Bien and R.C. McKinney
Cumberland Gap	Va.–Ky.	1882	Frank M. Pearson and Fred J. Knight
Jonesville	Va.–Ky.	1882	Frank M. Pearson and Fred J. Knight
Cranberry	N.C.	1882	Morris Bien, Chas. M. Yeates, and J.W. Hays
Nantahalah	N.C.	1884	Chas. M. Yeates
Knoxville	Tenn.	1884	Frank M. Pearson and C.G. Van Hook
Maynardville	Tenn.	1883	Frank M. Pearson and C.G. Van Hook
Murphy	Tenn.	1884	Frank M. Pearson amd Chas. M. Yeates

These quadrangles were the first to be mapped in the east when it was necessary for the Survey to produce maps in the shortest possible time at a minimum cost, at the risk of frequently rough and sketchy work. The scale of the surveys was small, often too small for certain areas, but the small scale, less accurate and less expensive maps, enabled the Survey to make a showing that won the support of the public and convinced Congress that continuing appropriations were justified.

At the end of 10 years, the requirements of the geologists and general public had increased to force the adoption of larger scales for the maps in many localities. The scale of 1:250,000, or about 4 miles to 1 inch, had been found not suitable for the areas under survey, and was practically abandoned. At the same time an increasing amount of work was being done on the mile scale (1:62,500) and a diminishing amount on the 2-mile scale (1:125,000). Thus while the number of maps produced per year was maintained, the area surveyed diminished. In other words, a smaller area was being mapped in greater detail. The trend is shown in table 10.

Table 10.   Mapping trends.

[--, no data]

Year ending June 30	Percent of total area on scale of:			Square miles mapped	Appropriations
	1:62,500	1:125,000	1:250,000		
1883	--	27	73	15,000	70,700
1884	1	46	53	45,150	[a] 100,150
1885	4	63	33	57,500	[b] 160,865
1886	5	72	23	81,829	[b] 184,510
1887	9	66	25	55,684	[b] 204,656
1888	7	67	26	52,062	202,311
1889	10	70	20	64,472	[b] 215,200
1890	17	83	--	67,650	[b] 216,200
1891	26	74	--	67,575	[b] 345,773
1892	15	85	--	53,000	[b] 266,200
1893	73	27	--	26,000	[b] 249,200
1894	66	31	3	35,650	[b] 209,200

[a] Includes $16,200 for Scientific Assistants, a continuing item, but after 1892 the amount was reduced to $9,200.

[b] These amounts were increased by State cooperative funds.

Manual of Topographic Methods

A "Manual of Topographic Methods," 300 pages, by Henry Gannett, appeared in early 1894 as Monograph XXII, intended primarily for the information of the men engaged upon that work. The manual opened with an account of the various surveys made under the general and state governments and by private parties, followed by a statement of the plan of the atlas made by the Geological Survey, its scales, contour intervals, size of sheets, contents, etc.

The work of map making is classified into two parts, the geometric control and the sketching, and the principal instruments and methods used in the various steps are fully treated. The chapter devoted to sketching contains a brief outline of the origin of topographic forms.

Following the text is a body of tables for use in the various computations required in the work, and the book is illustrated with cuts of instruments and with specimen maps.[21]

---

[21] U.S. Geological Survey Annual Report No. 15, 1893–94, p. 95.

Walcott Becomes Director

Major Powell resigned in May 1894, to take effect at the end of the fiscal year, thus terminating a service of 13 years as Director. His successor, taking office on July 1, 1894, was Charles D. Walcott, a geologist with the Survey since its organization.

As Director, Mr. Walcott proposed to make some readjustments in Survey policy in order to meet new conditions and bring the Survey more in touch with the economic and educational interests of the country. These adjustments, briefly outlined, were: (1) raising of the standard of quality of the topographic maps and the appointment of an editor, Marcus Baker; (2) the representation on all future maps in the public land states of the land subdivision lines, and, when possible, the location and elevation of township and section corners; (3) the resurvey of areas on existing maps that have proven defective or inadequate for the proper representation of a real and economic geology; (4) the planning of a system of triangulation to effectively supplement what had already been accomplished; (5) the placing of the entire topographic force within the classified service; (6) the obtaining of authority from Congress to print and sell topographic maps, with texts, for educational purpose, and to print more scientific reports for wider distribution of valuable information; (7) the enlargement and increase of the work of the Division of Hydrography and of the Division of Mineral Resources; and (8) the making of reconnaissance surveys, such as those made on the scale of 1:250,000 (restored) of regions supposed to contain important economic resources, in order to obtain information more quickly.

To emphasize the urgency and importance of geodetic control, the topographic work was reorganized. The two independent divisions under which it had been carried on were consolidated, retaining the name Topographic Branch and Henry Gannett as Geographer-in-Charge. The branch was organized by class of work into two divisions, Triangulation and Topography, and by locality of work into four sections as shown in table 11.

Table 11.   Topographic Branch Organization.

Location	Geographer-in-charge
Atlantic Section	Herbert M. Wilson
Central Section	John H. Renshawe
Rocky Mountain Section	Edward M. Douglas
Pacific Section	Richard U. Goode

Division of Triangulation

George T. Hawkins extended triangulation for control of quadrangles in Maine and Vermont, and under detail to the General Land Office, reran and double chained the north boundary of Nebraska as an excellent means of control of the system of land surveys in its vicinity, since the township corners were connected to the horizontal control.

Sam S. Gannett established, by triangulation, control points for quadrangles in New York and Pennsylvania, and, by primary traverse, control of the Monroe Quadrangle in Georgia. He also established an astronomic station in Eugene, Oreg., determining the longitude by telegraphic exchanges with the U.S. Coast and Geodetic Survey in San Francisco.

Professor Almon H. Thompson established 24 triangulation stations north and south of the Arkansas River east of the mountains in Colorado.

William T. Griswold measured a baseline in the neighborhood of Eugene, Oreg., 18,630 feet in length, and expanded southward to control 3 quadrangles by occupying 14 stations.

Division of Topography

In the Atlantic Section during the year ending June 30, 1895, 11 topographers with assistants, mapped 7,000 square miles; in the Central Section, 6 topographers mapped 10,475 square miles; in the Rocky Mountain Section, 9 topographers mapped 5,963 square miles; and in the Pacific Section, 11 topographers mapped 6,975 square miles.

The following year Mr. Gannett was assigned to other duties and Director Walcott took active charge of the Topographic Branch, continuing as chairman of the topographic committee, composed of the chiefs of the four topographic sections and of the new Indian Territory Section.[22]

Indian Territory Surveys (1896)

Indian Territory had been attracting more attention for a decade than any other western state or territory when an appropriation of $200,000 was made by act of Congress, approved March 2, 1895, for the survey and subdivision of the lands of the Indian Territory with the provision that the Secretary of the Interior might entrust this work to the Director of the Geological Survey, instead of giving it out by contract as had heretofore been done.

In order to allow for expansion westward, southeastern states prevailed on the Government to enter into treaties with the Cherokee Indians (1817) and the Choctaw Indians (1820) to vacate their lands and migrate across the Mississippi River to new lands. Similar treaties with other tribes were drawn up, until by 1840, the Five Civilized Tribes (Cherokee, Choctaw, Chickasaw, Creek, and Seminole) were established west of the state of Arkansas. Tribes of the Plains Indians were allotted lands farther westward and an effort was made to create an Indian State.

However, the Indians were not populous and their broad undulating prairies, used for little else than hunting grounds, were coveted by frontiersmen and home seekers. The most

---

[22] See #17 for Permanent Bench Marks.

vociferous of these were known as "Boomers." After a commission had studied the situation and made recommendations, the Congress passed the Dawes Act in 1887, providing for breaking up the Indian Reservations into individual allotments and opening the surplus lands to white settlement.

There followed one of the most thrilling events in the history of the West. At noon on Monday, April 22, 1889, the signal was given and the "Run" for townlots and farms in the "Oklahoma lands," purchased by the government from the Creeks and Seminoles was made by thousands of home seekers and speculators who had been restrained for days and weeks behind cordons of soldiers. Those who got across the starting lines secretly before the signal were known as "Sooners." On that eventful day, many towns and cities were founded, including Oklahoma City, the future Capitol of Oklahoma.

Other runs occurred as fast as Indian tribes accepted allotments in severalty and the surplus lands were opened to settlement. The Iowa, Sac and Fox, and Shawnee-Potawatomi Reservations to the east of "Old Oklahoma" were opened on September 22, 1891, and the Cheyenne–Arapaho lands to the west in 1892.

The greatest run of all occurred on September 16, 1893, when the Cherokee Outlet, together with the Tonkawa and Pawnee Reservations, were thrown open. The small Kickapoo Reservation followed in 1895.

Six years later on July 4, 1901, the Osage, Kaw, and Comanche–Kica Lands were opened, this time by lottery instead of a run. Drawings were made for 6,500 claims. Settlements prospered, and on November 16, 1907, Indian Territory, the area occupied by the Five Civilized Tribes, and the territory to the west were admitted as Oklahoma, the 46th State.

Plans and organization for the survey of Indian Territory were completed by March 21, 1895. The project, possessing the pleasing promise of topographic mapping as an inexpensive corollary to the cadastral surveys, was given to the Topographic Branch of the Geological Survey. A new unit was organized as a division with Charles H. Fitch in charge, and later Van H. Manning became second in charge.[23]

The requisite animals and camp equipage were purchased and shipped to South McAlester, in the Choctaw Nation, the headquarters selected for the work. Towards the close of March, two parties were placed in the field, one under George T. Hawkins to run the Indian Base Standard Parallel eastward 50 miles from the Chickasaw Nation to near the Arkansas state line, the other under Wm. J. Peters to run the Second Guide Meridian southward to near the Texas state line. Thereafter, these parties continued the work of running standard lines.

Triangulation parties, under Samuel S. Gannett and Charles F. Urquhart, were soon in the field to extend primary triangulation over the area subdivided, and to connect their stations, permanently marked, with section and township corners as a check upon the running of the lines and as a means

of recovering lost corners. Other parties were added for other places of the work until, by September 1st, a maximum force was engaged.

In addition to those mentioned above, two parties were engaged in surveying township exteriors. Sixteen other parties were engaged in running subdivision lines. These 16 parties, the surveyors of which for the most part were men temporarily employed for the purpose, were grouped by fours, each group being under the immediate supervision of a topographer, who supervised and examined their work and at the same time prepared a topographic map of the country subdivided. Thus the work was executed by three independent sets of surveyors, the work of each of them testing that of the others.

In accordance with the requirement, made in the act providing for the topographic surveys for the fiscal year 1896–97, that two permanent bench marks (one in forest-clad and mountain areas) be established in each township or equivalent areas, level parties were put to work extending lines of accurate elevation over all township exteriors and establishing permanent bench marks at the corners of townships.

The topographers who had charge of subdivision parties during the first year, in addition to Van H. Manning, were Charles E. Cooke, Wm. B. Morse, Robert A. Farmer, R.H. McKee, Dabney C. Harrison, Wm. S. Post, and R.M. Towson. They prepared a topographic map and had charge, also, of camps where the men were sheltered in tents, where cooks prepared and served food, and where teamsters took care of the mules.

The method of mapping the topography was, first, to prepare on drawing paper, 15 by 15 inches square, the plat of a township 6 miles by 6 miles, on the scale of 2 inches equal 1 mile, and enter upon the section lines all crossings of roads, streams, and ridges as described in the notes made by the line surveyors. This was done nightly by the two topographic assistants or aides assigned to the topographers in charge of each camp. With the prepared township drawing mounted on a 15 by 15 inch planetable, the aides, one usually assigned to the eastern one-half of the township, and the other assigned to the western one-half, astride mules, would ride along the newly-run and trimmed-out section lines, and add the necessary topographic details.

Section and half-mile corners recently established, were clearly visible and other details entered on the lines were usually easy to recognize. The aide determined contour crossings by reading an aneroid barometer that had been set in the morning to agree with the elevation established at camp by a line of Y-levels run from the nearest railroad, or by comparison with elevations marked up on the exterior lines of the township, likewise established by Y-levels.

The contour interval was 50 feet. Where the terrain had considerable relief, and there were many details to sketch, the aide dismounted and measured distances by pacing. He was warned at the outset that distances were not to be obtained by counting the paces of a mule because that cagey critter traveled away from camp with a reluctant, shortened step

---

[23] Seventeenth Annual Report (1895–96) of U.S. Geological Survey, General Land Office Report for 1895, p. 106.

and towards camp, wherever it might be, with a springy step several inches longer.

The topographer was constantly busy. With a similar township plat bearing the line crossings, he drove over the roads with a buckboard and team of mules. His equipment consisted of a planetable, tripod, aneroid barometer, and alidade, but no stadia rod. Setting up at the intersection of the road with the section line, he traversed parts of the road, measuring distances by counting the revolutions of a front buckboard wheel, to which a rag or twig had been tied, and obtaining elevations by aneroid or by reading a vertical angle to a chosen spot on the road, to which he had, or would have, the distance. He surveyed parts of section lines, measuring the distances by pacing and determining elevations by gradienter or vertical angles, as a check upon both the line surveys and the traverses made by the aides. Responsible for the execution of the work in his allotted territory, he looked to the line surveyors to set the pace, and when they had completed the subdivision of a township, or a convenient group of townships, he directed that camp be moved to a new, selected location. Unmapped sections left behind required longer rides morning and evening by the aides.

The running of lines of levels was not commenced until the appropriation for that work was made available on July 1, 1896, and the office work connected with the field surveys presented problems. The first draft of the line notes from the surveyor's rough notes was prepared at South McAlester, the field headquarters, and its receipt in Washington controlled the preparation of the three typewritten copies for filing with the Land and Indian offices. The execution of township plats and topographic maps was retarded while awaiting the adjustment of spirit level lines and their connection with railroad or control lines. Improved organization accelerated the office work and a further helpful step was taken when the work that was being done in Washington was transferred to South McAlester in December 1896.

The Indian Territory project was extended when an appropriation of $141,500 was provided in the Indian Act, approved June 6, 1897, for the resurvey of the lands of the Chickasaw Nation, which had been sectionized several years previously by private surveyors under contract. To be more conveniently located for this work, and for the completion of the Denison Quadrangle, the larger part of which lay south of the Red River, the general headquarters were moved about the middle of June, from South McAlester, Indian Territory, to Denison, Tex.

The field surveys of Indian Territory were completed by June 30, 1898, with the results accomplished shown in table 12.

The completion of the office work at Denison, Tex., on December 31, 1898, brought to a close the survey of the Indian Territory, a highly profitable project, for the land surveys had been made at an estimated savings of $85,000, as derived from the comparison with the probable cost of such surveys by private surveyors under the contract system. Indeed, the contract system was thereafter abandoned by the General Land

Table 12. Field surveys completed June 30, 1898.

[--; no data]

Activity	Indian Territory	Chickasaw Nation
Standard lines run	1,931	532
Township exteriors run	6,181	1,863
Section and meander lines	49,961	12,554
Retracement of lines	--	188
Number of townships subdivided	730	--
Topography mapped (square miles)	23,533	7,352
Number of 30-minute quadrangles mapped	26	--
Level lines run along railways and township lines	2,813	2,316
Permanent bench marks established	814	--
Vertical angle lines, miles	10,321	2447
Triangulation stations occupied	138	--

Office after it had been demonstrated that government personnel performed accurately and efficiently in running lines and marking corners.

Without additional cost, the Geological Survey established 138 geographical locations by triangulation and 814 permanent bench marks by Y-leveling, prepared 26 topographic maps for publication on a scale of 1:125,000, and acquired valuable camp equipage consisting of tents, wagons, mules, and sundries.

Had the project been one-half or one-quarter as large as it was it would not have proved a financial success for the factor of travel expense would then have become relatively larger. As it was, the large Indian Territory project met with considerable added expense for travel and long moves of parties in reorganization when the progress of the survey was interrupted and furloughs became necessary by the lapse of funds for the periods June 1 to June 10, 1896, and April 17 to June 6, 1897.

The Indian Territory Survey enrolled a number of young men who continued on careers of a lifetime with the Geological Survey:

John E. Blackburn—Teamster, chainman, topographer
Pearson Chapman—Draftsman, topographer
Robert Coe—Levelman, topographer
Acheson F. Hassan—Levelman, cartographer
Francois E. Matthes—Draftsman, topographer, geologist
Sledge Tatum—Surveyor, topographer, geographer

Leveling—Wilson–Douglas (1896)

An important change affecting the work of the Survey and precipitated by the surveys in Indian Territory was the enactment by Congress of legislation providing for the monumenting of the topographic surveys. The provision was embodied in the Sundry Civil Appropriation Act, approved June 11, 1896, and was in the following words:

"Provided, that hereafter in such surveys west of the 95th meridian elevations above a base level located in each area under survey shall be determined and marked on the ground by iron, or stone, posts, or permanent bench marks (fig. 14), at least two such posts or bench marks to be established in each township or equivalent area, except in the forest-clad and mountain areas, where at least one shall be established, and these shall be placed whenever practicable, near the township corners of the public land surveys; and in the areas east of the 95th meridian at least one such post or bench mark shall be similarly established, in each area equivalent to the area of a township of the public land surveys."

The enactment above quoted did not provide for the protection of the monuments thus to be erected from removal or defacement, but such legislation was incorporated in the Indian Appropriation Act, approved June 10, 1896, in which occurs the following clause:

"Provided further, that hereafter it shall be unlawful for any person to destroy, deface, change, or remove to another place any section corner, quarter-section corner, or meander post, on any Government line of survey, or to cut down any witness tree or any tree blazed to mark the line of a Government survey; that any person who shall offend against any of the provisions of this paragraph shall be deemed guilty of a misdemeanor and, upon conviction thereof in any court, shall be fined not exceeding $250, or be imprisoned not more than one hundred days."

The statutory authority for improving the topographic mapping as granted in the above legislation had been strongly advocated by Director Walcott. The original object of the topographic map was to serve as a basis for the geologic map of the United States, but, with the progress of the Survey from year to year, the public had become more and more acquainted with the maps, and a strong demand had developed for them. The value of the maps, steadily increasing as the quality of the surveys improved, might be enhanced 100 percent if permanent records were left on the ground, in the form of monuments to show the position of triangulation points and township corners, and their true elevation above sea level, or some fixed point. The utility and practical value of any survey is largely dependent upon the measures taken to preserve its results and to render possible the ready identification and practical use of these results.

The geologist often found it difficult to determine the exact location on his map of an important geologic feature. In the mountain regions, sharp summits of hills or mountains that have been used by the topographers as triangulation points could be used by a geologist to locate a given point, if a monument had been erected and was still standing. Less accurate locations could be obtained by the geologist from intersections of streams, valleys, and ridges, or of artificial features, such as roads, buildings, etc. On the Great Plains area his only

recourse might be to township or section corners, when they could be found. The stake, or stone, over a buried handful of charcoal, between two shallow holes had often been destroyed, obliterated, or moved, and had an unknown error of location, as it was not connected with any system of measurement.

In mining geology, a large part of the observations of the geologist are made underground, and their location is referred to a single point on the surface. The advantages would be tremendous for surface observations as well, if the point possessed exact geographical position and elevation and could be identified on the ground without any doubt. Such a point possessed, also, inestimable value to the surveyor and engineer as a starting or closing point for a project.

The extending level lines became, in 1896, an established phase in the basic control of topographic mapping. Permanent bench marks were set at stated intervals and, during 1896 and 1897, consisted of a copper bolt 4 inches long, and 1 inch in diameter fastened into masonry or solid rock, by driving it on a brass wedge placed in the bottom of a vertical hole, so that the top of the bolt was horizontal, and thus formed the bench mark. Triangulation stations were also marked in this manner.

In 1898, two new disks were adopted, each presenting a circular plate of bronze or aluminum, 3.5 inches in diameter, and 1/4-inch thick, appropriately lettered. One disk had a 3-inch stem for cementing in a drill hole, generally in the vertical walls of public buildings, bridge abutments, or other substantial masonry structures, the other type consisted of a hollow wrought iron post 4 feet in length and 3.5 inches in outer diameter, split at the bottom and expanded to 12 inches, for planting in the ground and resisting both easy subsidence and malicious extraction. The bronze cap was riveted to the top of the post (fig. 15).

On the tablet was stamped the elevation above sea level, to the nearest foot. Engineers and others finding these bench marks could obtain, if desired, the accepted elevation to the hundredth, or thousandth of a foot by communicating with the Director. In the Washington office were prepared lists of bench marks with their descriptions and elevations. As lines of levels were closed in circuits, adjustments were made in elevations. The starting points of the lines had been selected in the field as being nearest to the quadrangles to be mapped, in the hope that the elevations would be ready in advance of the topographers. To aid them in properly placing contours, temporary bench marks were set at 1 mile intervals between the permanent posts or tablets, which were set between 5 and 6 miles apart.

Instructions for spirit leveling were issued in 1896. The work was conducted as part of the topographic survey and was a part of the responsibility of the topographer in charge of the mapping. The work of the first year (July 1, 1896, to June 30, 1897) is summarized in table 13.

The results of spirit leveling first appeared as a part of Annual Report No. 18, 1896–97, and after 4 years of similar treatment, in 1901, were issued as Bulletin No. 185. Triangulation and leveling in Indian Territory 1895–98, were treated for publication as Bulletin No. 175, dated 1900, and authored by

Table 13.   Spirit leveling completed July 1, 1896, to June 30, 1897.

[BM, bench mark; NA, not applicable; average cost of the leveling was $4.75 per linear mile; average cost of the topography was $1.40 per square mile; percent of country considered covered by topographic surveys was 25 percent]

Section	Number of states	Number of level parties	Linear miles	Square miles topography	Number of BMs
Atlantic Section	9	9	2,245	7,456	338
Central Section	7	7	1,622	6,144	227
Rocky Mountain Section	5	6	2,005	3,120	386
Pacific Section	4	5	794	3,048	169
Subtotals	25	27	6,666	19,768	1,120
Indian Territory	NA	NA	4,174	7,678	700
Total			10,840	27,446	1,820

C.H. Fitch. In 1906, Bulletin No. 281 was published, containing results of spirit leveling in New York State for the years 1896 to 1905, inclusive, with values conforming to the readjustment, in 1903 of the precise level net in the eastern portion of the United States by the U.S. Coast and Geodetic Survey. The adjustment and readjustment of levels became a function of the Triangulation and Computing Section and was first performed by Harry L. Baldwin, Jr. and Clyde B. Kendall under the supervision of the chief, Sam S. Gannett. As levels were adjusted to conform to the 1903 readjustment of the precise level net the results were summarized by states, or groups of states, and published from time to time in bulletin form.

In 1878, the U.S. Coast and Geodetic Survey was authorized to run first order leveling, which was adjusted in 1929. Permission was given December 22, 1920, by the Treasury Department to put bench marks on Federal buildings. Hodgeson's letter about concrete collars for bench marks was issued on July 26, 1940.

The following facts and performances of Survey leveling are selected from the annual reports and field notebooks:

- The start of Survey leveling in 1896 was made by A.B. Pomme on June 2 near Moravia, N.Y.; followed by E.L. McNair on June 3 at Medina, N.Y.

- L.F. Biggs extended 20 miles in one 10-hour day in 1907 on the railroad between Las Vegas and Rhyolite, Nev. He used a prism level, and was assisted by B.A. Jenkins, recorder, Bayard Knock and Jake W. Muller, rodmen, and a man to set bench marks ahead. Mr. Biggs again extended 20 miles in one day in 1912 on the Norfolk and Western Railroad in Virginia.

- H.S. Senseney recommended, while at Ludowici, Ga., in 1917, that reference bench marks be set near every permanent bench mark, and the idea was incorporated into the regulations.

- R.G. Clinite extended 12.2 miles of levels, making 124 setups, in 5-3/4 hours on June 22, 1926, along the C&0 Railway, on the Coveville Quadrangle, Va.

- E.L. McNair, in 1925, leveling forward and backward over a course of 9.6 miles to the summit of Mount Washington, N.H., (elevation 6,288 feet) with 9 bench mark tablets that had been previously set, and establishing 34 temporary bench marks during the progress of the leveling, attained the record in table 14.

Some of the lines extended were so-called fly-levels, or fourth order levels. These were extended over the roads of a quadrangle for the use of the topographer. The levelman was allowed sights in excess of 300 feet and determined the elevation of road forks, railroad crossings, water surfaces, high and low points and contour crossings of roads, etc., painting such elevations with white paint on fence posts, telephone poles, and trees. The topographer, often with the roads already traversed and transferred to his map sheet, would drive over the roads with the planetable on his lap and sketch in the contours. He determined his locations and distances by counting the revolutions of a rag tied to a front wheel of the buggy. To fill in areas away from the road he had to walk, counting paces for distances and using a hand-level and aneroid for elevations.

Table 14.   Record set by E.L. McNair, 1925.

[Equipment: U.S. Coast and Geodetic Survey precision level and invar rods graduated in meters]

Total rise and fall	9,846 feet
Total setups	1,128
Total time	67 hours, 40 minutes
Average number set ups	16.7 per hour

The method of sketching contours while seated in a buggy did not possess the apparent luxurious ease. In front of the seated topographer was a harnessed horse who perhaps could not stand still, and in fly time, May to September, would keep up a continuous tap dance. When it came time to start off, the horse might have both reins clamped under his tail as tight as a vise.

Distances up to 15 miles were occasionally extended in a day by a fly-level man. Topographers sometimes sketched the contours on a 15-minute quadrangle in 2 days. Open, nearly flat country, such as found in the Mississippi Valley States, lent itself best to this method. In the summer of 1907, C.L. Sadler worked a quadrangle in Oklahoma on the fly-level, buggy-wheel-traverse method and H.H. Hodgeson mapped an adjoining quadrangle by the planetable stadia method. The relative merits and costs of the two methods were discussed that winter at a Triangle Club meeting.

Mapping Forest Reserves—Wilson–Douglas, 1897 (Henry Gannett)

With the Indian Territory project (1895–98) nearing a highly successful conclusion, the Geological Survey was soon to be engaged in another special task, the mapping of the forest reserves. The Sundry Civil Act, approved June 4, 1897, contained a provision for the survey of the forest reserves and the establishment of a forest policy on the part of the government. Since June 1883, the Geological Survey had been collecting forest statistics and overprints showing forests areas on every quadrangle.

Movements in favor of protection for the forests resulted in an act of Congress approved March 3, 1891, authorizing the President, by proclamation, to set aside and reserve any part of the public lands, wholly or in part, covered with timber or undergrowth. The first Forest Reserve (later to be called the Shoshone National Forest, in Wyoming) was established by President Harrison that same year, and by September 28, 1893, 17 forest reservations were established, aggregating 27,343 square miles in 7 western states and territories. Data accumulated by the Geological Survey helped to determine the areas and the boundaries of the new reserves. Their establishment attracted little attention and created little or no opposition, since no real protection was afforded and the wasteful methods of timber cutting and destructions by fires went on within their limits, as elsewhere.

The Secretary of the Interior, early in 1896, had requested the National Academy of Sciences to study forestry matters and to make recommendations. A committee of five, one of whom was Arnold Hague, geologist of the Geological Survey, was appointed and began work on July 2, 1896. In accordance with the recommendations of the committee, on February 22, 1897, 13 additional reserves were established by Executive Order, containing an aggregate area of 34,918 square miles.

As was anticipated, the establishment of these reserves produced a strong protest from the residents of the states interested, resulting in the provision incorporated in the Sundry Civil Act approved June 4, 1897, suspending until March 1, 1898, the Executive Orders and proclamations of February 22, 1897, and providing for the examination and survey of the reserves by the Geological Survey during the intervening time, $150,000 being appropriated for this purpose. The object of this was to obtain for the use of the Department, and the President, the necessary information for revising the boundaries of the reserves, and subtracting from them such areas as were found to be more valuable for agriculture or mining than for the timber they contained.

The plans for this work, as approved by the Director, were ready by July 1, 1897, and provided that the special survey of the forest areas by expert foresters to determine the size and density of the timber, the leading economic species, the damage inflicted by forest fires, the amount of dead timber, the extent of the timber already cut, and the effects of the deforesting upon the springs and flow of streams, etc. was assigned to Henry Gannett (fig. 16; the topographic surveys were assigned to the Rocky Mountain (Edward M. Douglas in charge) and Pacific (Richard U. Goode in charge) Topographic Sections of the Geological Survey.

The purposes of the topographic surveys were defined as:

1. The preparation of topographic maps, on a scale of 2 miles to 1 inch, with contour intervals of 100 feet, as base maps for the representation of forestry details; agricultural and mineral lands; and future geologic survey;

2. the establishment of bench marks indicating elevation above sea level, for vertical control in topographic mapping, and for all mining, engineering, and geologic work;

3. the subdivision of reserves, where necessary, by running township lines for the purpose of designating tracts of land;

4. the demarcation, by means of section lines, of tracts which are more valuable, as agricultural and mineral lands than for timber; and

5. the mapping by the topographer in charge of each party of the outlines of all wooded and forest areas.

It was anticipated that the 62,261 square miles of forests then included within the reserves could be thoroughly and economically surveyed within 5 years, provided adequate appropriations were made for the purpose.

Cadastral surveys were begun, or were contemplated for the ensuing year, in seven of the forest reserves. However, the Geological Survey was relieved of this class of work by the act of Congress approved March 3, 1899, which so changed the law authorizing surveys within forest reserves as to require township, and subdivisional surveys to be executed thereafter by the General Land Office.

The work of this nature already accomplished amounted to approximately 91 miles of standard lines, 43 miles of township lines, 101 miles of section lines, 69 miles of meander lines and 11 miles of retracements in five forest reserves in the Pacific Section; in the Rocky Mountain Section, four complete townships adjacent to the Lewis and Clark Reserve, Mont.; and all of the Black Hills Reserve, S. Dak.–Wyo., embracing 12 complete townships. The 16 completed townships were surveyed by Wm. H. Thorn, who had previously made similar surveys in Indian Territory and brought with him some of his trained assistants, among whom was John E. Blackburn, who later, as a topographer with unlimited enthusiasm for hard work, surveyed many quadrangles in the roughest areas of the country.

The expert foresters to make the special studies with relation to the economic distribution of the forests were selected by Mr. Gannett chiefly from the Division of Forestry,

Department of Agriculture, a unit that had existed in that Department as far back as 1881, and had been given permanent statutory rank in 1886. In 1898, the unit acquired Mr. Gifford Pinchot as Chief and became the Bureau of Forestry by the act of Congress, March 21, 1901. Mr. Pinchot was an outstanding champion of forest conservation, shared his views with another outstanding champion, President Theodore Roosevelt, and vigorously administered a policy shaped to draw private lumbering into conservation practices and to train the personnel in practical forest management.

As the Secretary of the Interior had been recommending for several years, in order to effect better administrative results, the Forest Reserves were transferred to the Department of Agriculture, by act of Congress, approved February 1, 1905. The Agriculture Appropriation Act of March 3, 1905, designated the former Bureau of Forestry as the Forest Service.

Geography and Forestry Survey

Henry Gannett, as geographer of the Survey, continued in charge of the revision of the large map of the United States known as the "nine-sheet map." The preparation of a Physical Atlas of the United States, upon which much work had been done in former years, was continued. The office collection of railroad profiles was also overhauled and an index prepared, and copies of them were made for publication.

As president of the Board on Geographic Names, Mr. Gannett was obliged to give much time to its deliberations. In order to secure information in relation to certain topographic features of which little is known, Mr. Gannett proceeded, under instructions, to the West and Southwest. He visited the Blue Mountains in northeastern Oregon for the purpose of obtaining a general idea of their structure, extent, and altitude. He also visited Mount Hood, to study the surrounding topography and its glaciers. On his return, by the southern route, he made investigations of the San Bernardino Mountains and the valley lying south of them, and of the Staked Plains of Texas.

Plans for the special survey of the forest areas with relation to the economic distribution of the forests were completed by Mr. Gannett in June 1897, preparatory to starting the field work in July.

During fiscal year 1898, Henry Gannett had compiled, from all available sources, a map of western United States showing on a scale of 40 miles to 1 inch, the distribution of woodland and of what was thought to be merchantable timber. For most of this area, the information obtainable was amply accurate for this representation, consisting as it did of maps prepared by the Geological Survey and by the Hayden, Wheeler, and Powell Surveys, together with much unpublished information.

In general geographic work, a folio on physiographic types, being folio 1 of the Topographic Atlas of the United States, was completed and published. A Gazetteer of Kansas, accompanied by a map of the States, on a scale of 1:750,000, with a contour interval of 100 feet, was completed to be published as Bulletin No. 154.

Forestry work consisted principally in the superintendence of the examination of seven of the forest reserves established by order of President Cleveland February 22, 1897. The work was placed in charge of Mr. Gannett June 14, 1897, and under his instructions, the following field assistants were employed: E.Z. Graves, for the examination of the Black Hills, S. Dak.; F.S. Town, for the Bighorn Reserve, Wyo.; Dr. T.S. Brandegee, for the Teton Reserve and the southern portion of the Yellowstone Reserve; J.B. Leiberg, for the Priest River Reserve and the eastern portion of the Bitterroot Reserve; W.G. Steele and N.W. Gorman, for the eastern portion of the Washington Reserve, and H.B. Ayres, for the western portion, the line of division between them being the summit of the Cascade Divide.

Mr. Gannett's movements during the field season were directed toward familiarizing himself, so far as possible, with the areas and the forest conditions of the regions under examination. For this purpose, he proceeded directly to the Pacific Coast in July and, after a short trip in the Mount Rainier Reserve, went up Lake Chelan to the upper waters of Stehekin River, in the Cascade Range. Here he spent a number of days before crossing the range. Going down to the coast by way of the Cascade and Skagit Rivers, he came east to the Black Hills. After visiting them, several days were spent on the examination of the Bighorn Mountains; then 3 days were devoted to the accessible portion of the Bitterroot Reserve, after which he returned to the Pacific Coast, reentering the Washington Reserve at Monte Cristo. Thence he proceeded to the San Jacinto Reserve, in Southern California, and spent several days in the examination of that and the other reserves in that part of the State. On his way, east a stop of several days was made at Las Vegas for the purpose of examining a proposed addition to that reserve in the neighborhood. Reports on the examinations mentioned have been prepared and published as part of the 19th Annual Report, 1897–98.

Division of Geography and Forestry

Henry Gannett, geographer of the Survey, continued in charge of the revision of the large map of the United States, and also gave consideration to such geographic matters as were referred to him.

Forestry examinations of the western reserves were continued during the field season and, during the winter and spring, reports were prepared on the results of the work done. These reports are presented in Part V of the 21st Annual Report, 1899 to 1900. During the progress of the surveys of Indian Territory, the woodlands were mapped with great accuracy, and notes were made by the surveyors concerning the character and quality of the timber. A map of the Territory on a small scale, with contours, showing the woodland, was prepared and issued, with an abstract of the surveyor's notes, as a reconnaissance of the region. It was intended to supplement this as soon as means were provided, by an examination of the forests by experts.

During fiscal years 1901 and 1902, Henry Gannett continued in charge of the Division of Geography and Forestry, and the examination of the forest reserves.

Mr. Gannett, during the summer of 1902, was personally engaged in examining conditions in Central Utah, from which region many requests for the establishment of small reserves had been made for the protection of the farming industry against overgrazing by sheep. An agreement with the sheep grazers and farmers was reached by which the entire mountain region of Utah, which constituted the summer range for sheep, would be reserved. It stipulated that in such portions of these reserves as contributed to the water supply of the agricultural settlements sheep grazing was to be prohibited; that the remaining portions of the reserves would be allotted to the various sheep owners for extended periods, and that the number of sheep to be grazed upon a unit of area would be restricted far below the usual number.

In the fall of 1902, Mr. Gannett went to the Philippines, where he was engaged in taking the census, and the work of the Division of Geography and Forestry was temporarily transferred to the Hydrographic Branch, in charge of F.H. Newell. On Mr. Gannett's return, late in September 1903, he resumed charge of the Division.

During the spring of 1905, the work of examining forest reserves, etc., together with the men employed upon it, was transferred to the Bureau of Forestry of the Department of Agriculture.

As this closes the work of the Geological Survey in the examination of forest lands, it may be well to make a brief resume of the results accomplished. The work was committed to the Survey by act of Congress in 1897, and has therefore been carried on for 8 years. During this time the reserves in table 15 have been examined and reported on.

The names listed in table 15 are in most cases those which were in use at the time of the examination. Many of them have since been changed through consolidation.

Besides these reserves, large areas were examined with a view to the formation of new reserves, or their inclusion in existing reserves. The total area examined during the 8 years amounted to 110,000 square miles.

Henry Gannett continued in charge of the Division of Geography and Forestry during fiscal year 1906, and was occupied mainly with the following duties:

- Assisting the Forest Service, particularly in matters relating to the geography and topography and in the organization of its reserve force.

- His work as chairman of the Board on Geographic Names took considerable time, especially after the Presidential order of January 23, 1906, which extended materially the powers the Board.

- Much time was devoted to assisting the committee on Departmental methods, first in the investigation of the Bureau of Statistics of the Department of Agriculture, and second in the investigation in the departments and

Table 15.   Reserves examined and reported on, 1897–1905.

Arizona	Colorado—Continued	Oregon
Black Mesa	South Platte	Ashland
San Francisco Mountain	White River	Cascade Range
California	*Idaho*	*South Dakota*
Lake Tahoe	Priest River	Black Hills
Plumas	*Idaho and Montana*	*Washington:*
San Bernardino	Bitterroot	Mount Rainier
San Gabriel	*Montana*	Olympic
San Jacinto	Flathead	Washington
Santa Barbara	Lewis and Clark	*Wyoming*
Sierra	Little Belt	Bighorn
Stanislaus	Yellowstone (part)	Teton
Yosemite Park	*New Mexico*	Yellowstone (part)
Colorado	Gila River	
Battlement Mesa	Lincoln	
Pikes Peak	*Oklahoma*	
Plum Creek	Wichita	

bureaus of the government of the organization for carrying on scientific and routine processes.

- The editing of the report of the Eighth International Geographic Congress, published by the Public Printer, required considerable time during the first 3 months of the year.

- An outline map of North America, on a scale of 1:5,000,000, was completed and engraved. All the atlas sheets published by the Survey were reduced and made ready for the engraver. On the unsurveyed areas, the county lines and railroads were revised and are also ready for engraving.

The Division of Geography and Forestry was continued under the direction of Henry Gannett during fiscal year 1907. The principal results were:

- The revision of the three-sheet map of the United States on a scale of 1:2,500,000 was continued and good progress was made.

- The preparation of sheets, as part of the World-map scheme, on a scale of 1:1,000,000 was continued. Seven such sheets have been projected, and all available surveyed material has been added to them. Separate maps of Wyoming, Utah, Colorado, Arizona, New Mexico, West Virginia, Virginia, Maryland, and Delaware have been drawn, embodying all available surveyed material.

- Preparation of maps of the national forests has been continued in the form of township plats. These plats were at first made on a scale of 2 inches to 1 mile, and each township was published as an individual map. In order to better meet the needs of the Forest Service, the scale was changed afterwards to 1 inch to 1 mile, and six townships are published together on an atlas sheet, on which is shown the land, timber, and alienation classification by color and by symbols.

- Additions to national forests that had been completed necessitated many changes in the manuscripts.

- The preparation of "The Areas of the United States, the States, and the Territories," published as Bulletin No. 302, which is the result of cooperation between the Census, the General Land Office, and the Geological Survey, is an attempt to bring the figures of areas into accord.

- A list and descriptions of 100 atlas sheets illustrating topographic forms was prepared and published.

Mr. Gannett continued as chairman of the Geographic Board. Much work was done for this organization in the preparation of lists of all towns in the United States where the post office name differs from the town name, a matter that has been taken up with the Post Office Department, and in the preparation of sheets showing different conventional symbols used by Government offices for the representation of the same object. An effort will be made to have one set of symbols adopted for Government use.

From June 1907 to October 1908, Henry Gannett was furloughed for work on a census of Cuba. Upon his return to Washington, he was selected as geographer to the National Conservation Commission. As editor of the report of the National Conservation Commission, Mr. Gannett had charge of the most important stock-taking of natural resources ever undertaken by any government. It was a "doomsday book" of an entirely new order. The three volumes of this report furnished the basis for the formulation along constructive lines of public policies of far reaching value, and the report became, and continues to be, a veritable arsenal of information on the topics covered.

Thus, within less than half a century after Henry Gannett began his topographic work on the millions of unsurveyed acres of our national domain, he was called to act as statistician of a Conservation Commission dedicated to the work of saving from spoilation the seemingly inexhaustible natural resources which his surveys opened up to settlement.[24]

From June 1909, to his death on November 5, 1914, Henry Gannet was occupied with his duties as chairman of the U.S. Geographic Board. He also handled any general geographic work that was referred to him.

Idaho–Montana Boundary Line

The various activities of the Astronomic and Computing Section were increased when the Sundry Civil Bill was approved June 4, 1897, providing for a survey of the northern portion of the boundary line between Idaho and Montana. The line was to be the 39th meridian west of Washington, D.C., or 116 degrees, 3 minutes, and 2.30 seconds west of Greenwich, northward from the crest of the Bitterroot Mountains to the boundary between the United States and British possessions, a distance of approximately 72 miles.[25]

To establish an initial point in the boundary, a belt of triangulation was extended from the Spokane base eastward over a longitudinal distance of about 70 miles. In spite of adverse weather conditions and smoke alternating with storms throughout the field seasons, E.T. Perkins, Jr., established 15 triangulation stations, of which one was a little over a mile east, and the other about the same distance west of the boundary line, the two being about 16 miles apart. On the basis of computations that winter, it was necessary the following summer to traverse along the crest of the mountains a distance corresponding to 6,072 feet in longitude in order to locate the initial point on the meridian line. This traversing was done by Samuel S. Gannett assisted by Dewitt L. Reaburn; distances being measured by chain and checked by stadia, and directions being controlled by azimuth observations at the beginning and ending of the traverse line. (In 1897, Mr. Gannett had determined, by astronomic observations, the intersection of the 107th meridian with the Colorado–New Mexico boundary line.)

From the initial point, Mr. Reaburn traversed the random line northward. Horizontal and vertical distances were checked by the triangulation and the differences of elevation by the profiles of the Northern Pacific and the Great Northern Railroads. The distance run was about 72 miles, the number of transit stations 1,051, and number of azimuth stations 17. The entire line was in very rough country and snow was encountered on the high ridges. The survey of the random line was completed on October 31, 1898, in snow 2.5 feet deep.

An adjustment of the random line to the triangulation was made during the winter office season. Tables were prepared showing the exact offsets to be applied at each transit station to locate the true line, and all available data were plotted on sheets to be used in sketching the topography. It then remained to establish the monuments on the true line, to cut out and blaze the line, to secure data for the preparation of the final map, and also to remeasure with the steel tape the northern section of the line, a distance of about 6 miles. Mr. Reaburn organized a party at Leonia, Idaho, June 15, and completed the work on October 5, 1899. That winter, in Washington, the map was drawn to the scale of 1 mile to 1 inch with 100-foot contours, and four copies of the notes and plats were prepared for filing; one with the General Land Office, one each in the offices of the Surveyor General of Idaho and of Montana, and one to be retained by the U.S. Geological Survey.

---

[24] "Henry Gannett, President, National Geographic Society," by S.N. North, 1915.

[25] Survey Bulletins on boundaries no. 13 (1885); no. 171 (1900); no. 226 (1904); no. 689 (1923); no. 817 (1930).

Northwest Boundary Survey

At the request of the Department of State, E.C. Barnard, topographer, was detailed in 1901 to make certain surveys and investigations of the boundary between the United States and Canada, extending westward from the summit of the Rocky Mountains, and he was to cooperate with C.H. Sinclair of the Coast and Geodetic Survey. The object of the work of the joint parties being (1) to clearly define and temporarily mark the boundary in three localities where there was considerable uncertainty as to the line; and (2) to make an examination of the whole line and report upon its existing condition in conjunction with three geologic parties.

Having agreed that work should start in the westernmost locality, (the Mount Baker mining district) Messrs. Barnard and Sinclair, with an outfit using 20 pack horses, approached the line from Chilliwack, British Columbia, a station on the Canadian Pacific, on May 22, 1901, and from a camp on Silica Creek found and identified an old astronomic observatory post used by the Northwest Boundary Commission of 1856 to 1860. From this point a transit line was extended carefully southwest a distance of about three-quarters of a mile to the supposed position of the boundary line, and after diligent search, a mound of stones was discovered, which was identified as a boundary monument by a cut in a nearby rock, corresponding in distance and bearing with those given in the records of the Boundary Commission.

Accepting the latitude of the astronomic point as determined by the Commission and verified by Mr. Sinclair, the measurements made by Messrs. Barnard and Sinclair gave the position of the rock mound as about 7 feet north of the 49th parallel. From points on the east and west tangent through the rock mound, as determined by an observed azimuth, offsets were made to the boundary line. Vistas were cut and iron posts were set over a distance of more than 2 miles. This was the locality in which there were many disputes, and it was impracticable to extend the line any farther as snow lay deep on the summits to the east and west.

Messrs. Barnard and Sinclair arrived, via the Canadian Pacific Railroad, at Midway, the second locality, on June 27. The next day they visited a Canadian boundary party camped in the vicinity, under Mr. O'Hara, who stated that he had been instructed to open the boundary line between existing monuments from Midway westward and also to make a topographic map of a strip of territory 10 miles in width on the Canadian side of the line. Messrs. Barnard and Sinclair surveyed and cut out the boundary line from Midway to Grand Forks, a distance of about 16 miles, and inspected the monuments for an additional 6 miles farther eastward to Cascade. The uncertainty in the stretch between Midway and Grand Forks was due to the existence of two lines; the old astronomic line that was cut out, but on which the monuments had been torn down, though not removed and the mean parallel, which was defined by monuments but only partially cut out. The uncertainty was removed by the Barnard-Sinclair party clearly and distinctly cutting out the line defined by the monuments on the mean parallel.

The third locality was north of Montana and the combined party reached Phillips, British Columbia, a point on the boundary line, on August 14, traveling by wagon from Elko on the Canadian Pacific Railroad. From a monument near Phillips, the line was surveyed and cut out eastward to Wigwam, another station on the old boundary survey, a distance of 13 miles, and westward to a point about 4 miles west of the Kootenai River. The field work was completed on October 14.

Photographs were taken of all old marks and all new monuments and a topographic map was made on a scale of 1:45,000 with 100-foot contours. The total of the work in the three localities was: the determination of the latitude of 9 stations; the clearing out and surveying of 42 miles of line; the rebuilding of 5 monuments and setting of 9 iron posts; and the mapping of 164 square miles along the axis of the line.

River Surveys

In March 1902, arrangements were made whereby surveys of certain rivers upon which hydrographic investigations were in progress were to be made under the direction of Herbert M. Wilson, Geographer-in-Charge of the Eastern Section of topography. The Survey, since April, 1891, had been making measurements of stream flow of eastern rivers. The measurements included profiles based on the annual reports of the Chief of Engineers, U.S. Army, as well as railroad profiles and Geological Survey maps. The profiles differed from one another greatly in accuracy, as the elevations used were derived from level lines and barometric readings, and were shown on the very small scales of 100 miles to 1 inch for horizontal distance and 2,000 feet to 1 inch for vertical distance.[26] These scales were maintained for low country and mountainous streams alike, in order that comparisons between the different rivers might be made directly.

Approval being given to the operations of the Hydrographic Division by the 100 percent increase in its appropriation for the fiscal year 1903, the Division was expanded into the Hydrographic Branch. One of its earliest endeavors was to improve the river surveys so they might render a greater service. The arrangement in 1903 provided that the Topographic Branch would have entire charge of the surveys, their supervision and administration; be charged only with one-half of the expense of surveying the areas not yet mapped topographically; and retain custody of the notes and maps.

Mr. Wilson designated Wm. O. Tufts, topographic aid, to organize the first party for surveys of certain rivers in Georgia, and J.R. Eakin, topographic aid, to organize a party for surveys of certain rivers in Georgia, North Carolina, and Tennessee. After several weeks of training, these parties were turned over to Fred A. Franck and Carroll Caldwell. Field and office supervision was assigned to W. Carval Hall, assisted by G.H. Guerdru, both topographers.

River surveys totaled 733 miles in Georgia, Tennessee, North and South Carolina; in Wisconsin, 42 miles of the

---

[26] U.S.Geological Survey Water-Supply Paper 44, Henry Gannett, 1901.

Chippewa River, surveyed by F.T. Fitch, field assistant, after James R. Ellis had extended level lines to supply elevations in Maine; 135 miles of the Kennebec River, surveyed by Alvah T. Fowler, topographic aid, assisted by Frank J. MacMaugh and F.T. Fitch, field assistants, after Mr. Fowler had extended the line of levels. After the field season, Messrs. Franck, Caldwell, and Fitch were retained in the Washington office for several months, inking the maps, plotting profiles, and writing reports.

The field manuscripts were plotted on the scale of 1:22,500 and the profiles were drawn on the horizontal scale of 1 inch to 1 mile, and vertical scale of 1 inch to 100 feet. The maps were not published, but the profiles and tables of elevations and distances were published in Water-Supply Paper 115. The cooperation was discontinued at the end of 1903, as the surveys, made primarily as parts of topographic maps, did not present all of the engineering features of the water power sites desired by the Survey hydrographers and cooperating state engineers.

Grand Canyon Maps, Arizona

With the completion, in 1901, by the Santa Fe System of the 67-mile spur railroad from Williams, Ariz., to Grand Canyon, replacing the stage lines from Williams and Flagstaff, an increase in visitors to the Canyon was assured, and a demand was foreseen by the U.S. Geological Survey for a modern detailed map of the Grand Canyon for the information of the public and scientists. The map in existence had been made in the early 1880's, under the direction of Major Powell, and was of the reconnaissance type, on a small scale.

The Grand Canyon of the Colorado River is a complex chasm 225 miles long as measured along the course of the river, from 4,000 feet to 6,000 feet in depth, and from 6 miles to more than 12 miles in width. Between its outer rims is a bewildering array of immense buttes or "temples", colored cliffs, pinnacles, domes, amphitheaters, and many partly hidden side canyons. The outer walls are, in reality, a series of walls and terraces, the walls of distinctive coloring and extending unbroken for many miles. In a few places, where walls had been shattered by geologic faulting, trails had been built from the south rim down to the river. Only one trail, the Bass-Shinumo Trail, led from the river to the north rim. Thus, the Grand Canyon offered a greater obstacle to travel than any mountain range in the United States.

Two standard 15-minute quadrangles, Bright Angel and Vishnu, were to be mapped. F.E. Matthes was assigned the topographic mapping, and horizontal and vertical control were ready for him.

Early in 1902, H.L. Baldwin, Jr., extended triangulation northward from the primary triangulation net already existing in Central Arizona, occupying Bill Williams Mountain, several peaks north of the Santa Fe Railway, Red Butte on the Coconino Plateau, and projecting capes on the south rim of the canyon. In 1903, J.T. Stewart strengthened the system,

occupied 10 stations on the north rim and controlled, in all, seven 15-minute quadrangles. Timber along much of the canyon rims made it necessary to resort to heliotroping for sighting from the peaks south of the canyon. Stations along the canyon rims were marked, if in timber areas, by tripods 12 feet high, if outside timber areas, by rock cairns 6 feet in diameter and 7 feet high, both kinds of signals covered with white cloth and displaying a white flag.

Starting in 1902, and continuing in 1903, J.T. Stewart extended a line of primary levels from a bench mark at Williams on the precise level line along the Santa Fe Railway and the Grand Canyon Railroad to the triangulation station on Hopi Point, down the Cameron Trail to the river, eastward by road through the woods to Grand View Point, for the control of the Bright Angel Quadrangle; farther eastward by road to the triangulation station on Comanche Point, down to the river along the Hance Trail, for the control of the Vishnu Quadrangle.

In leveling down the steep canyon trails, levelmen found it necessary, for long stretches, to set up the instrument in the trail and take back sights of 6 feet or less; and the canyon heat in summer, showing a minimum of 96 degrees some days, shrunk the level bubble to the vanishing point in mid-day and made the instrument and the ground too hot to touch.

The mapping required two radically different topographic methods. The plateaus back from the canyon rims, with an undulating surface of slight relief and devoid of commanding view points, lent themselves to the planetable traverse method. Control traverses were extended along the rims, over the few wagon roads and trails, in circuits of a dozen miles or less. Then these circuits were crossed by stadia traverses and to fill in the small unmapped areas mostly in densely wooded areas. Traverses were executed by pacing and observing elevations on an aneroid.

The Grand Canyon was filled with sharply sculptured buttes and walls, and since these were inaccessible, and the side canyons between them so remote as to render travel to them impracticable, mapping by long distance intersection was necessary. From the easily reached stations along the south rim, hundreds of square miles were mapped at distances between 2 and 6 miles. Defiladed portions here and there remained to be completed by planetable stations and traverses in the lower parts of the chasm.

The large size planetable, 24 by 31 inches, was required for each quadrangle and the intersection work was executed on the large covering main sheet. Sight lines were drawn to all salient features as it was found necessary to locate 40 to 135 points in a square mile. Elevations for one-third of them were determined by vertical angles with a possible error of 5 feet. Sight lines were not numbered with descriptions in a notebook, but were tagged on the map sheet with brief form lines which were approximately in place and could be shifted into position when the intersections were completed by "cuts" from a second table station.

Hundreds of sight lines were made from a single station; from Grand View Point, for instance, they numbered 1,214.

The method of marking the sight lines ensured their identification and was a guarantee against loss, for some were not cut in for months, even a year, afterwards. All vertical angles, on the other hand, were recorded in a notebook and the elevations copied on an oversheet, an 8 by 10-1/2 ruled tablet paper, or a piece of tracing paper a little larger than a 5-minute block, each elevation exactly over the needle-sized plot marking the intersected point to which it belonged. The map sheet thus was not cluttered up with figures of elevations, and the oversheet was a complete record of elevations.

A short time after starting the mapping, in the summer of 1902, Mr. Matthes experimented with the delineation of the Zoroaster Temple and sent a tracing of it to Washington with recommendations, which were approved. Instead of 100-foot contour lines, which would not show some of the lesser cliffs, 50-foot contours would be used to adequately express all the topographic elements of consequence. However, the resulting map would be so crowded with lines that it could not be reduced to the scale of publication (1:62,500); therefore, an exception was made for the Grand Canyon maps and they would be published on the field scale (1:46,000).

Further, two established customs in cartography were changed. Because of the prevailing parallelism of the cliff patterns, in each of which a number of contours merge to form a band, the fifth contour would not be accentuated, but would be distinguished by a broken line. Also, because the fifth lines often are concealed for considerable distance in cliff bands, elevation figures would be put on intermediate contour lines, wherever necessary for clarity. These innovations compelled the topographers to furnish field maps of sufficient accuracy neatness to serve, unreduced, as copy for the engraver, and involved the most painstaking kind of drafting, first in pencil, and later in ink.

In the late summer of 1902, the survey had to be moved across to the north rim. The unfordable Colorado River was unbridged for a stretch of 800 miles, from Grand Junction on the Denver and Rio Grande to the Needles on the Santa Fe line. To be sure, there was Lee's Ferry, situated at the head of the Marble Gorge, but to reach it would require slow travel across about 100 miles of the Painted Desert of the Navajo's, and beyond the ferry another 100 miles of rough and waterless wagon trails.

Bright Angel Canyon, in full view from Grand Canyon Station, seemed to offer a direct route for travel to the north rim, but old timers declared it to be impassable for animals. It was decided, therefore, to try the Bass Trail, about 30 miles to the west, although it had been constructed as an aid to mining and was not intended for pack animals larger than a burro. Mr. Bass had a boat at the river and offered it's use. A member of the party was familiar with the trail as he had already traveled over it afoot to erect a signal on Point Sublime.

By removing the packs and saddles from the horses and mules the party managed to get them by the overhanging cliffs on the trail. Then, arriving at the foot of the trail, they succeeded in leading them, again unloaded and unsaddled, down the remaining 100 feet of rock slide to the river, pushing them

one at a time head first into the water and towing them safely across, all 10 of them. All the equipment and supplies were carried by the men themselves down the rock slide and ferried across the river in the small boat, which had been moored on the north side and was brought across by two members of the party who swam across to get it.

For two long weary days the party followed a little-used and partly washed out trail out of the heated chasm to the rim of the Kaibab Plateau (about 7,900 feet), which was cool and covered with a beautiful forest. This isolated plateau in the northwest corner of Arizona contained, in summer, one forest ranger and his small family, and herds of cattle, which were driven in by cowboys in the spring and out in the fall, and also herds of deer, which in the fall were hunted for buckskin by Paiute Indians. In the winter, the plateau was vacant, as the snow was deep. It took the party 5 days to make the journey from Grand Canyon Station to Point Sublime, an airline distance of 11.5 miles. It was decided that the packer would go for mail and supplies to Kanab, Utah, 80 miles to the north, and he made that trip on horseback with one pack mule twice each month.

Extending the work eastward, the first of November found the party at the head of Bright Angel Canyon, when two haggard prospectors and a worn out burro climbed out of the canyon. They allowed that a pack train might get through, so on Sunday, November 9, a day before the first heavy snow as it turned out, the Survey party broke camp and started on its adventurous return trip. In the milder climate below, the party tarried several days in order to map the canyon. In the descent to the river, where fortunately a boat was found, Bright Angel Creek had to be crossed 106 times by wading to knee depth. On the south side of the river, the dangerous climb of 1,500 feet up over the burro trail to the Tonto Terrace was accomplished without losing an animal.

Mr. Matthes was assigned the mapping of the Jerome, Ariz., Quadrangle that winter and returned to the Canyon in the summer of 1903. That fall R.T. Evans, who had been associated with Mr. Matthes on the Jerome work, upon the completion of his mapping at Cripple Creek, Colo., was sent to the Canyon to help finish the two quadrangles. He accompanied Mr. Matthes to the north rim via Bright Angel Creek, two members of the party first packing down a light steel boat, specially made in two parts for packing on muleback, by which to make the crossings on the river. The stay on the north rim was for the month of October only. From camps on Walhalla Plateau, Mr. Matthes continued intersection work from points on the rim, while Mr. Evans mapped the park-like plateau itself by extending numerous stadia traverses across it.

Particularly interesting was the successful heliotroping on October 25 from 1 to 2 p.m. from the triangulation station on Cap Royal to Mount Kendrick, one of an array of peaks across the Canyon to the south and distance about 50 miles. J.T. Stewart, triangulator, had requested that this be done when, more than a month before, he sent a schedule of his movements.

On the return trip the steep climb up from the river to the Tonto Terrace was marred by tragedy. One pack mule, crowded by another on a narrow zigzag of the burro trail, slipped off and rolled over and over, pack and all, down several hundred feet to a rocky ledge, barely missing other animals, and men too, in the descent.

On the south rim, a few remaining blank spaces on the map in the neighborhood of Grandview, were filled in spite of the first snowfalls of winter. Then came the long-planned expedition to the northeast corner of the Vishnu Quadrangle, an isolated section north of the Little Colorado River with elevation a thousand feet lower than the south rim, and consequently with milder weather in November. Five men made up the party and they traveled with a camp wagon drawn by four animals, the men walking part of the way and later working on foot.

The morning of the third day, the Little Colorado River was forded without much trouble at Cameron's Crossing and, by the end of the fourth day, a Navajo Indian had guided the party to the "Hogan", the Indian name for the rock cairn that stood over the triangulation point "6141." Camp was set up and the teamster, "Pa" Johnson, unhitched the animals from the wagon and, riding one, took them back to the Willow Springs Trading Post, the nearest water and feed.

The work was completed in the estimated time, as the weather did not interfere and, with the animals returned as prearranged, the return journey was begun. Camp was set up the first night at Willow Springs Trading Post, and the next day being Thanksgiving, it was celebrated by a visit to Tuba City Indian Agency, the wagon going around by road while Evans, Elliott, and Oliver took a trail that was several miles shorter, but through sand that was ankle deep.

There still remained enough field work for the month of December, sections of the Tonto Terrace and most of the granite gorge from the foot of the tourist trail downstream to the west edge of the Bright Angel Quadrangle were incomplete. The gorge is 900 to 1,300 feet deep and from 1,300 to 2,600 feet wide at the top, which culminates in the 200-foot Tonto Cliff. From stations along this cliff (the rim of the Tonto Terrace) most of the river and gorge can be readily mapped by intersections.

Indian Garden and Hermit Creek offered convenient camp sites and the weather was mild. Snow rarely reaches into the depths of the Grand Canyon at the Tonto Terrace, nor does ice form. One-half square mile of the Tonto Terrace on the north side of the river between Crystal and Tuna Creeks, abutting the west edge of the quadrangle, could not be satisfactorily obtained by intersection. It was necessary to cover it with stadia traverses. So, when light enough to see, Evans and Philips, a rodman, left the tent of Louis Boucher, prospector, hiked the 1-1/4 miles down the boulder strewn bed of Boucher Creek, and then in Boucher's crude bateau glided down the river 1-3/4 miles to Crystal Creek. A break in the Tonto Cliff enabled them to complete the 900-foot climb up from Crystal Creek to the Tonto Terrace; the mapping was completed in the afternoon, then the return trip was made before dark.

Reorganization

The Topographic Branch, on May 1, 1903, the date on which the new manual of "Instructions" took effect, was reorganized for administrative purposes into two divisions (table 16).

Table 16.   Reorganization of the Topographic Branch, May 1, 1903.

Division of topography	Geographer-in-charge
Atlantic Section	H.M. Wilson
Central Section	J.H. Renshawe
Rocky Mountain Section	E.M. Douglas
Pacific Section	R.U. Goode
Triangulation and Computing Section	S.S. Gannett
Division of Geography and Forestry	Henry Gannett

Mr. Goode died on June 9, 1903 and, as a consequence, several changes occurred, effective July 1. The Director then resumed the active chairmanship of the Topographic Committee. Consolidation resulted in the reorganization show in table 17.

Table 17.   Reorganization of the Topographic Branch, July 1, 1903.

Division of topography	Geographer-in-charge
Eastern	H.M. Wilson
Western	E.M. Douglas
Geography and Forestry	Henry Gannett

The two divisions of topography included jointly three sections, the chiefs of which reported to the Topographic Committee (table 18).

Table 18.   Topographic divisions.

Division	Personnel in charge
Section of Triangulation and Computing	S.S. Gannett, Geographer
Section of Inspection of Topographic Surveying and Mapping	J.H. Renshawe, Geographer
Section of Instruments and Topographic Records	S.A. Aplin, Topographer

In addition to the above, the eastern and western divisions of topography were each subdivided into several sections composed of one or more states, each section in charge of topographers acting as section chiefs. This organization with slight changes became permanent April 27, 1904 (table 19).

Table 19. Organization and number of persons employed on March 3, 1904.

[--, unknown]

Branch	Division/Section	Persons	Appropriations
Administrative	Executive/Correspondence, records, supplies, etc.	14	--
	Steam engineer, mechanics, messengers etc.	43	--
	Disbursements and Accounts	14	--
	Library	9	$94,640
Geologic	Geological and Paleontological Surveys	67	163,700
	Chemical, Physical, etc. researches	49	37,000
	Alaskan Mineral investigations,	--	60,000
Topographic	Eastern, Western, Triangulation	--	309,200
	Inspection and Geography and Forestry	92	--
	Surveying Forest Reservations	5	130,000
Hydrographic	Hydrography–Hydrology	46	--
	Reclamation Service (16 arid states and territories)	187	200,000
Publication	Editorial Texts, Maps, Illustrations,	35	68,280
	Engraving and Printing	85	100,000
	Documents (disbursed by Public Printer)	12	215,000
Total		658	1,377,820

Survey Pennant and Button

R.H. Chapman designed the Survey flag in 1903. On March 1, 1904, E.M. Douglas, Geographer-in-Charge of the Western Division, requested F.E. Matthes to design a Survey button that would be worn on coats used by topographers on field duty as a mark of identification and distinction.

The buttons were soon available to topographers, as well as a Survey pennant bearing the same design, for display over all Survey camps, beneath the American flag. This practice was prescribed by Col. H.C. Rizer, Chief Clerk as Acting Director, in a memorandum which read in part:

"With reference to a suitable flag for the use of the U.S. Geological Survey, I have approved, in accordance with the recommendations of a majority of the members of the Survey, a design consisting of a blue rectangular flag with white triangle, cross hammers and thirteen stars, in which the hammers extend beyond the sides of the triangle (fig. 17).

"It is hereby directed that the chiefs of all Survey camping parties shall fly the U.S. flag and the Survey ensign from their camps at all times."

The Survey buttons were worn by field men until they wore the uniform of the U.S. Army in 1917 and 1918.

Twenty-Fifth Anniversary

The 25th anniversary of the Survey happened to fall near the date set for the opening of the Louisiana Purchase Exposition at St. Louis, and it was deemed desirable to include in the Survey's exhibit, for the information of the visiting public, a small bulletin (No. 227) setting forth an account of the organization and work of the Survey, and the results it had achieved. The following items are selected from the bulletin:

The first appropriation made March 3, 1879.................................$106,000

The appropriation for fiscal year ending June 30, 1904...............1,377,820

The permanent force of the Survey in 1879–80........................39 persons

The field force of the Survey was increased by seasonal employees in each of the three Branches: Geologic by 50, Topographic by 200, and Hydrographic by 250, thus bringing the total number of employees up to about 1,200 during the field season of 1903.

To the Federal appropriation for topographic mapping for the fiscal year ending June 30, 1904, was added $105,750 by 11 states with the areas mapped by April 30, 1904 (table 20).

On Saturday evening, April 2, 1904, the Survey celebrated it's 25th anniversary with a banquet at Rauscher's, and the Secretary of the Interior, Ethan Allen Hitchcock, the Speaker of the U.S. House of Representatives, Director Walcott, and the branch chiefs were the speakers.

Speaker Joseph G. Cannon said, among other things:

"I came here 30 years ago and there were the great geologists and scientists—King, Hayden, and Powell—that man of courage, wisdom, and endurance. I cannot speak of Major Powell without a tear coming to my eye * * *. With all these needs in front of it, the Geological Survey has its hands full. So buckle on your armor and go forth to further work, for there is plenty to do, and you are the men to do it, especially when you are backed up by such women as I see before me."

The menu was an attractive feature, depending upon geologic terms to convey the sense of the feast that was offered. The walls of the banquet hall were decorated with American flags interspersed with humorous colored drawings of the principal officers of the Survey.

Table 20.   Appropriations and areas mapped, April 30, 1904.

State	State cooperating official or agency	Allotment	Square miles mapped in year	Area of state mapped	
				Square miles	Percent
Alabama	Geologist	$1,000	90	17,534	33
California	Engineering Department	10,000	2,532	67,148	42.4
Kentucky	Geological Department	5,500	824	12,727	30
Louisiana	Experiment Station	1,500	486	7,003	14
Maryland	Geologist	750	53	9,835	80
Maine	Survey Commission	2,500	221	6,122	18
Michigan	Geologist	1,700	550	3,014	5
Ohio	Governor	28,800	2,942	9,529	23
New York	Engineer and Surveyor	19,000	2,747	33,871	69
Pennsylvania	Survey Commission	15,000	1,052	15,422	34
West Virginia	Geologist	20,000	1,394	20,945	84
Area in 20 other states mapped			13,818		
Total		105,750	26,709		

Area of United States mapped (27 percent including Alaska)................................31

The rate of mapping of the country, on steadily larger scales, by the Topographic Branch had progressed:

During first year (1879–80) on scale of 1:250,000..3,400 square miles

During last full year (1902–03) on scales of 1:125,000 and 1:62,500..............31,000 square miles

Suggestions and Awards

The following circular letter, dated December 8, 1905, was received by Survey employees:

> "The Director has approved a recommendation that hereafter credit be given in some formal manner to employees of the Survey for suggestions or inventions which tend to improve the work, lessen its cost, or otherwise benefit the service. These suggestions may concern field or office work, forms of accounting, or clerical features.

> "In a few instances in the past, valuable suggestions have been given recognition through more rapid promotion, and it is hoped that in the future many suggestions or inventions may be offered of such merit as to deserve similar action.

> "It is now proposed to give credit by honorable mention in the Annual Reports of the Survey to those who have, during the year, proposed changes of sufficient value be adopted, in the judgment of a competent committee."

Annual Report No. 28 (1906–07) gave honorable mention to:

Lewis F. Biggs, Levelman—for a useful method for checking rod readings, adapted to the yard rods used with precise level lines, consisting of a series of graduations in feet on the back of the rods.

Albert O. Burkland, assistant topographer—a combination of the short-sight alidade with the ordinary boxwood plotting scale, the resulting device being a light alidade at one-half the cost of the all-metal instrument, and also a convenient plotting scale.

Harry L. Baldwin, Jr., topographer—to plot primary and secondary traverse lines on 10 times the desired scale on a separate piece of paper, and then to connect the terminal or controlling points by a straight line that can be readily transferred to the proper place on the map by means of two triangles, and at the same time reduced in length by 10.

Glenn S. Smith, topographer—a movable arm for holding the compass on a traverse planetable board, which permits easy adjustment for changed declinations.

The Beaman Arc

The Beaman Arc (fig. 18) was developed by W.M. Beaman, topographer, from 1896 to 1904, and it greatly facilitated the execution of stadia surveys.

It was one of the earliest and most useful improvements to the alidade and measured vertical angles in steps instead of degrees, simplifying the computation of elevation differences when used with stadia distances. The steps were graduated in sequence above and below the level line, and varied in magnitude from 34.38 minutes for level sights to 69.77 minutes for sights 30 degrees above or below the horizon. When the index was set exactly on a Beaman Arc division, the elevation difference along the line of sight was obtained by multiplying the stadia intercept on a vertical rod by the number of steps above or below the horizontal, a simple and easy field computation. The size of the steps compensated for the distance discrepancies because of the inclination of the line of sight, and the fact that the stadia rod was not at right angles to the line of sight. The Beaman Arc included a scale expressed in percent for converting stadia distances on inclined sights to corresponding true horizontal distances.

Although it has been modified considerably in details, the Beaman Arc is still used in 1954 on alidades and on many engineers transits. To the alidade, an auxiliary bubble has been added to level the arc assembly without leveling the telescope, and the arc was changed from the drum type to the conventional flat form, although the scales remained the same. Later, a small magnifying glass was attached for quicker and more accurate reading. The last modification provided a separate bearing for the arc, independent of the telescope, to eliminate motion of the arc when the telescope was raised or lowered in sighting.[27]

---

[27] "An Alidade Development in G.S.," by C.S. Maltby and J.L. Buckmaster, Topographic Division Bulletin, March 1954.

The Baldwin Solar Chart

Early in the mapping operations of the Geological Survey, an urgent need developed for some means to orient the planetable, an instrument used in graphic traverses, when used through areas where the compass needle does not serve. To meet this need, the "Solar Azimuth Chart" was designed by D.H. Baldwin, topographer, in 1908; it worked as a sun dial with a movable vertical gnomon.

During the years of practical application of the chart to work in the field, experience has indicated some desirable changes in the original design. From time to time, these changes have been incorporated into subsequent editions of the chart, but the principle upon which it works remains the same.

The primary purpose of the Solar Chart is to mark direction; therefore, it is logical to use a ruler with a vertical shadow-casting gnomon, because angles of direction are measured around the vertical plumb line as an axis. For an observation at a given time and place, two points must be found on the chart—the sun-time point, and the pivot point. These two points are to be set in line with the shadow of the gnomon and the chart will then be oriented.

The 1943 edition of the original solar chart contains new features providing for its use at latitudes numerically greater than 30 degrees in Northern and Southern Hemispheres. The Tropic Supplement, which was made available in 1943, is for use between latitudes 30 degrees north and 30 degrees south.[28]

Government Employees Mutual Benefit Association

Topographic aids, as well as other Civil Service appointees, often took the advice of older associates and bought insurance policies covering life, health, and accident. The 20-year-pay life insurance policy was considered a good investment. Temporary employees, such as recorders, rodmen, teamsters, and cooks, as a rule, did not carry any form of insurance.

Government regulations required that the pay of a civilian employee stop immediately if he became incapacitated for work by reason of death, injury, or sickness. As the outdoor work was healthful, sickness and accident, in spite of hazards, rarely marred a field season for many survey parties.

However in 1904, in West Virginia, a levelman fell ill with typhoid fever and died. The topographer in charge of field parties in the states, Albert M. Walker, was so moved by the evident financial distress in the case that in the following winter in Washington he campaigned among his associates for some form of protection for the temporary field employees that would compensate them for expenses and loss of time, and would be more effective than the limited assistance that the chief and other members of the party might be able to

render. His associates lined up in support, for they wanted such protection for themselves as well. The result was the organization of the Government Employees Mutual Benefit Association by members of the Geological Survey, the Forest Service, and the Reclamation Service.

The business of the Association was placed in the hands of a Governing Committee. The committee prepared application blanks for membership and had them ready a month in advance of the opening date, June 1, 1905. Any permanent or temporary employee could join the association by paying an admission fee of $1.00 and a premium of $1.00 per month or semi-annually in advance. The policy provided for:

- Loss of time not to exceed $150 in any 12-month period.

- Doctor bills, medicines, hospital expenses, $100 in any 12-month period.

- In case of death, not to exceed $600; if at home a cash payment of $200 for funeral expenses.

Chiefs of parties were instructed to urge all temporary assistants to apply for the insurance and to explain to them that if disabled by sickness or injury, certainly if for as long as a week, their pay would stop, and that the chief of party would not feel called upon to assume any responsibility of their support or care in view of the opportunity so cheaply offered through an organization created by the Survey for their protection.

The first annual report of the Committee, dated January 6, 1906, stated that the Association had 177 members, and that a credit dividend would be available at the end of 1906 to those who remained in good standing. One member died, Herbert B. Blair, topographer, on December 28, 1905, and three members received indemnities.

The Association was 3-years old when Congress passed the Act of May 30, 1908, (35 Stat. 556), granting compensation, if injured in the course of employment, to any person employed by the United States as an artisan or laborer in various construction work of a hazardous nature, particularly under the Isthmian Canal Commission, or other employee engaged in any hazardous work under the Bureau of Mines or the Forest Service.

Survey Order No. 13, dated June 29, 1912, directed that all accidents to employees be reported on Department of Commerce and Labor Form CA–21, even though Survey employees were not entitled to compensation under the Act of May 30, 1908.

Protection against loss of pay was extended to all civil employees, permanent and temporary, of the United States by the Federal Compensation Act of September 7, 1916, which provided compensation for injuries sustained in the performance of duty that cause disability to perform work for a period of more than 3 days. The act also provided for the payment of certain expenses incurred for medical and hospital attention resulting from such injuries.

---

[28] "The Baldwin Solar Chart," by D.H. Baldwin, published in Photogrammetric Engineering Journal, v. X, no. 2, April, May, June 1944.

Chiefs of parties were supplied with a copy of "Regulations concerning the duties of employees, official supervisors, medical officers, and others", a printed list showing the location of all designated medical offices, hospitals, and dispensaries, and blank forms for reporting injuries and filing claims for compensation. In case an employee was injured in a section of the country where no U.S. medical officer or hospital was located, and there was no physician designated in the printed list for that vicinity, a local physician might be employed. In the latter case, full explanation was exacted.

The Act of September 7, 1916, provided no benefits for sickness not caused by injuries. Such cases continued to look for relief to the Mutual Benefit Association.

It was not until June 6, 1924, that an amendment to the act was approved, providing that compensation shall be paid for occupational diseases or "any disease proximately caused by the employment."

The Harrison Letter

Making the field surveys and drafting a topographic map in the office did not always add up to a 12-month's cycle of a happy, creative occupation. There were hardships and disappointments. A recital of some of these was contained in a topographer's letter and cited as grievances against his Chief, Herbert M. Wilson, Geographer-in-Charge of the Eastern Division.

The topographer, Dabney C. Harrison, a Virginian with an A.B. from Hampden Sidney College, with 2 years of teaching school and 3 years with the Mexican National Railroad, had been appointed to the Survey on June 27, 1885, and had been chief of party in topographic surveys since 1889. A nervous condition led to an extended absence from April 1902 to April 1904 for treatments in a sanitarium. Upon recovery, he was reemployed in the Eastern Division. Soon he developed a dislike for Mr. Wilson that became an obsession; he blamed him for the misfortunes, real and fancied, that seemed to plague him.

Mr. Harrison's grievances included: A last minute change in his field assignment to a less important quadrangle where the living conditions were more difficult, the change necessitating a hurried shift in the disposition of his family; a transfer late in the field season from his almost finished quadrangle to another one, causing him embarrassment and extra personal expense; ambiguous and contradictory instructions, at one time to hurry the field surveys so as to make them cost as little as possible, at another time to get all the topographic detail carefully and accurately; unfair treatment in not being made a strawboss; and sundry acts of broken faith and persecution. The letter was presented to the Director on Saturday, December 15, 1906.

Director Walcott, after deep consideration, set up a committee of three as a Trial Board to hear and weigh the charges: Henry Gannett, Colonel Rizer, Chief Clerk, and George Otis Smith, geologist, as chairman. A few short afternoon sessions were held. Mr. Harrison for months had recorded in a black Survey notebook the uncomplimentary remarks and criticisms that any of his colleagues had made against Mr. Wilson and depended on those colleagues to substantiate their remarks. The existence of this book created a panic among Harrison's acquaintances. No one was quite sure he had not spoken disparagingly of "the boss" at some time or other. The notebook proved of little value as only two or three topographers admitted the criticisms ascribed to them.

The Director announced on Friday, March 8, 1907, his acceptance of the findings of the Trial Board. Not one of Mr. Harrison's charges was sustained. Mistakes made in administering the far flung field work are recognized as inherent in the office of the chief. Some partiality may have existed. At any rate the hearings brought out the fact that Mr. Wilson was too unpopular to continue as Chief of the Eastern Division. Too valuable a man to be lost to the Survey, he was transferred to the Survey's new Technologic Branch, effective April 5, as Chief Engineer. Mr. Harrison, whose satisfactory mapping had been supplemented by his design of Harrison Percenter Level (made by the Eugene Dietzgen Company) was transferred to the Bureau of Forestry, Department of Agriculture.

The Harrison letter seemed to be the spark that set off a chain of events that kept the Survey personnel in Washington stirred up to a high degree of excitement throughout the spring of 1907.

Director Walcott, whose broad interest in science was evidenced by his connection with the National Museum as its head from January 1897 to July 1898 with the title of Acting Assistant Secretary of the Smithsonian Institution; with the Carnegie Institution in Washington, D.C., 1902–05 as Secretary; with the Washington Academy of Sciences as its President since 1899; was appointed Secretary of the Smithsonian Institution January 23, 1907. The position was one of eminence in the scientific world.

Dr. Walcott was tendered a farewell banquet at Rauscher's on March 13. Several topographers, whose feelings had been ruffled by recent events, did not attend. Dr. Walcott's resignation from the Survey became effective on April 30.

First Booklet of Field Assignments

The localities of field work and assignments of employees of the Geologic and Topographic Branches of the Survey for the season of 1906 were contained in a pocket size (4 by 6 inches) paper covered booklet of 24 pages. It was printed by the Government Printing Office and dated August 1, 1906. It listed 43 states and territories in which geologic investigations and topographic surveys were being carried on by 90 geologists, 16 of whom were working in Alaska, and by 109 topographers, 2 of whom were working in Alaska.

Printed Bulletins

As a means of keeping the personnel informed of addresses and to facilitate the handling of mail and express, there was issued, on September 9, 1912, a multigraphed list of "1912 Assignments of Scientific Appointees, Topographic Branch."

One hundred and sixty-two (162) names appear on the list and mapping operations were carried on in 25 states, the District of Columbia, and the Territory of Hawaii. Four topographers were on furloughs.

Clarence L. Nelson and Washington B. Lewis were still in Argentina. Thomas H. Moncure was with Arctic Coal Company, at Tromso, Norway; William J. Peters with the Bureau of Terrestrial Magnetism, Carnegie Institute, Washington D.C.

In 1913, a little booklet, 6.25 by 4.25 inches, containing 56 pages, and printed at the Government Printing Office, titled "Service Bulletin, Personnel and Field Assignments," was distributed.

This bulletin, giving the personnel of the entire Survey (786 in number) and program of field assignments for 1913, was issued for the use of members of the Survey and cooperating bureaus. It gave the organization of the Survey, and of the field forces, the supervision and scope of the work of each branch, division, and section, and a summary of the work by states. Similar booklets were issued for 1914, 1915, and 1916. (Beginning in 1911, the Survey issued a Press Bulletin for release weekly to newspapers for the purpose of apprising the public of its activities).

Geological and water supply investigations were in progress in 46 states, Alaska, and the Territory of Hawaii; topographic surveys were being made in 32 states, Alaska, and Hawaii. The field force of the Topographic Branch consisted of 8 geographers, 19 topographic engineers, 30 topographers, 32 assistant topographers, 43 junior topographers, and 3 topographic aids, a total of 167.

Swamp Lands in Minnesota

An appropriation of $15,000 was made by Congress in the Indian Act for fiscal year 1906–07 for the survey of the swamp lands in the ceded lands of the Chippewa's in northwest Minnesota. A.P. Meade, topographer, was detailed for this purpose. About 600 square miles were covered with a network of levels. In addition, a line of precise levels, 71 miles in length, was run by C.H. Semper from Crookston northeastward. With sufficient data thus obtained, estimates were made by A.P. Meade and W.C. Hall for the drainage of the Mud Lake District and submitted as a report to Congress.

The Congress made a further appropriation of $10,000 for the next fiscal year for a continuation of the survey into five adjoining counties. The work was conducted by A.P. Meade and E.L. McNair. Approximately 1,800 square miles were covered by a network of levels, 188 miles of precise levels and 272 miles of primary levels being run, in connection with which 79 permanent bench marks were established.

Earthquake

The city of San Francisco was hit by the most destructive earthquake that ever occurred in the United States at 5:30 a.m. Wednesday, April 18, 1906. With terrific rumble and roar, shocks lasting almost a minute, wrecked buildings, and ruptured electric power and gas mains. Lesser tremors followed all day. The conflagration that raged for 3 days and nights multiplied the tragedy and misery of the inhabitants. Five hundred lives were lost and property damage reached $350,000,000. San Jose, Santa Rosa, Petaluma, and four other California communities suffered similar devastation. Sacramento, on the east edge of the affected area, was shaken but not damaged. Its citizens soon learned of San Francisco's catastrophe and organization was promptly effected for the relief of refugees and the dispatch of supplies to the stricken city.

Director Walcott telegraphed Robert B. Marshall, Geographer-in-Charge of the Pacific Division, to assemble camp equipment from the Survey storehouses in the Sacramento Valley and prepare for shipment to San Francisco. Before the end of the day, Mr. Marshall was designated head of the purchasing unit for relief supplies by the General Relief Committee. Donations exceeding $150,000 were received by the Committee in a few days. On the first trip of the *S.S. Joaquin*, Mr. Marshall was aboard in charge of the cargo, and among his Survey helpers were George R. Davis, Jake W. Muller, C.L. Nelson, A. Benson Searle, Sidney N. Stoner, and Albert H. Sylvester. During the night, trip stops were made to take on additional supplies brought to the wharves by farmers who were eager to help. The boat docked at the Presidio and volunteer stevedores, citizens, and soldiers assisted the Survey men in unloading the cargo so the boat could return to Sacramento for more cargo. Gen. Frederick Funston, in command of the San Francisco District, had established military rule over the city, prohibiting persons from sleeping indoors and making fires for cooking, except in the streets, and restricting visitors from outside.

There have always been earthquakes caused by strains within the earth's crust. The line of parallel valleys in California are due in part at least to faulting that took place long ago. These associated natural processes have generally occurred beneath the surface of the earth. But on April 18, 1906, a crack reached the surface for a distance of 192 miles in almost a straight line from Point Arena in Mendocino County to San Juan in San Benito County. The crack followed the well known San Andreas Rift, an ancient fracture line of the earth's crust that may be traced southeastward beyond San Juan, past San Bernardino, to the vicinity of Yuma. A belt of peculiar topography marks the San Andreas Rift, in places by ponds or lakes in straight rows; in other places by long lines of very straight cliffs. These have been clearly shown on topographic maps of a scale as large as 1:24,000. The crack made in 1906, though disturbing the ground surface for a width of about 22 feet and causing offsets in roads, fence lines, and rows of trees, was too small a feature to be shown on standard topographic maps other than by a single line. The principal concern was to determine the amount of change that occurred in the basic vertical and horizontal control.

Marshall, Chief Geographer

For several years the demands for topographic surveys had become more insistent. Appropriations for topographic mapping had been increasing in spite of some setbacks and undoubtedly would continue to increase. Director Walcott, before leaving the Survey, had appointed a committee to work out a plan for the reorganization of the Topographic Branch, and Robert B. Marshall was recalled from his headquarters at Sacramento, Calif., to be chairman of the committee.

The committee, on March 19, 1907, submitted a plan that provided for five divisions instead of two; but, because of the impending transfer of H.M. Wilson, this plan was modified on April 6 by the consolidation of the Northeastern (Mr. Wilson) and Southeastern (Mr. Sutton) Divisions into the Atlantic Division. The four divisions and their selected chiefs are shown in table 21.

George Otis Smith, geologist, succeeded Mr. Wilson as chairman of the Topographic Committee.

The appointment of a Chief Geographer-in-Charge of the Topographic Branch was recommended.

Mr. Marshall was virtually Chief Geographer from the date of his selection as chairman of the Reorganization Committee, though his official appointment was dated January 23, 1908. His successful management of the California Section was known to all, and almost equally well known was Dr. Smith's earlier declaration that should he become Director, he would appoint Mr. Marshall as Chief Geographer. (This promise, and its fulfillment assured both of them a measure of unpopularity in the Eastern Division, where several topographers were willing candidates for the position.)

Mr. Marshall took charge of the Topographic Branch with some positive ideas, several of which stirred and shocked the members of the old Eastern Division. The sketching of contours by chiefs of parties while seated in a buggy would

be supplanted by the more accurate method of locating them by planetable and stadia. To spread knowledge of the latter method there would an interchange of field personnel between the East and the West.

During the winter months in Washington, social meetings would be held and Survey work discussed, and the field men would become better acquainted and their ideas about their work could be exchanged and properly appraised. With Mr. Marshall's approval, and in some cases at his suggestion, clubs were formed and during 4 months, January to April, hardly a week passed without two or more gatherings, and the Topographic Branch became a family.

Plans were developed also for winter field work, both around Washington where little or no expense would be added to salaries, and in distant southern states where cooperation might be obtained.

Special Forest Maps

In the spring of 1907, the Division of Geography and Forestry was terminated by the transfer of forestry personnel, and their work, to the Department of Agriculture. Retained by the Survey was the compilation of special National Forest maps, and this was placed in the charge of Arthur C. Roberts. These maps were made on a scale of 2 inches to 1 mile and each township was published as an individual map in order to meet the needs of the Forest Service. The scale was changed to 1 inch to a mile and six townships were published together as an atlas sheet, on which was shown the land, timber, and alienation classification by color and by symbols. All the field work performed by the Geological Survey, the General Land Office, the Hayden, Wheeler, and Transcontinental Surveys, as well as by private surveys and by the Forest Service, was incorporated in the maps. The maps that together covered a National Forest constituted a folio, and folios were to be completed as fast as possible. The final completion of a folio was often retarded by changes of boundaries and redistricting of National Forests.

Topographers with first hand knowledge of forest conditions were sometimes assigned to Mr. Robert's unit. Once, Joseph F. McBeth was in Union Station en route to a summer's field assignment when, one-half hour before his train was to leave, a Survey messenger delivered a change of orders, that kept him in Washington throughout the summer, for office work with Mr. Roberts.

Transfers to General Land Office

Shortly after the reorganization in 1907, three of the older topographers with broad experience in triangulation, leveling, topographic mapping, and surveying boundary lines, chiefly in the rugged west, were transferred to the General Land Office.

Abner F. Dunnington, on May 1, had been appointed assistant topographer of the Survey on

Table 21.   Divisions and Chiefs, April 6, 1907.

Atlantic Division
Comprising the 21 states of the Appalachian and Atlantic Coastal areas, Frank Sutton, Geographer-in-Charge, assisted by James H. Jennings, geographer and Van H. Manning, topographer.

Central Division
The 10 states of the Mississippi Valley, Wm. H. Herron, Geographer-in-Charge.

Rocky Mountain Division
The 10 states and territories of the Rocky Mountain Region, Edward C. Barnard, Geographer-in-Charge.

Pacific Division
The seven far western states, Robert B. Marshall, Geographer-in-Charge. Geographer-in-Charge of office: Edward M. Douglas. Geographer-in-Charge of map editing: Henry Gannett. Section of Inspection: John H. Renshawe, geographer; Wm. M. Beaman and Francois E. Matthes, topographers.

August 16, 1882. He was Chief of the Drafting Section of the GLO for many years before his retirement, at the age of 70, on October 31, 1932. He died August 15, 1938.

Harry L. Baldwin, Jr. was appointed assistant topographer on July 23, 1884. A graduate of Princeton and a dynamo of energy, he pitched into the mapping of the United States as if he could complete it in a few years. He was nicknamed "Bee-line" and was said to get off the train running and keep on running until the field work was done and he caught the train back to Washington. He should have had a second platoon of assistants and animals.

Robert A. Farmer, on May 29 was appointed assistant topographer. He joined the Survey an October 24, 1888, as a station assistant. He retired on May 31, 1932, at the age of 70, and died on November 23, 1934.

The survey and resurvey of the public lands, by act of Congress approved June 25, 1910, provided that the Secretary of the Interior select competent surveyors for the work to be done by the GLO, in effect bringing to a close the contract system and authorizing the employment by a permanent corps of U.S. surveyors.

Alaska Mapping

A brief outline of the Survey's work in Alaska is appropriate at this time. An epic of adventure and accomplishment, it warranted the creation of a separate branch, the Alaskan Branch, and it deserves a history of its own.

In the summer of 1889, the Survey was given the opportunity to send Israel C. Russell, a geologist, to the Yukon Valley as an attaché to a U.S. Coast and Geodetic Survey party sent out to determine some positions on the 141st meridian, the boundary between Alaska and Canada, for about 100 miles. The following two summers, the same geologist was able to combine geologic reconnaissance in Alaska with an expedition under the joint sponsorship of the National Geographic Society and the Survey for the purpose of exploring a portion of the St. Elias Range. On the first of the expeditions, the geologist, as Chief, was accompanied by Mark B. Kerr, a topographer, as Assistant Chief.

Systematic studies were launched in 1895 when Congress made its first appropriation of $5,000 for the investigation of the gold and coal resources of Alaska. The annual amount was increased to $20,000 in the urgent deficiency bill early in 1898, as a recognition of the importance given to discovery of rich gold deposits in the Klondike Region, in the Canadian Territory. In 1901, the Survey received $60,000 for its work in Alaska and, in succeeding years, the appropriation often reached $75,000.

Field surveys in the almost unconquerable remoteness of Alaska had to be especially well planned, prepared, and executed. A typical season was that of 1898 when four parties were outfitted in Seattle and sailed northward on the U.S. Gunboat *Wheeling* on April 5. Upon reaching Skagway, the parties[29] were detached to cross by the Chilkoot Pass to the Yukon River, which they were to descend together as far as the White River. At Cook Inlet, the other two parties[30] disembarked to ascend the Susitna River and cross over the divide into the drainage system of the Kuskokwim River. All were to rendezvous at St. Michael on the coast some 50 miles northeast of the Yukon Delta on September 15. The first two parties traveled more than 1,300 miles over rugged country before reaching St. Michael.

Of primary importance has been the personnel of parties. Necessary qualifications were sturdy physique and experience in packing loads on the back, and familiarity with boats and with general habits of rough camp life. Travel was variously by dog team, canoe, raft, pack train, and by foot. Of volunteers among the Survey's topographers and geologists, only those were chosen who had demonstrated their special fitness during several field seasons in the States. They could be expected to cope with every problem encountered and bring the information collected. The chief of party was a geologist with a topographer as assistant chief, or the other way around, and the instructions to the assistant contained the clause "in case the chief is disabled, you are herewith authorized to take charge of the party and to carry out the plans as far as possible."

The Survey work in Alaska possessed great attractions. The field season, though strenuous, was short, usually about 3 months (June 15 to September 15); the comfortable office work in Washington lasted for 6 or 7 months; the railroad journeys between Washington and Seattle in the spring and fall were restful; the voyages north of Seattle among the wooded islands of southeast Alaska and along the coast where glaciers come down from the mountains to the sea furnished entertainment and thrills; and the 3 months in the silent rugged wilderness offered high adventure and successive successful seasons fitted geologists and topographers for greater responsibilities as chiefs of sections, divisions, and branches of the Survey.

The 586,400 square miles of Alaska presented problems analogous to those that the vast western mountains and deserts presented to the newly created Survey in 1879, and during the early 1880's. Similar methods of work were adopted. Direction of the work was given to Dr. Alfred H. Brooks, a geologist, in 1901, and his unit was given the title of "Division of Alaskan Mineral Resources" in 1907. On April 1, 1922, the division became the Alaskan Branch, which operated as a small Geological Survey as it embraced most of the functions of the larger organization, though in lesser degree.

The Alaskan Branch was dissolved on July 1, 1946, by Survey Order No. 150, dated May 17, 1946, and its topographic personnel, equipment, and functions were transferred to the Topographic Branch, to be attached in part, to the new Rocky Mountain Division.

Dr. Philip S. Smith, who became the Chief Alaskan Geologist after the death of Dr. Alfred H. Brooks on November 22, 1924, retired March 31, 1946, and died May 10, 1949.

---

[29] Edward C. Barnard, Topographer, Chief, Wm. J. Peters.

[30] Robert Muldrow, Topographer, William S. Post.

The short field season in Alaska, further abbreviated by a too abundant rainfall in many sections and by unexpected difficulties of travel, compelled constant study toward speeding the surveys and devising shortcuts.

As early as 1904, C.W. Wright and P.E. Wright, two geologists, built a panoramic camera from a commercial instrument and fitted it with level bubbles, a plate for screwing it on a planetable tripod, and internal scales that formed shadowgraphs on the negatives. The photographs taken at control stations and covering a complete horizon, when placed in a specially designed photoalidade, were converted into many square miles of topographic maps in the Washington office under comfortable conditions. Improvements in this method of mapping were made. In 1911, James W. Bagley, a topographer, was instructed to return from Alaska by way of Ottawa "to investigate the method of photo-topographic surveying in use by the Canadian government, with a view of obtaining information which will be useful in introducing these methods for some of the Alaskan surveys."

That winter, studies by Jim Bagley, F.M. Moffit, and J.B. Mertie, two Alaskan geologists, with the idea of photographing the ground from an airplane, led in 1916 to the beginning of their construction of a tri-lens camera and transformer. In 1917, Bagley was commissioned a major in the Engineer Officers Reserve Corps and was able to improve the camera with the aid of personnel and funds supplied by the Corps of Engineers and Army Air Service. The first photographs with the tri-lens camera were taken in the winter of 1917–18.

The lack of sufficient funds, however, prevented the adoption of the airplane and aerial camera for mapping operations. In 1926,[31] the Survey profited by an arrangement with the U.S. Navy, whereby the Alaska Aerial Survey Expedition photographed 10,000 square miles of southeastern Alaska, using the Army T–1 cameras (the Bagley tri-lens) mounted in two amphibian planes. Three years later, a second expedition photographed 12,750 square miles farther to the north, using the Army T–2 cameras (Bagley 4-lens).

In 1936, funds permitted the photographing of the entire Tanana River Basin, the Survey plotting the flight lines and furnishing the camera, a commercial company supplying the plane and pilot, the Hydrographic Office of the Navy assigning Lt. Cdr. John M. Hayne, an expert photographer.

Prior to 1926, ground surveys had accounted for the mapping of about 40 percent of Alaska on reconnaissance standards (1:250,000 scale with 200-foot contours), with surveys on scales of 1 inch to the mile or larger for practically all mining camps. In all, more than 400 maps and 400 geological reports had been produced. The small permanent force, usually four geologists and two topographers, engaged in this work was recognized, with the coming of World War II, to possess a wealth of information concerning the strategically important Territory and the equally important process of making reconnaissance maps from aerial pictures. The Alaskan Branch

became the school for the Aeronautical Chart Service and helped to develop the trimetrogon method of making topographic maps, expanding to a force of 250 persons. The senior topographic engineer of the Alaskan Branch, Gerald FitzGerald, was rewarded for his services by being commissioned a colonel and placed in charge of the Aeronautical Chart Service of the Army Air Forces. He became Chief, Topographic Branch, on May 22, 1947.

Immediately after World War II, the Geological Survey greatly increased its topographic mapping force in Alaska. Work was undertaken in many parts of the Territory, but the heaviest concentration of field men was in southeastern and central Alaska. This accelerated mapping program was aided by photographic units of the Navy who provided excellent single-lens coverage of all southeastern Alaska, as well as of extensive areas in western and northern Alaska. In addition to "Trimetrogon" photography, units of the Air Force completed standard mapping photography covering extensive areas in central Alaska. Airborne Shoran observations were made to provide horizontal control for much of this photography and a Geological Survey project established vertical control for a considerable part of the same area through the use of airborne altimetry.

Preparation of the Alaska 1:250,000-scale series (about 4 miles to 1 inch) consists principally of recompilation and readjustment of the older published maps with the new mapping where available. Because of military priorities, most of the topographic mapping is being done at a scale of 1:63,360 (1 mile to 1 inch) and larger. Contour interval is 100 feet with a 50-foot supplemental contour for the valleys and flatter areas.

Field parties were concerned principally with obtaining horizontal and vertical control (triangulation and elevations over the areas photographed) to provide a framework for stereo-compilation of topographic maps. The latter operation was done in Denver at the headquarters of the Rocky Mountain Region of the Geological Survey. Under ordinary conditions, field parties would leave Denver for Alaska early in May and return to headquarters in October. Depending on the location and the type of terrain, about six field parties of 10 men each were assigned to Alaska each year and operated under the direction of a District engineer, two project engineers, and a Resident Engineer. Field transportation was handled mainly by helicopter that carried surveyors and their equipment to observation points throughout the area to be mapped. This method of rapid transportation revolutionized mapping in Alaska.

The first experimental work was carried on with a chartered helicopter by the Rocky Mountain Region Engineer at Canon City, Colo. During the 10 days of this experimental project, 24 round trips were made to mountain-top stations with a payload on each of these trips of from 200 to 250 pounds. The helicopter was flown 16 hours and 45 minutes and covered a distance of about 650 miles. As the result of this experiment, three helicopters were chartered for work in Alaska during the summer of 1948. One ship assigned to southern Alaskan parties made 260 mountain-top landings in

---

[31] "Aerial Surveys in Southeastern Alaska," by R.H. Sargent, The Military Engineer, May–June, 1928.

142 hours of flying and covered an area of about 1,000 square miles. The stations occupied ranged from 3,500 to 4,500 feet above sea level. In the meantime, the two helicopters assigned to the vicinity of Fairbanks in the interior of Alaska were used for successful survey operations in establishing control for 2,500 square miles. Operational flights on this project were made to peaks at an elevation of nearly 9,000 feet above sea level. This work was so satisfactory that helicopters have been used in Alaska for survey operations with outstanding success during field seasons since 1949 by the Geological Survey, the Corps of Engineers, and the Coast and Geodetic Survey. The successful use of the helicopter for survey operations depends on careful planning of the project, ground and air communication, proper mechanical maintenance of the helicopter, and experienced and capable pilots.

Since the end of World War II, the Geological Survey has recompiled an up-to-date small scale map of Alaska which been published in three editions as Standard Base Map "E" scale 1:2,500,000; Map "A", scale 1:500,000, and a large wall map, scale 1:1,584,000. In addition, a new general purpose series covering the entire territory at a scale of 1:250,000, consisting of 153 sheets, has been designed, compiled, and reproduced. A large number of these new maps are being published in shaded relief and are considered to be outstanding examples of this type of general purpose map.[32]

In November 1953, Reynold E. Isto became Resident Engineer with headquarters at Fairbanks and he would direct the activities of the field parties and serve as contact representative for all activities of the Topographic Division in Alaska.

The Annual Report for 1954 states that the accelerated mile-to-the-inch mapping program was continued, but the scale of operations was reduced somewhat. Mapping on this scale (1:63,360) was continued on 409 Alaskan quadrangles covering an area of about 80,000 square miles. Actual accomplishment, in terms of stereo-compilation, was about 20,000 square miles, and 90 quadrangle maps were published, bringing the total of published maps of this series to 233.

George Otis Smith, Director

George Otis Smith was announced as the new Director on April 5, 1907, and he took office on May 1. Dr. Smith was a native of Hodgdon, Maine, and his determined pursuit of an education attained for him several degrees, an A.B. from Colby College in 1893; an A.M. in 1896; and a PhD. from Johns Hopkins University in 1896. After three summer seasons of geologic studies in several states he received an appointment with the Geological Survey in 1896 as assistant geologist.

---

[32] "Surveying and Mapping in Alaska," by Gerald FitzGerald.

Section of Cartography

The Section of Cartography, with Acheson F. Hassan in charge, began work in July 1908 on the 1:1,000,000 scale map of the United States. The meridian of Greenwich was adopted as the initial point in the numbering of the sheets, but at a meeting of the International Geographic Society, held in London in 1909, the initial point was changed to the meridian of 180 degrees from Greenwich. The sheets now are numbered from 1 to 60 and proceed eastward around the world. Each sheet comprises an area of 4 degrees between parallels and 6 degrees between meridians.

Mr. Hassan's assistant was Frederick M. Hart, a Washingtonian who came to the section in 1907 with government service as a draftsman dating from July 2, 1900.

Idaho–Washington Boundary Line

An appropriation of $25,000 was made by Congress for the survey of the Idaho--Washington boundary line, and the work was assigned to the Geological Survey by the Secretary of the Interior. The line, originally run from 1874 to 1875, was the 117 degree, 2 minute, 54 second meridian, extending northward from the center of the Snake River at the junction of the Clearwater River, a short distance west of Lewiston.

The work, undertaken in the summer of 1908 by S.S. Gannett, C.L. Nelson, and S.G. Lunde, was completed by the latter two men during the following summer. The line, a total length of 177 miles, was marked by 157 iron monuments, 6 feet long, 4 inches in diameter, filled with concrete; by 23 granite monuments, 6 feet long and 10 inches square; and by 229 iron monuments 4 feet long and 3 inches in diameter, filled with concrete, placed at the closing corners of the public land surveys. The topography of an area one-half mile in width on each side of the line was mapped by Mr. Nelson, assisted by Bayard Knock, for publication at the scale of 1:62,500 with a contour interval of 50 feet.

Foreign Assignment, Canada

Upon the request of the government of Canada for a topographer to help start a topographical survey of Canada, Robert H. Chapman was loaned to the Canadian Geological Survey from May 1, 1909, to December 31, 1911. He had direct charge of the triangulation and topography on Vancouver Island, British Columbia, and the instruction of assistants in those phases of mapping.

In 1912, for use by the 12th International Geological Congress held in Ottawa in 1913, an informal cooperative arrangement was made with the Canadian Geological Survey for a special resurvey of the Niagara Gorge, the work on both sides being done by members of the Topographic Branch, but the entire expense of the work, 11 square miles in the United States and 17 square miles in Canada, being borne by that Government. The field surveys and resulting superb topographic map, on the scale of 1:12,000, with contours at internals of 10 feet, were the work of Charles E. Cooke.

The map depicts an electric railway on each side of the river below the Rainbow Bridge, the one on the east side being within the gorge a few feet up from the water's edge, affording an exciting scenic ride. Both railways were discontinued in 1932.

The partial Niagara Falls Quadrangle, scale 1:62,500, and contour interval 20 feet, had been published by the Survey in May 1901. The mapping had been done in 1893 and 1900 in cooperation with the state of New York.

Triangulation was by the U.S. Lake Survey with topography by Frank Sutton, J.H. Wheat, and W.W. Gilbert of the U.S. Lake Survey. International boundary information was furnished by the International Waterways Commission.

Mapping the Hawaiian Islands

A few years after the Hawaiian Islands became a territory of the United States, the American planters, many of them descendants of New England missionaries, realized that further agriculture development depended on definite knowledge of the available water supply. In 1909, it was realized that comparative little could be done without adequate topographic maps. In a conference with the Chief Geographer of the Geological Survey, it was decided to start a program of topographic mapping on a field scale of 1:31,680, or 2 inches to 1 mile, with contour intervals of 10 and 50 feet. The island of Kauai was to be surveyed first, and work was begun early in 1910, under the jurisdiction of the Department of Public Works. Until June 30, 1910, the expense of this work was borne entirely by the Territory, and about $27,000 was expended previous to that date. The Geological Survey cooperated to the extent of loaning men and instruments to do the work.

On July 1, 1910, an agreement was entered into whereby the territory allotted $15,000, and the Geological Survey $5,000, for the continuation of the work, and cooperation on this basis was continued each year until the latter part of 1914, when the island of Kauai, containing 555 square miles, and 871 square miles of the island of Hawaii, had been mapped at a cost of about $111,000, $90,000 of which had been expended by the Territory. Using these figures, which include the original $27,000 expended by the Territory, the total cost of the work amounted to about $78 per square mile, but included in this is the cost of considerable leveling and triangulation, and about $2,000 that was used in making a map of the island of Oahu on a scale of 1:62,500 from data supplied by the War Department. This map was engraved and published on one sheet. The Kauai work was printed at 1:31,680 scale in seven photolithographs and was later engraved at the scale of 1:62,500 and published in one sheet. The work in Hawaii was covered by 13 photolithographs and was later engraved and published at 1:62,500 scale in five 15-minute quadrangles. C.H. Birdseye was topographer in charge of this work and was

assisted by N.E. Ballmer, W.H. Barringer, L.F. Biggs, S.H. Birdseye, A.O. Burkland, G.R. Davis, A.T. Fowler, J.L. Lewis, H.L. McDonald, T.H. Moncure, A.J. Ogle, J.M. Rawls, Olinus Smith, and O.G. Taylor.

On May 12, 1920, an agreement was made between the Director of the Geological Survey and the Governor of Hawaii for the continuation of the cooperative topographic survey of the Territory of Hawaii, as provided for in the allotment of $25,000 made by the Governor from his Contingent Fund on the same date. This money was to be under the jurisdiction of the Commissioner of Public Lands and immediately available. It was understood that it would be met with a like amount of Federal funds on July 1, 1920. After this date, similar cooperative agreements were signed for each fiscal year and a total of $237,500 had been allotted by the Geological Survey to meet the Territorial appropriations made to June 30, 1930. For the fiscal year beginning July 1, 1930, only $3,000 was allotted, as this was deemed sufficient to complete the cooperative work, but was later supplemented by special allotments of $1,000 each to ensure the completion of the work, and to reprint the Kauai and old Hawaii photolithographs, making a total of $242,500 allotted by the Geological Survey to meet territorial appropriations to the close of June 1932.

This work began in June 1920, and by the end of 1927, field work had been completed on all the remaining islands except Oahu. The total area mapped during this period, on a field scale of 1:31,680 was 4,393 square miles. An additional 22 square miles, constituting a part of the island of Oahu were mapped by the Coast and Geodetic Survey and incorporated in the Geological Survey map of that island. The cost of the work including photo lithography was approximately $358,000. During the 7 years, an average yearly output of 627 square miles was maintained at a cost of about $81.50 per square mile.

The mapping of Oahu was begun in July 1927, and the field work was completed in a little less than 3 years. At the request of the War Department, the mapping was executed on a scale of 1:20,000 with contour intervals of 10 and 50 feet, depending upon the slope. The cost of the work was borne equally by the Territory of Hawaii and the Geological Survey while the War Department cooperated by furnishing aerial photographs of the entire island, as well as pack horses and equipment when needed. The work on this island represented a much greater state of refinement than that on the other islands. Including Pearl Harbor, a total of 604 square miles was mapped at an expense of about $135,000. This represented an average cost of $224 per square mile. The island was divided into 16 quadrangles and, where possible, these were made to extend 6 minutes in latitude by 8 minutes in longitude. These 16 engraved maps will be combined on one sheet for the island of Oahu, and maps of all of the islands have now been engraved.

A.O. Burkland was in charge of this topographic mapping and was assisted by F.A. Danforth, G.S Druhot, Adolph Fankhauser, Helen M. Frye, Clerk, Max J. Gleissner, J.B. Leavitt, J.F. McCook, M. Madeleine M'Grath, Clerk, R.R.

Monbeck, Bishop Moorhead, Kostka Mudd, E.S. Richard, R.G. Stevenson, Arline G. Weeks, Clerk, R.M. Wilson, and E.G. Wingate.

A second group, consisting of H.B. Smith, J.O. Kilmartin, V.S. Seward, A.K. Andrews, and L.B. Lint spanned the years 1920–30.

In 1948 and 1949, the Engineering Association of Hawaii requested Governor Stainback to ask the Geological Survey to send representatives to the Territory of Hawaii to investigate the need for revision of topographic maps, and Gerald FitzGerald, Chief Topographic Engineer, arrived in Honolulu on June 13, 1949, for conferences. It was agreed that as the Territory of Hawaii had no money for cooperation, the Geological Survey would start a revision program. Accordingly, on October 30, 1949, Adolph Fankhauser, Resident Project Engineer, left for Honolulu. There were difficulties in obtaining aerial photographs from the Navy, Army, or commercial firms, but in fiscal year 1951, field operations were started for the revision of the obsolete map series covering the Hawaiian Islands. According to the 1954 Annual Report, mapping in the Territory of Hawaii was continued for the publication of topographic maps at the scale of 1:24,000; this scale being preferred by local engineers and surveyors. Operations were conducted on the islands of Oahu, Molokai, and Maui on a total of 31 quadrangles covering an area of 1,360 square miles. Four of the five quadrangles covering the island of Molokai were published, and the fifth was nearly completed. It was planned to extend this mapping program to the island of Hawaii at an early date. Mr. Fankhauser was assisted by Karl Thornton and Melville Baker, Jr.

National Park Service

The Topographic Branch was asked to help solve some of the problems of the new Bureau in the Department of the Interior, the National Park Service. Beginning with the creation of the Yellowstone National Park in Wyoming, by act of Congress on March 1, 1872, unusual tracts of land in various sections of the United States, from time to time, were set aside as pleasure grounds for all people and their natural wonders preserved from desecration. With the establishment of Glacier National Park in Montana on May 11, 1910, 12 of these reservations were under the general jurisdiction of the Department of the Interior, and were administered by a small unit in the Secretary's office.

The Secretary of the Interior, in his report for the fiscal year ending June 30, 1910, recommended that a bureau be created to have sole charge of the National Parks, and bill S.826 was presented to the session of the 63d Congress. The bill failed of passage repeatedly until August 25, 1916, when it became a law, establishing the National Park Service as the ninth bureau within the Department of the Interior. In the fall of 1914, a general superintendent, with headquarters at San Francisco was appointed. This arrangement proved unsatis-

factory and on December 10, 1915, Robert B. Marshall was appointed general superintendent at Washington, D.C.

In 1894, Mr. Marshall had mapped the Yosemite 30-minute quadrangle. On June 14, 1904, he was appointed a member of the Yosemite National Park Commission, created by the Secretary of the Interior for the purpose of changing the park boundary as authorized on August 31, 1940. As a member of the first conference of departmental officials, park superintendents, representatives of various railroads, concessionaires in the various reservations, and others, called and presided over by the Secretary of the Interior, in Yellowstone National Park in September 1911. Mr. Marshall presented a paper on the general administration of the national parks and their particular needs. As a member of the second superintendents conference held in the Yosemite National Park in October 1912, under similar auspices, he presented a paper on the important subject of the admission to national parks of automobiles and their restricted use to selected roads. Later that fall, he was asked to report on the advisability of creating a national park in the vicinity of Estes Park, Colo., and he submitted his report on January 9, 1917, when he returned to the Topographic Branch as its Chief Geographer, and Stephen T. Mather was appointed the first Director of the National Park Service. Isabelle F. Story, Arthur E. Demaray, and James V. Lloyd, who had transferred with Mr. Marshall, remained with the National Park Service. Members of the Topographic Branch that were transferred or detailed to the National Park Service are listed in table 22.

Table 22. Members of the Topographic Branch who were transferred or detailed to the National Park Service.

R.H. Chapman detailed as Acting Superintendent of Glacier National Park from May 15 to December 1, 1912.
W.O. Tufts detailed for special surveys in the parks during 1916.
W.B. Lewis transferred as Superintendent of Yosemite National Park on March 4, 1916.
O.G. Taylor transferred as Resident engineer of Yosemite National Park on March 2, 1920.
J. Ross Eakin transferred as Superintendent of Glacier National Park on May 12, 1921.
C.L. Nelson detailed as Acting Superintendent of Mount Rainier National Park from June 11, 1922, to November 8, 1923.
R.T. Evans detailed as Acting Superintendent of Zion National Park during tourist seasons of 1925 and 1926. Transferred as Superintendent of Hawaii National Park on January 1, 1927, and transferred back to the Survey on November 21, 1928.
M. Madeleine M'Grath transferred on November 16, 1925.
Olinus Smith transferred on November 1, 1933, and his task became the survey, and purchase, or transfer, of tracts adjacent to park boundaries.

Maryland–West Virginia Boundary Line

S.S. Gannett was the technical member of the commission appointed in 1910 by the U.S. Supreme Court to resurvey and mark the comparatively short west boundary of Maryland. This was known as the "Deakins" or Old State Line, and disputes over its proper location concerned the ownership of about 40 square miles of land. The commission established the initial point, or the southwest corner of Maryland, 3,989 feet almost due north from the Fairfax Stone. As re-established, the "Deakins line" was a jogging line with a general bearing a little east of north. There were five offsets in the line, which extended nearly east and west, and ranged in length from 54 to 971 feet. A large concrete monument was erected at each angle and many at intermediate points, 60 in all, on the line, which extended nearly 36 miles to the Mason and Dixon Line.

North Carolina–Tennessee Boundary Lines

In 1910, at the request of the Attorney General of North Carolina, S.S. Gannett was designated by the Director of the Geological Survey to investigate and, if possible, to find the line as marked by blazed trees in 1821 along the disputed part of the North Carolina–Tennessee boundary line in the Slick Rock Creek and Tellico River region. As a result of these investigations, he appeared as an expert witness in the suit: No. 4 Original, the state of North Carolina, Complainant vs. the state of Tennessee, 1914, of the Supreme Court of the United States, which suit was decided in favor of the Complainant, by the Court.

Mapping Mount Rainier, Washington

The topographic survey of Mount Rainier National Park, completed in 1914, provided an adequate map for the Park Administration. Mount Rainier, about 42 airline miles southeast of the city of Tacoma, Wash., is the dominating feature of the park. It is the loftiest and most magnificent volcanic peak in the United States, rising 8,000 feet above the skyline of the Cascade Range, and is doubly impressive by the vast extent of its snow mantle, consisting of 28 glaciers that cover an area of 45 square miles. Each glacier is the source of a river that cuts a deep trench through the surrounding rugged mountains. The area was set aside in 1899 as the sixth national park and contains 325 square miles.

The topographic mapping was begun in 1910 when E.M. Bandli extended 45 miles of primary levels and established 12 permanent bench marks, the highest (5,557) at the end of the road in Paradise Valley.

C.F. Urquhart occupied 7 triangulation stations; G.R. Davis and F.E. Matthes mapped 36 square miles for publication on the scale of 1:62,500 with a contour interval of 100 feet. They returned in 1911 to map 141 square miles. The work was suspended in 1912, but resumed in 1913, by C.H. Birdseye, as supervisor, with W.O. Tufts, O.G. Taylor, and S.E. Taylor, as assistants, and completed in 1914.

To settle a controversy, which had arisen as to the actual height of Mount Rainier, it was decided that a set of reciprocal angles should be observed from the summit. To reach the summit was an adventurous ascent that called for hours of foot climbing on snow ice, with the prospect of a storm rendering the effort futile. It was like Birdseye to claim the adventure for himself and his two field assistants, C.B. Harmon and Frank Krogh.

On the first ascent they were turned back short of the Summit by a violent storm that promised to hold out all day. A couple of days later, on August 15, they started as before from the Camp Muir shelter cabin (10,000 feet) at 4 a.m., and reached the summit in 4 and one-half hours of climbing. Again a storm compelled them to retreat. This time the driving snow obliterated the marked route and, blinded by the storm, and fearful of the many crevasses likewise concealed by the new snow, the three men made their way back to the summit and sought refuge in one of the tiny ice caves among the steam vents in the large crater; there they spent the night. Encompassed by ice, they were kept from freezing by the steam issuing from the vent which also kept their clothes wet and clammy. They took turns in maintaining a watch that none should fall asleep. A little comfort was derived from a light lunch and a cup of hot tea from a thermos bottle. The next day, with the storm considerably abated, they descended the mountain; they were half starved and almost exhausted, and Birdseye and Krogh had varying degrees of frostbite.

Their third attempt, on August 19, was successful. Leaving the hotel in Paradise Valley (5,400) at 1 a.m., accompanied by eight tourists, they climbed at a brisk pace, passing Camp Muir (10,000) at 4 a.m., Gibraltar Rock (12,679) at 6 a.m., and reached Register Rock on top at 8 a.m., a climb of nearly 9,000 feet in 7 hours. The day was clear. The greater portion of the state of Washington and the northern part of Oregon lay unfolded before them. The necessary vertical angles were observed and the mapping of the entire summit was completed by 1 p.m., as the tourists carried the rod and instruments from point to point.

Columbia Crest is a huge dome-covered summit with perpetual snow, measuring about 20 feet in thickness. This does not vary much, as wind accompanies snowfall and prevents new snow from settling. The vertical angles established a height of 14,408 feet, displacing the former supposed height of 14,363 feet, and making Mount Rainier the third highest mountain in the United States, being overtopped 87 feet by Mount Whitney (14,495) in California and 23 feet by Mount Elbert (14,431) in Colorado[33].

Mapping Costs

The relative costs of field surveys and of office work were discussed in the various meetings of topographers during the winters in Washington. In January, 1911, analysis was made of seven selected 15-minute quadrangles, worked on the 1:48,000

---

[33] See "Surveying our Greatest Volcano," by Claude H. Birdseye, Travel Magazine, May 1915.

field scale and recently completed, as follows in table 23 and funds appropriated are listed in table 24.

Table 23. Selected 15-minute quads completed in January 1911.

Quadrangle name and year mapped	Contour interval (feet)	Field surveys	Office work	Total	Topography by
Blaine, Wash. (1905)	20	$23.00	$1.66	$24.66	Pearson Chapman and L.E. Tucker
Hagerstown, Md. (1908–10)	20	$19.17	$2.53	$21.70	Jos. H. Wheat and R.L. Harrison
Laurelville, Ohio (1910)	20	$15.22	$2.76	$17.98	Jesse A. Duck
Mason, Mich. (1908–09)	20	$8.10	$1.54	$9.64	C.D.S. Clarkson
Moeteetee, Wyo. (1911)	25	$11.09	$2.32	$13.41	C.C. Gardner and C.C. Holder
Quincy, Wash. (1909)	25	$17.01	$1.57	$18.58	C.F. Eberly
Rolla, Mo. (1908–09)	20	$27.80	$3.40	$31.20	C.G. Anderson, H.H. Hodgeson, and F.W. Hughes
Average cost per square mile	$17.34	$2.25	$19.59		
For 1:96,000 quadrangles	9.22	0.88	10.12		
For 1:92,000 quadrangles	2.34	0.22	2.56		

Potomac River Survey

At the request of the Commissioners of the District of Columbia, the Chief Hydrographer of the Geological Survey, Marshall O. Leighton, made a study of the available water supply at Great Falls to determine whether it could be increased to meet the growing daily consumption of water in the District of Columbia, and whether it could be harnessed to produce light and power. Mr. Leighton's report, dated January 14, 1911, while favorable, pointed to the necessity of more detailed studies based upon accurate topographic information. Before these studies were made by the Corps of Engineers, U.S. Army, as was authorized in the District bill approved June 26, 1912, a topographic survey was undertaken of the Potomac River from Seneca, Md., to the Chain Bridge above Washington.

Table 24. Funds appropriated for topographic mapping for the fiscal year ending June 20, 1911.

Activity	Funds appropriated
Topographic surveys—Federal	$359,200
Surveying national forests	75,000
Cooperative allotments by 18 states and Hawaii	154,383
Total	588,583
For this funding, the square miles mapped were:	29,956
Average cost per square mile:	19.71

A small project, it was performed by topographers in the winter, (1910–11), so as not to interfere with their regular summer assignments, and to be at minimum expense by having topographers fill out the parties as recorders and rodmen. Assignments were made in groups, leaving to each group the selection of the instrumentman. The instrumentmen were D.H. Baldwin and J.R. Ellis, on control; and R.T. Evans, E.P. Davis, Fred Graff, Jr., R.L. Harrison, C.E. Cooke, and H.W. Peabody, on topography. The mapping was on the scale of 1:12,000 with a contour interval of 5 feet. The work was begun in November by the two topographers who were first arrivals in Washington, both starting up river above Seneca; one following down the Maryland side of the river, the other, the Virginia side. Pleasant fall weather prevailed until they were in the neighborhood of Great Falls, about the time the other parties were organized and ready to start.

The winter of 1910–11 was one of Washington's severest. An early snow fell all day Monday, December 5, 1910, to a depth of 11 inches in the suburbs and was followed by a series of cold waves that sent night time temperatures as low as 10 degrees and continued until near the end of January. The snow did not melt during those 7 weeks and stayed about 15 inches deep, for a second storm on December 10 had added 6 inches to the snow already on the ground.

The field work coincided with the seven wintry weeks. The daily routine began with a street car ride from home to Georgetown in the dark, a ride in a farm wagon with the beginning of dawn to the neighborhood of the work, standing at the instrument, or tramping in the snow all day, and the return trip, reaching Georgetown at dark. By way of variety, some parties found places to stop overnight, one such place being the Great Falls Inn at Lock 20, which ordinarily took in travelers only in the summertime, but agreed to accommodate the survey parties. The meals were good, but the unheated bedrooms were like refrigerators!

The mapping progressed in spite of the snow and cold. Everyone was furnished a pair of rubber boots, a bonfire was kept going at every instrument station, the absence of leaves permitted longer sights, and the ice on quiet water was thick enough for planetable stations from which sights were taken on the rod at the water's edge. This simplified the work among the wooded islands. Only once did the ice give way and then with the tallest rodman, John H. Wilson. He was pulled out and dried at a bonfire.

The report of the Army engineers confirmed the belief that there was sufficient water in the Potomac River to supply the needs of Washington by building a second conduit from Great Falls. For power purposes, two high dams were considered: one just above Chain Bridge to create a pool with a surface elevation of 115 feet; the other, at Great Falls, to create a pool with an elevation of 215 feet and extending 39 miles to Brunswick, Md. The two pools would develop a surplus amount of hydroelectric power after supplying light and power for uses of the United States and of the District of Columbia; would require one or more storage reservoirs in the upper reaches of the Potomac to guarantee a calculated minimum flow of 2,000 cubic feel per second during the low flow stages of the river, as the observations of the Survey hydrographers since 1891 had indicated a fluctuation between 653 and 390,000 cubic feet per second; would submerge the Chesapeake and Ohio Canal and the Great Falls, and utterly destroy the magnificent natural scenery of the Potomac River gorges so near to the capital city of the United States.

Washington and Vicinity Map

The Potomac River survey incited many inquires for map information concerning the city of Washington, and its mapping was begun by the Survey in the spring of 1912. Topographers were urged to hasten with their office work and were assigned to various areas in and about the city, working from their homes, without travel allowances, and without horizontal and vertical control until triangulators and levelmen could later reach them (table 25). It was decided to make the map as large as could be printed by the Survey's largest press; that is, about 50 inches by 40 inches, which would permit the mapping of a broad suburban area on a scale of 1:31,680 (2 inches equal 1 mile). Then 10 feet was chosen as the contour interval. The map extended from 38 degrees 45 minutes to 39 degrees 5 minutes in latitude and from 76 degrees 54 minutes to 77 degrees 15 minutes in longitude.The mapping was not completed until the close of 1915.

Table 25. Field engineers who worked on the Washington, D.C., and vicinity map.

C.G. Anderson	W.H. Griffin	F.E. Matthes
C.W. Arnold	R.L. Harrison	W.H.S. Morey
N.E. Ballmer	E.I. Ireland	Robert Muldrow
J.E. Blackburn	B.A. Jenkins	A.J. Ogle
F.W. Crisp	Oscar Jones	H.W. Peabody
J.A. Duck	R.A. Kiger	S.T. Penick
J.R. Eakin	J.H. LeFeaver	Albert Pike
C.F. Eberly	L.L. Lee	Roscoe Reeves
C.A. Ecklund	J.L. Lewis	H.S. Senseney
R.T. Evans	S.G. Lunde	Olinus Smith
S.P. Floore	J.F. McBeth	W.O. Tufts
J.I. Gayetty	H.L. McDonald	J.M. Whitman
C.E. Griffin	Fred McLaughlin	Fred Graff, Jr.

Survey Order No. 5

Survey Order No. 5, dated December 19, 1911, reads as follows:

"Every request for manuscript map data in the possession of the Topographic Branch must hereafter be made through the Chief Geographer. Requests for advance photolithographic sheets, or for special photographs of partial sheets, are to be submitted in the same way, and all business pertaining to the Millionth-Map work must be conducted with, or through the Chief Geographer, who is Chairman of the Committee in charge of that work.

"The purpose of this order is two-fold, to protect manuscripts from loss or damage, and to prevent the misunderstanding that so often results from informal procedure. The regulations, page 128, that correspondence between branches should pass through the Director's office, need not apply to such routine matters as those mentioned, but inter-branch correspondence must pass over the desks of the branch chiefs."

Survey Order No. 11

Survey Order No. 11, Rearrangement of Divisions in the Topographic Branch, dated June 25, 1912, effective July 1, 1912, was as follows:

"Effective July 1, 1912, a rearrangement of the divisions in the Topographic Branch will be made, as follows:

"Atlantic Division (unchanged)
In charge: Frank Sutton, Geographer.

"Central Division (unchanged)
In charge: William H. Herron, Geographer.

"Rocky Mountain Division (unchanged)
In charge: Sledge Tatum, Geographer.

"Pacific Division, consisting of the States of California, Nevada, Utah, Arizona, and the Territory of Hawaii
In charge: George R. Davis, Geographer.

"Northwestern Division, consisting of the States of Idaho, Oregon, and Washington
In charge: Thomas G. Gerdine, Geographer."

Winter Work in the White Mountains

Discussions and debates by earnest people in the States of the Appalachian Region culminated in the Weeks Act (John W. Weeks, M.C., from Massachusetts), approved March 1, 1911. Sections of the act provided for the acquisition of lands for the protection of the watersheds of navigable streams; the creation of a commission to administer the act (the commission to be composed of the Secretaries of War, Interior, and Agriculture; two U.S. Senators; and two U.S. Representatives); and perform the examination of lands, by the U.S. Geological Survey, before selection and purchase by the National Forest Reservation Commission. Topographers assigned to the White Mountains are listed in table 26.

Table 26. Topographers assigned to the winter work in the White Mountains.

E.I. Ireland	W.H.S. Morey	Hersey Munroe
Roscoe Reeves	Karl E. Schlachter	J.H. LeFeaver
E.E. Witherspoon	Robert L. McCammon	F.L. Shalibo, Rodman

The Geological Survey, in undertaking the examination of lands, accepted the demonstrated facts that forests prevented or retarded erosion that caused soil and rocks in times of heavy rains to be washed down to silt streams and form sandbars in the navigable portions, but recognized the need for particular studies in order to establish the effect of vegetation in upland areas on the stage of flow in navigable streams. Besides vegetation, the quantity of water in a stream was dependent upon as many as five other factors: (1) rainfall and climate; (2) geology in respect to rock formations, thickness of overlying rock debris and soil, and underground flow; (3) topography in respect to surface irregularities and degree of slope; (4) geographic position; and (5) works of man, such as storage reservoirs.

In the White Mountains of New Hampshire, the Geological Survey found but slight evidence of erosion and, therefore, conducted investigations for the purpose of determining the regulative effect of the forest upon the stream flow. Two areas were selected with great care for the study. The Shoal Pond Brook and Burnt Brook were similar in every way, except that the first was forested whereas the second had been burned over, and the mat of vegetal soil depleted by successive fires. Observations, carried on during the winter months of 1911–12, demonstrated that the rate of runoff from the forested basin was from 50 to 67 percent of that from the denuded basin.

The National Forest Reservation Commission wanted the facts before July 1, 1912, which meant field work would be done before spring opened up. Hydrographers of the Water Resources had established gaging stations on Shoal Pond, Burnt, and five other brooks and in March, 1912, set up in the basins of the two selected brooks many snow scales and rain gages for the observation and recording of precipitation and runoff.

In March, also, topographic parties were sent into the snow-covered basins. The mapping was on scales adapted to the size of the several basins, and varied from 1 inch to 2,000 feet, or 1:24,000, and included the careful location of all gaging stations, snow scales, and rain gages.

Elevations began with 1,500 feet and ascended to 3,000-6,000 feet along the crests of the different basins. Throughout the month of March, planetable work was difficult, in snow that varied from a depth of 2 feet to 20 feet, where the snow had drifted, and was performed with the aid of snowshoes. Vertical control was executed by Reuben A. Kiger, and included the extension of a line of levels along the cog railroad to the summit of Mount Washington, at 6,284 feet.

The topographic mapping was executed in cooperation with the National Forest Reservation Commission and was completed in time for the Survey's report, the determination of boundaries of purchase tracts by the Forest Service, and the approval for purchase by the Commission of three tracts containing a total of 70,370 acres in the White Mountains before July 1, 1912. Approved for purchase also were 15 other tracts aggregating 643,045 acres in the States of Maryland, Virginia, West Virginia, North Carolina, Tennessee, and Georgia. When purchased, the tracts, consisting of forested, cutover, and denuded lands, became National Forests, and were transferred to the Forest Service for administration and protection from illegal timber cutting, over grazing, and fires. Not long after July 1, Congress amended the Weeks Act by making available, until expended, the entire amounts appropriated for the four fiscal years from 1912 to 1915, thereby making available $8,000,000 for the purchase of land, and for the expenses of examination and survey.

Operations of the National Forest Reservation Commission continued without interruption through the years. It was estimated that about one-third of the area of the United States would better serve economic and social-purposes by being set aside for the growth of timber. Proof of the soundness of the program was shown by the multiplying number of states that enacted the requisite acts of consent to the initiation of the program within their borders. Examination of new watershed areas by the U.S. Geological Survey continued to show that forests have a favorable affect upon the regimen of navigable streams both in minimizing erosion and the silting of stream channels, and in tending to equalize stream flow. The greatest progress that was made in carrying to fruition of the purchase program of the Commission was during the interval from 1933 to 1936, when President Roosevelt allotted $46,712,150 for land acquisition funds, which made possible the purchase of more than 11,000,000 acres of land.

Motorcycles and Motorcars

Mr. Marshall instructed Walter N. Vance to try out a motorcycle on traverses of roads on the Shauck Quadrangle, Ohio, in 1912. Road surfaces were rough in those days and riders of motorcycles were bumped and frequently spilled. The rodman, Rhea B. Steele, rode a bicycle. To make better time to and from work in the morning and evening, Mr. Vance contrived to tie the stadia rod to his motorcycle and to tow Mr. Steele on the bicycle at the end of a 15-foot rope. Mr. Steele held the rope after a couple of turns around the handle bars. Even so, there were times when he couldn't release the rope quickly enough to avoid a spill. Also, the rod was coming loose and tangling with the rear cyclist. It was not long before Mr. Steele also acquired a motorcycle; however, this mode of transportation was not adopted by the Survey.

The Topographic Branch then experimented with automobiles (fig. 19). Albert M. Walker used one in North Dakota in 1914. The Geologic Branch tried one out in California. When not in use, it was loaned to the Topographic Branch. In 1916, the two topographers mapping the Turner Ranch Quadrangle west of Merced in the San Joaquin Valley used it for a couple of weeks and were delighted with its greater speed than that of a pair of mules and buckboard. It was a Ford touring car and for some reason, the gauge was a couple of inches wider than standard, preventing the wheels from fitting into the ruts of private and mountain roads. To drive it was a new adventure for the topographer, and occasionally required all his skill and courage when necessary to steer the wheels on the two planks set at standard gauge, which substituted for bridges across irrigation and drainage ditches.

The Pacific Division, early in 1917, acquired five new Ford touring cars, and in March they were driven, with tops down and a small American flag flying on each, by a group of topographers from Sacramento to King City, Calif., where a base camp was established. The famous Tie Sing was cook. In a day or two parties were organized and departed for the several quadrangles to be mapped. There were not sufficient cars to go around, leaving to some topographers the usual buckboards and mules. The fortunate topographers, however, found that the water boiled in the radiators on long mountain grades and that extra water had to be carried in large canteens for the purpose of refilling the radiators. Further, to add to the transportation difficulties, the cars were unable to climb the steepest mountain roads.

In the spring of 1918, field parties that could not be equipped with the usual camp layout were permitted to live off the country and rent automobiles for use in the daily work. The cars were usually the touring type and the monthly rental was between $100 and $150, the Survey paying for all running expenses in addition. Improved roads were surfaced with gravel or crushed rock and sharp fragments punctured the tires all too often. Hardly a day passed without at least one puncture. If the tires were over inflated so as to avoid punctures, a sharper bump in the road resulted in a broken spring.

New Building Authorized

In the winter of 1912 to 1913, Congress authorized the expenditure of $2,596,000 for a new building for the Geological Survey on the block covering 18th and 19th Streets, E and F Streets, NW. It was to be a fireproof building "of modern office building type of architecture" and would accommodate, besides the Geological Survey, the Reclamation Service, the Indian Office and the Bureau of Mines, and all Bureaus of the Interior Department whose work was closely related to that of the Survey.

The authorization was the culmination of 26 years of effort. In the early spring of 1885, the Survey offices were moved from the National Museum to the Hooe Building, 1330 F Street, NW. It was the first building in Washington of iron construction. The Survey laboratories would remain in the National Museum. As early as 1886, the Survey called attention to the crowded working conditions, with attendant inadequate ventilation, and lighting, and the impossibility of systematic arrangement of office and of working materials. Some relief was found by the rental of nearby office space: in the early 1890's, the top floor of the Adams Building, 1333 to 1335 P Street NW.; some rooms in the old Post Office Building on F Street between Seventh and Eighth Streets, NW. In 1899 to 1900, paleontologists often were quartered in the National Museum. In 1906, the basement floor of the Hooe Building was secured. In 1913–14, the Rocky Mountain Division was accommodated for its winter office work on an upper floor of the Munsey Building and in 1914–15, on the fifth floor of the Adams Building.

The crowding of rooms with furniture and materials, and some hallways with cases of records created a real fire hazard, of special concern to geologists, hydrographers, and topographers whose original notes of investigations and map manuscripts, the fruits of their labors during one or more field seasons, were placed in constant jeopardy. Some fires did occur but the most disastrous one was in the front basement on Sunday afternoon, May 18, 1913. The fire damage was estimated at approximately $100,000. About 175,000 copies of the latest topographic maps were destroyed; 200,000 copies of geologic folios were destroyed or seriously damaged; and about 60,000 book publications, most of which were reserve copies, including the older and rarer publications, were partly or wholly destroyed.

The Survey had many suggestions as to desirable features for the new building, and in 1913, an advisory committee was organized with A.H. Brooks as chairman, and Sledge Tatum and Herman Stabler as members. The Survey moved into the new Interior Building in May 1917, and after the South Interior Building was completed, it was known as the North Interior Building. In August 1943, it was designated the Federal Works Building, and in September 1949, the name was changed to General Services Building. Again, the Geological Survey had outgrown its quarters and a new, larger building was contemplated.

Shaded Relief Maps

In 1913, John H. Renshawe, inspector of maps and accomplished watercolor artist, at the request of the Secretary of the Interior, prepared shaded relief maps (fig. 20) of the western national parks. Shading was applied to a special printing of the topographic base in a manner to emphasize the slopes of the mountains and hills, thus creating the effect of the third dimension for the map and rendering its larger features more understandable; especially to the layman. The Survey had experimented with relief shading in the early years, but the cost of reproduction by the lithographic process was then prohibitive.

In 1914, A.M. Walker was appointed inspector to take over Mr. Renshawe's duties, as his time was fully occupied with relief mapping. Shaded relief maps of Crater Lake, Yosemite, Glacier, Yellowstone, and Mount Rainier National Parks were prepared by Mr. Renshawe. These maps were so popular with the public that during the following years, he prepared relief maps of many areas: Mesa Verde and Rocky Mountain National Parks; San Francisco and vicinity, California; also a model showing the Survey's work for the Exposition at San Francisco; the United States; Salem and vicinity, Oregon; state of Ohio; Southern California; Alaska with periodic revisions, for the Division of Alaskan Mineral Resources; aviation route between San Diego and Los Angeles, etc.

From July 1, 1920, to June 30, 1925, when he retired, Mr. Renshawe was Geographer-in-Charge of the Section of Relief Maps. Ralph W. Berry, who had been selected by Colonel Birdseye in 1921, as the engineer with the most artistic ability, to train with Mr. Renshawe and continue shaded relief mapping after Mr. Renshawe's retirement, was Engineer-in-Charge of the Section of Relief Maps from July 1, 1925, until the section was discontinued on June 30, 1928. Mr. Berry resigned on May 31, 1930, but returned to the Survey on September 29, 1933, and was assigned to the Cartographic Section. The map of the Monument Springs Quadrangle, Tex., was the first shaded relief map published (1921) by the Geological Survey and was followed by a series of maps, most of which were topographic quadrangle maps of 1:62,500 scale. The relief shading on these maps was done with charcoal, pencil, crayon, chalk, and in some instances, with a combination of two or more of these media. The copy was then photo processed, transferred to stone, and touched up by lithographic artists for printing.

In World War II, the U.S. Army, recognizing the advantages of relief shading in teaching map reading, and for briefing purposes, requested the Geological Survey to reactivate a relief shading unit to prepare small scale maps of the Far Eastern Theater. This unit, under Mr. Berry's direction, prepared a series of maps of China and Japan, which proved to be very popular. Mr. Berry received a meritorious award from the Department of the Interior for excellent relief shading.

In 1946, the Shaded Relief Unit was moved to the Rocky Mountain Region in Denver, Colo., and continued as a part of the Region's Cartographic Section. Under the direction of Hal Shelton, shading was done by airbrush, and after Mr. Shelton's resignation on April 15, 1949, was continued by Gene Harbert. Shading by airbrush, together with improved reproduction techniques, resulted in speedier output and better tone qualities in the end products. An Eastman contact magenta halftone screen was used in the preparation of reproduction negatives. The resulting maps were a great improvement over those prepared by the old methods, but some still lacked the desired contrast and depth in color.

Subsequent experiments developed the present process of preparing two drawings for each map; a relief drawing and a green tone drawing. From the relief drawings two negatives are made. One at normal exposure, and the other at an exposure just long enough to photograph the darkest parts of the copy. Three press runs are required in printing. The plate made from the normal negative is run first, in full tone color (brown) so that the greatest possible contrast can be obtained; followed in order by the plate made from the second or shadow negative, which accentuates the darkest parts of the map, and then the plate made from the green tone drawing, which further accentuates the valley depths. This procedure is now in use, except that the standard interposing halftone screen is used instead of the magenta screen. More study and experimentation is necessary before the maximum effectiveness in making relief drawings can be reached; the search for improvements in shading methods, equipment, and reproduction processes will be continued.

In 1930, the new shaded relief map of the Crawford Notch Quadrangle, N.H., was sent out for comment and Albert Pike, former Atlantic Division Engineer, wrote as follows:

> "Many thanks for the shaded map of the Crawford Notch Quadrangle, N.H.. It is most effective, very interesting, and what a fine exhibit could be made; sample of maps, one in each 48 States.

> "Many years ago, Mr. Renshawe made the first one I had ever seen, the best I have ever seen, the old man was a real artist. I thought of the same thing at that time. We ought to have an exhibit, and too, why don't the cooperating states place a topographic map in each of its high schools. Maybe since the Survey has a more liberal director we might do a bit of educational work.

> "I am still interested in the Topographic Branch, and its progress toward an accurate map is most gratifying. You know that R.B. Marshall in changing from aneroid and buggy wheel to stadia did a fine thing. The pioneer often gets cussed. In my office work my first thought was to get a more accurate map. The airplane photo was a help but the pioneer Multiplex machine in the T.V.A. was a Godsend, and while they did well, it was hard to convince them that they could do better. Any photo seemed to be good

enough and it took the Clarendon office to prove that photos with full foliage was a mistake and too little control was a mistake. Now you are getting more control and cheaper control with the barometric instruments; also better pictures.

"I believe the Survey is making a wonderful map and it deserves plenty of support. More power to you and with best wishes."

In fiscal year 1954, relief shading was completed for 10 quadrangles and 3 state maps. (It is estimated that relief shading adds about 5 percent to the total cost of the average map.)

The Panama–Pacific International Exposition and The Panama–California Exposition

The Panama–Pacific International Exposition at San Francisco and the Panama–California Exposition at San Diego were opened in February 1915, to celebrate the completion of the Panama Canal (opened informally on August 15, 1914, and formally by presidential proclamation on July 12, 1915) and the quatercentenary of the discovery of the Pacific. The Geological Survey, for the use of visitors to these west coast extraordinary attractions, issued new printings of two quadrangle maps and a set of four guidebooks of the western United States.

The San Francisco 15-minute quadrangle with 24-foot contours had been mapped in 1892–94, the topography compiled from previous surveys by the U.S. Coast and Geodetic Survey, the U.S. Engineer Corps and by the City, and by original surveys by R.H. Chapman, R.B. Marshall, and W.H. Otis. In 1914, B.A. Jenkins was assigned to the work of revising the culture, which included particularly the addition of the splendid new exposition palaces. They were conspicuous on the map in the exposition grounds on the north edge of the city and fronting on San Francisco Bay for nearly 2.5 miles.

The San Diego 15-minute quadrangle with 25-foot contours had been mapped in 1902; the triangulation by A.P. Davis and the U.S. Coast and Geodetic Survey, the topography by J.E. Rockhold, and the U.S. Coast and Geodetic Survey; it was reprinted without revision.

The guidebooks, published as Survey Bulletins, covered four railroad routes west of the Mississippi River. Railroad travel was being stimulated by the two expositions, by the mounting popularity of the national parks, and by the diversion to the west of vacation travel usually directed eastward to Europe, then in the throes of war.

The Survey considered it an opportunity to present to the traveler-reader much geographic and geologic data, which it had accumulated in decades of pioneering investigation.

The guidebooks were illustrated by photographs, geologic diagrams, and geologic-topographic strip maps covering the routes traversed.

The maps were prepared with a degree of accuracy probably never before attained in a guidebook. Projections were made to fit 15 by 15 inch boards, as the bases for the compilation of all available map materials, published Survey maps, and maps yet in the process of being published. The recorded alignments of the four railroads in the files of the Interstate Commerce Commission, and the track profiles with elevations at mileposts obtained in the general railroad offices were printed on the bases. The gaps between the maps were filled in by original field surveys, made with a planetable. One or more topographers were assigned to each railroad, in the spring before starting a regular assignment and in the late fall after the regular summer assignment had been completed.

With the railroad alignment, and the mileposts plotted on a field board, it was easy for the topographer to set up the planetable at a milepost, orient by the compass or by sighting along the railroad alignment, and to sketch the drainage and contours after locating, by intersection, the necessary control points and determining their elevations by vertical angles. In flat deserts and open prairie country, topographic features as distant as 50 miles were located, and as many as 200 square miles were sketched in one day. The scale was small, for the guide maps were to be printed on the scale of 1 inch equals 8 miles with a contour interval of 200 feet.

The railroads concerned were very cooperative, furnishing the topographer a gasoline speeder and operator, (a railroad man who was familiar with the schedule of trains.) The speeder was light enough to permit two men to lift it off the track when a train was due, and speedy enough to permit a drive of 50 miles, if necessary, to a place to stay overnight.

For the convenience of the traveler, the sheets of the route map, varying in number from 19 to 29 in the four guidebooks, were so arranged that he could unfold them one by one and keep each one in view while he was reading the text relating to it. Railroad crossties were shown 1 mile apart and mileage figures were given at 10-mile intervals, counting continuously from the starting point.

Boundaries of published quadrangle maps were indicated and their names given. The geologic information on the maps, and the descriptive text, were prepared by a Survey geologist who supplemented his material derived from many sources by making a field examination of the entire route in 1914.

Descriptions of towns and excerpts from American history as it was enacted in the various localities added to the wealth of information in the text, and rendered it instructive and entertaining.

Each guidebook contains a list of recent publications describing geology along the route, a glossary of geologic terms, and an index of the railroad stations.

Part A (Bulletin 611).—The Northern Pacific Route, with a side trip to Yellowstone National Park. 1,910 miles, 27 map sheets. Gap topography by J.G. Staack, C.L. Sadler, J.L. Lewis, N.E. Ballmer, and W.O. Tufts.

Part B (Bulletin 612).—The Overland Routes with a side trip to Yellowstone National Park. 1,782 miles, 29 map sheets. Gap topography by C.H. Birdseye and J.L. Lewis.

Part C (Bulletin 613).—The Santa Fe Route, with a side trip to the Grand Canyon. 1,810 miles, 26 map sheets. Gap topography by W.O. Tufts.

Part D (Bulletin 614).—The Shasta Route and Coast Line—Seattle to Los Angeles. 1,427 miles, 19 map sheets. Gap topography by B.A. Jenkins and J.E. Blackburn.

Mr. John H. Renshawe did a water-color painting depicting field work of Topographic, Geologic, and Water Resources work for this Exposition as part of the U.S. Geological Survey exhibits. A.C. Roberts, in Washington, had charge of the drafting of the maps.

With a small allotment from the congressional appropriation for the Government exhibit at the Exposition, the Survey set up a creditable and instructive exhibit. Ralph W. Stone, geologist, was in charge, and gave illustrated lectures daily, describing the work of the Survey. The Survey was awarded one grand prize, three medals of honor, three gold medals, two silver medals, and one bronze medal.

At the time Lassen Peak created a great deal of interest. A volcanic peak (10,453 feet), about 200 miles north-northeast of San Francisco, and supposedly extinct, on May 30, 1914, emitted clouds of steam and ash, and on the following June 8, and July 18, threw up a column of steam that rose to a height of 10,000 feet above the crest.

Temporary Field Employment

The first Topographic Instructions Manual, edition of 1915, states:

"Persons desiring temporary employment on the field force in grade 1 must file application, on the form provided for that purpose, with the chief geographer in Washington. Laborers (grade 2) will be employed in the field by chiefs of parties as necessity may arise, and no applications for such employment will be received or filed in Washington. All persons are ineligible for employment if they are suffering from any contagious, or other disease, and, in grade 1, if they are under 20, or more than 40 years of age, or if they are students pursuing a college course, unless they bind themselves to remain with the party until their services are no longer needed. Preference is given to applicants from states in which the work is to be done, and in case of residents assigned to parties working in western states, the age limit for rodmen, recorders, etc., may be waived, provided that no such employee is to be less than 18 years of age (fig. 21)."

Each year, there were many more applications than there were positions and it was sometimes hard to convince the eager, bright-eyed boys, seeking adventure in the West, that there were no positions for them. It still warms our hearts though, when some of the previous employees write that they were with a topographic field party in the 1890's and had never forgotten it, and could we possibly find a place for their grandsons.

Mr. Marshall gave instructions that applications from sons of Topographic Branch employees should be brought to his immediate attention, and he would see that they obtained the first vacancies. He added that with limited appropriations he could not adequately compensate topographers for their good work and long hours, but he would make the sons realize that their father's name would obtain for them one of the coveted field positions. This policy is still being carried out in the Topographic Branch, although applications are no longer filed in the Chief Topographic Engineers office, but in the four Regional offices.

In the spring of 1943, an extensive mapping program made it necessary for the Survey to obtain authority for the employment of 16-year old boys, and also women in positions heretofore filled by men. The lowering from 18 to 17 years of age for seasonal assistants in field parties by the Secretary's memorandum of May 29, 1942, improved the situation for awhile, but conditions again became acute, largely because of the lowering of draft age to 18 years. An effort was made to employ men 50 years and older. One topographic field party at Media, Pa., had three rodmen whose ages were 61, 67, and 71 years. They had been coaxed to take jobs but could not be coaxed to work any faster. E.E. Harris, levelman in California, was the first to give rod-girls a tryout. He found them willing workers and got along fairly well, but a 12-foot invar rod was quite a burden for a 5-foot girl. Recording was much easier for them.

In the Chattanooga Section, the loss of personnel to the armed forces was made up by women trainees. In January 1943, the Section reported that its Multiplex output was about 30 square miles per operator, per month, on topographic mapping at a scale of 1:10,000, whereas the planimetric work was about 100 square miles, per operator at a scale of 1:12,000. The Alaskan Branch in September 1943, numbered 100 women among their mapping staff of about 270, who were engaged upon the trimetrogon process of compiling aeronautical charts for the Army Air Forces in various parts of the world. Final publication scale usually was 1:1,000,000 with a 1,000-foot contour interval. The output per month was 300,000 square miles of planimetry and about 50 square miles of topography.

To further meet the war demands, the Survey obtained authority to recall retired men for temporary regional office work.

Ohio–Michigan Boundary Line

Ohio was the first state formed from the original "Territory Northwest of the River Ohio", and its western boundary, a line drawn due north from the mouth of the Great Miami River, was surveyed and marked in 1799 as the first principal meridian of the General Land Office. When admitted as the 17th State on November 29, 1802, its northern boundary was the projection of the east-west line from the southerly extreme of Lake Michigan to the most northerly cape of Maumee Bay on Lake Erie. Lines were run in 1817, and established the position of the line between the two geographical features as 5 to 7 miles north of the due east line from the southern extremity of Lake Michigan. As the distance was about 70 miles, the strip between the two lines amounted to about 520 square miles, which remained in dispute until January 26, 1837, when Michigan was admitted as the 26th State and relinquished her claim. Parts of the line, as marked in 1817, were retraced and remarked in 1837 by the General Land Office.

In 1915, the Legislatures of Ohio and Michigan authorized a resurveying and monumenting of the line; all existing marks of previous surveys to be recovered and, where none existed, straight lines to be extended between known points. S.S. Gannett was in charge of this work and set 71 posts at 1-mile intervals.

Survey Order No. 67

Survey Order No 67, Designation of Acting Chief Geographer, dated December 11, 1915, was as follows:

"Secretary Lane has ordered the transfer of Mr. Marshall to the position of Superintendent of the National Parks, effective at once, and to continue during the balance of the fiscal year.

"Mr. Sledge Tatum is hereby designated Acting Chief Geographer and will continue to serve as Geographer-in-Charge of the Rocky Mountain Division."

Survey Order No. 69

Survey Order No. 69, Assignments in the Topographic Branch, dated January 22, 1916, was as follows:

"The great loss which the whole Survey has suffered through the death of Mr. Tatum necessitates the following assignments in the Topographic Branch:

"William H. Herron is hereby designated Acting Chief Geographer to serve for the balance of the fiscal year, and Glenn S. Smith will act as Topographic

Engineer-in-Charge of the Central Division for the same period.

"Claude H. Birdseye is designated Topographic Engineer-in-Charge of the Rocky Mountain Division.

"In case of the temporary absence of the Branch Chief, the division chiefs will act for him with full authority, in the following order of seniority: Mr. Frank Sutton, Mr. T.G. Gerdine, Mr. G.R. Davis, Mr. C.H. Birdseye, Mr. G.S. Smith."

Survey Order No. 70

Survey Order No. 70, Rearrangement of Divisions in the Topographic Branch, effective April 1, 1916, states:

"Effective April 1, 1916, a rearrangement of the divisions in the Topographic Branch will be made, as follows:

Atlantic Division (unchanged)
In charge of Frank Sutton, Geographer.

Central Division (unchanged)
In charge of Glenn S. Smith, Topographic Engineer.

Rocky Mountain Division, consisting of the States of North Dakota, South Dakota, Nebraska, Kansas, Oklahoma, Texas, Wyoming, Colorado, and New Mexico
In charge of Claude H. Birdseye, Topographic Engineer.

Northwestern Division, consisting of the States of Washington, Oregon, Idaho, and Montana
In charge of Thomas G. Gerdine, Geographer.

Pacific Division (unchanged)
In charge of George R. Davis, Geographer."

Survey Order No. 77

Survey Order No. 77, Reassignments in Topographic Branch, dated December 30, 1916, reads as follows:

"Mr. R.B. Marshall, having resigned from the Bureau of National Parks, will resume the duties of Chief Geographer.

"Mr. W.H. Herron will resume the duties of Geographer-in-Charge of Central Division.

"Mr. Glenn S. Smith will resume duties as Topographic Engineer.

"Effective January 1, 1917."

World War I

Involvement of the United States in the European conflict became inevitable during 1916 by reason of the depredations of U-boats upon merchant ships. On February 1, 1917, the German government announced unrestricted submarine warfare, and 2 days later the United States broke off diplomatic relations with Germany. On April 6, the United States declared a state of war existed with Germany.

For the topographic map so essential for military operations, the War Department looked to the Geological Survey to supply the needed experts for service at home and in France. To insure the closest cooperation, the Director, on August 18, 1916, proposed to Gen. William B. Black, Chief of Engineers, U.S.A., that Glenn S. Smith be given a commission and be put in charge of a new Survey division, Division of Military Surveys. The proposal was approved by General Black on January 19, 1917. G.S Smith had attended the Cayuga Lake Military Academy, N.Y., and the U.S. Naval Academy, Annapolis, Md., for 1 year before coming to the Survey in 1886. While making the topographic surveys at Fort Leavenworth, Kans., in 1906–08, he cultivated the friendship of officers on duty there and kept up his contacts and interest in military matters.

Topographers and other Survey technical experts were urged to accept commissions in the Army's new Engineer Officer Reserve Corps, the rank determined by salary grade and age limitations. The response was nearly 100 percent. Assignment to active duty followed in June and July. R.B. Marshall and G.S. Smith were commissioned as major, both later promoted to the rank of lieutenant colonel.

After March 26, 1917, most of the regular topographic work was confined to the general military surveys along the borders of the United States. More and more, however, it became necessary to make special surveys of sites for munitions plants, testing ground, artillery ranges, airplane and balloon fields, cantonments, and other military reservations.

Field surveys on the 15-minute quadrangles were conducted as usual though party chiefs now had additional details to handle. Military information had to be gathered, certain categories to be assembled on oversheets, and others to be embodied in written reports. Military regulations had to be studied; new forms to be filled out. To some field parties were assigned for training, one or two second lieutenants, young engineers newly commissioned from private life. Most instrument men now were army officers. It took some of them longer than it did others to find out the field surveys could not be performed in most types of country in expensive tailor-made uniforms.

The parties constantly were losing commissioned instrument men by military order and experienced field assistants by voluntary enlistment in topographic companies being organized at Camp Devens (Ayer, Mass.) for service overseas. This continued even after the registration of men (ages 21–30)

became effective on June 5, by authority of the Selective Military Conscription Bill, which was signed by President Wilson on May 18. To keep the parties supplied with field assistants Secretary Lane, on July 13, 1917, granted Mr. Marshall's request that the 20-year minimum age limit for rodmen, recorders, etc., established by order of April 9, 1904, be changed to 18 years.

In Washington D.C. in May 1917, the Survey moved into the North Interior Building, now called the GSA Building, begun in July 1915, and located at 18th and F Street NW., The Survey was housed in the center wing. The Topographic Branch was allotted the sixth floor, and to the front of the building where the north light was considered best for drafting. In addition to its regular activities on July 1, a training school for topographers was established for the purpose of training a force of field assistants, sent by engineering colleges throughout the country, in the rudiments of topographic engineering. About 75 candidates received instructions for a month from Maj. A.M. Walker and others.

General Pershing, with Maj. G.S. Smith and his staff of 53 officers and 146 men, arrived in England on June 8, 1917, and at Boulogne, France, on June 13. By the preliminary organization, announced on board the boat, and later confirmed, Col. Roger G. Alexander, U.S. Army, was designated as Chief G–2–C (Topography), and in this capacity, he exercised general supervision over the work of the 29th Engineers, at the Base Printing Plant, and allied activities.

In France, Major Smith and other officers were surprised to learn that only a strip of country about 20 miles wide along the northern border of France from Belgium to Switzerland had been mapped on a scale sufficiently large for military operations. As this strip had been captured by the Germans early in the war, the Allies were left in a region that had been mapped more than 50 years before the war on a scale of 1:80,000, with the relief shown by hachures, the maps being thus wholly inadequate for use in modern warfare. Topographic units of the French and English armies had remapped this region as rapidly as possible. Plans directeurs (battle maps) on scales of 1:10,000 and 1:20,000 had been prepared in areas where they were most needed to guide the Allies in repelling the advance of the Germans.

The 29th Engineers found the required maps already prepared, and only once was given the opportunity to demonstrate their capabilities. It was after the disaster to the Allied Army in the vicinity of Amiens, in March 1918, that the French called upon General Pershing to furnish topographic engineers to assist in the surveys needed to provide several battle maps of areas in the vicinity of Amiens and Chateau Thierry. The first operation covered 225 square kilometers to the westward of Creteuil, and the other was a regular 1:20,000 scale plan directeur of St. Auld.

After a thorough search in November 1917 for a good location for a Base Printing Plant, a choice was made and arrangements perfected with the French authorities for the occupation and use of 14 buildings of Tureene Barracks at the south edge of Langres, Haute Marne. In spite of difficulties, the necessary accessories of water, sewage, light, and heat were installed and, as they arrived from the States, printing presses, machines, cameras, and miscellaneous equipment, so that by the time the American Army was organized and in the field, the Base Printing Plant was ready to supply the needed maps and printed matter. Major Smith was Director of the base, which also was the headquarters, and home of the 29th Engineers. Capt. (later Maj.) Harold R. Richards, U.S.A., was Commanding Officer and Capt. L.L. Lee (Survey) was adjutant. The troops numbered about 100 officers and 1,200 enlisted men, though only a part of this force was at Langres at any one time.

The 29th Engineers had its beginning at Camp Devens, Mass. There Maj. Frank Sutton, assigned to duty with troops and relinquishing his Atlantic Division to Major Herron and Gerdine, gathered the required specialists from among officers and men and organized them into companies for the surveying, mapping, and printing work to be performed overseas. Company A was ready on October 28. Embarkation was from Hoboken, N.J., whence, via Brest, they arrived at Langres on December 3, 1917.

Company B was commanded by Capt. H.H. Hodgeson and other officers from the Survey were Oscar H. Nelson, 1st Lt.; Charles H. Davey, 2d Lt.; Howard Clark, 2d Lt.; Leo B. Roberts, 2d Lt.; Fred W. Crisp, 2d Lt.

The Company was ready to move at Christmas time, but an epidemic of measles delayed its embarkation at Hoboken on the Transport *Calamares* until January 30, 1918. Other companies followed at intervals.

In August, Maj. Frank Sutton with Capt. E.I. Ireland as adjutant, who had been assisting with the organization work, moved a battalion of 1,100 men from Camp Devens to Washington Barracks, Washington, D.C., and after full preparation, embarked the battalion from Newport News, Va.

In addition to the Survey personnel assigned to the 29th Engineers, a map organization, 13 Survey men were detailed as orienteur officers to the Coast Artillery of the Army. The 13 volunteers, selected by the Survey in June 1917, sailed in July for France via England, with the 6th, 7th, and 8th Regiments of the 1st Brigade of Coast Artillery. The three regiments were later designated as the 51st, 52d, and 53d Regiments.

When the 2d Brigade of Coast Artillery was organized, and prepared to sail for France, a like request on the Survey was made by the Coast Artillery for orienteur officers, but denied by General Black, Chief of Engineers, as Survey personnel were required for the Corps of Engineers assignments.

The Survey engineers who were detailed to the Coast Artillery under the command of Maj. C.H. Birdseye were Capt. B.A. Jenkins; Capt. C.B. Kendall; Capt. L. Nelson; 1st Lt. Kostka Mudd; 1st Lt. A.J. Ogle; 1st Lt. J. Ross Eakin; 1st Lt. J.H. Wilson; 1st Lt. Roscoe Reeves; 1st Lt. D.H. Watson; 1st Lt. Charles Hartmann; 1st Lt. C.R. Fisher; and 1st Lt. Thomas H. Moncure.

Upon arrival in France, the majority of the Survey personnel attended a 3-months course in the Artillery School for young Artillery officers. The knowledge gained in this school was invaluable to the orienteur officers in knowing the performance of each type of guns, firing data, rifling, drift, sighting instruments, bore sighting, and other pertinent data characteristic to heavy guns.

At the close of the Artillery School, an orienteur officers course was inaugurated for the artillery officers to learn the duties of the engineer officers, and the relation of duties of those involved in orienteur work, and artillery officer's responsibilities. The Survey personnel of the brigade served as instructors under the direction of Maj. C.H. Birdseye, both in the classroom and field training in the use of Survey instruments and computations.

The first engagement in 1918, in which the Survey personnel served as orienteur officers, was the St. Michael Sector. Several types of heavy guns were involved, including 10-inch tractor mounts, railroad mounts, and mobile GPF, or 155 mm. Lt. Kostka Mudd was the only Survey orienteur officer wounded in the early days of the bombardment, which resulted in a fractured leg, for which he received the Croix de Guerre. (Mr. Mudd's wound resulted in a permanent limp.)

At the close of hostilities, all returned to civilian life, and back to the Survey, with the exception of Lt. J.H. Wilson, who remained in the Service and reached the rank of brigadier general during World War II.

Survey officers stationed at Langres constantly were being sent out to make surveys of small areas designed for such use as hospital sites, warehouses, aviation fields, and training camps. One assignment mapped by personnel of Company A, shortly after its arrival in France, was a detailed survey of the Loire River flat at St. Nazaire with a contour interval of one-half meter, the basis for some 200 miles of railroad track built during 1918, for warehouses, and for ship docks at the river edge.

A school for officer candidates was located at Langres, and Survey officers conducted classes in topographic mapping and in the interpretation and restitution of airplane photographs. Here, too, before the end of hostilities, was established a department for the production of relief models, which were in demand for certain visibility studies.

Prior to Armistice Day, the output of the Base Printing Plant, often operating day and night, amounted to 3.5 million maps, or plans directeurs, 3 million impressions of type, 6,000 contact prints, 500 enlargements, 800 blueprints, etc. Miniature base plants added to these totals. They were established at the headquarters of the 1st Army in Neufchateau, the 2d Army in Toul, and later, the Army of Occupation at Coblenz. Lesser units were set up in Paris, and at G.H.Q. in Chaumont. The same methods were employed in making field surveys, and in reproducing maps used in the United States. They were acknowledged to be superior by the French and English, who frequently sent their topographic engineers to observe and study operations.

With the signing of the Armistice on November 11, 1918, the eyes of all Americans were turned towards home across the Atlantic, and the next few months were charged with impatience and restlessness, ameliorated to some extent in the 29th Engineers by the printing of many technical and historical reports, and by the performance of certain interesting tasks. One unit of 30 men made a careful large-scale survey of the hidden concrete dugouts extending for 2 kilometers east and west of the Mon Plasir Ferme, one of the strong points in the enemy defenses between Verdun and Metz. Several small parties sketched and described the most likely sites for monuments that might be erected in commemoration of the best advance against the enemy that was made by each American Division. Individuals were selected for convoy duty with the trucks that took supplies twice a month to the Intelligence Unit with the Army of Occupation at Coblenz, in Germany.

A group of surveyors, in conjunction with other detachments to form the Graves Registration Service, made search, surveys, notes, and plans throughout the battle areas until all remains of fallen American soldiers were found and identified. For their final resting place, Lt. R.M. Wilson helped to select a site near Romagne, and then had charge of six transit parties that staked out this entire cemetery, involving setting about 6,000 stakes for a cemetery designed to hold 24,900 graves. For this work, the planetable and stadia were substituted for the transit and tape, after trial, as affording more rapid progress. Subsequently, he selected and staked out a smaller cemetery at Theaucort, France.

For the purposes of experiment and record, a Mobile Topographical Unit was organized, and three field problems of 10 days each was devised. A large amount of technical equipment was loaded on great trucks, a complement of officers and men selected, and the unit began its movement on March 10, to terminate at Sedan on April 14, 1919. Wartime conditions were simulated, and the battle areas from Mont Sec to Sedan were scouted, surveyed, photographed, drafted, restituted, summarized, and printed, just as they might have been done in conjunction with the Army in full operation. The area mapped, corresponding to the width of two Army corps, varying from 20 to 30 kilometers, was in ruins, deserted, and either under a blanket of snow or very muddy.

When the orders came in June to return, the 29th Engineers marched out of the Tureen Barracks with flags flying and it's fine band playing, paraded through the main street of Langres and down the winding road to the railroad, and entrained for the ports of St. Nazaire and Brest. Only one American remained in Langres; Lt. Thomas H. Moncure, who was to hear and record claims for damages incurred by the French townspeople during the 8-months stay of the 29th Engineers in Langres. (December 3, 1917, to June 20, 1918.)

That the 29th Engineers fulfilled their important purposes is attested by the following letter of commendation from the Commander-in-Chief, American Expeditionary Forces (A.E.F.):

"American Expeditionary Forces
Office of the Commander-in-Chief
France, April 28, 1919

Colonel Roger G. Alexander, General Staff, G–2–C
Commanding 29th Engineers,
Chaumont

My dear Colonel:

In considering the work of the various special troops in the American Expeditionary Forces, there is none that has been of more vital importance to successful operations than the work of the 29th Engineers, Survey and Printing. The adequate supply of accurate maps was a vital necessity. The field survey and reproduction work necessary for such maps under war conditions called for technical ability of the highest order, for foresight resourcefulness and good organization.

That our map supply was entirely adequate is due to the efforts of the officers and men of the 29th Engineers, at General Headquarters, at the Base Printing Plant, and with the Armies, Corps, and Divisions. They gave to their work at all times their best efforts, with the same zeal and self-sacrificing devotion to duty that has characterized the American fighting man at the front. For myself, personally, and in the name of their comrades in the Forces, it gives me the utmost pleasure to express to you, and through you to them, our heartiest congratulations and thanks.

Sincerely yours,
s/ John J. Pershing"

The following members of the Topographic Branch received medals, citations, or personal letters of commendation for meritorious services:

- French Croix de Guerre:

  - Lt. Kostka Mudd, with 8th Army Artillery Staff—Wounded by shell.

  - Maj. Luria L. Lee, during Chateau Thierry offensive, July 1918.

  - Capt. Alvah T. Fowler, under fire in Picardy, May to July 1918.

- Citations by Commander-in-Chief of the A.E.F.:

  - Lt. Col. Glenn S. Smith, as Director, Base Printing Plant.

  - Maj. Joseph H. Wheat, as instructor Army Intelligence School.

- Maj. William O. Tufts, in topographical surveying with 1st Army.

- Maj. Herbert H. Hodgeson, in charge of triangulation at the front.

- 1st Lt. James B. Leavitt, as topographic officer for 6th American Corps.

- Mr. Louis H. Gott, in drafting section, General Staff.

- Maj. Luria L. Lee, as Adjutant 1st Battalion, 29th Engineers.

- Capt. Albert O. Burkland, as topographic officer, 1st Division.

- 1st Lt. Charles H. Davey, as instructor, Army Intelligence School.

- 1st Lt. Reuben A. Kiger, as topographic officer, 5th Corps.

- Capt. Oliver G. Taylor, as topographic officer, 1st Corps.

- Maj. Leo B. Roberts, as topographic officer, G.H.Q.

- Personal letters from General Pershing :

  - Lt. Col. Alfred H. Brooks, as chief geologist of the A.E.F.

  - Lt. Col. Glenn S. Smith, as Director, Base Printing Plant.

- Citations and medals of the Order of University Palm from the French Republic (Ministere de l'Instruction publique et des Beaux-Arts):

  - Lt. Col. Claude H. Birdseye, Officier de l'Instruction Publique—Gold Palms

  - Lieutenant Colonel Smith, Major Wheat, and Major Roberts—Officier d'Academie—Silver Palms

Maj. J.W. Bagley, assisted by F.M. Moffit, and for a part of the time by J.B. Mertie, (all of the Division of Alaskan Mineral Resources) pursued experiments towards designing and constructing a tri-lens camera for taking aerial photographs on a continuous roll of film, and a rectifying printer for restoring the scale of the wing exposures. This work was executed under the direction of the Chief of Engineers, U.S. Army, and the Director of the Geological Survey. Sixteen engineer officers, most of whom were sent to different aviation fields in the country, were trained in the use of these instruments, and about 40 enlisted men assisted them in mapping areas, which included among others, the airplane mail

route from Washington to New York, and the airplane route from Washington to Langley Field, Va.

Major Bagley and contingent reached France too late for active operations.

The Geological Survey had a war service record probably unequalled by any other bureau in the Federal Civil Service. All employees, including clerks, draftsmen, map engravers, and printers were engaged in assisting the War and Navy Departments on work connected with the war. The number of men who went into the military or naval service during the war by branches is shown in table 27.

Table 27.   U.S. Geological Survey employees who served in the military.

Branch/Division	Enlisted	Officers
Topographic Branch	305	110
Water Resources Branch	78	28
Administrative Branch	41	3
Geologic Branch	28	17
Engraving Division	21	3
Land Classification Board	6	2
Total	479	163

The Topographic Branch was proud of its share in the services and Chief Geographer Marshall had all the names carefully drafted, framed, and hung on a wall in his office (table 28).

The Director, on December 18, 1918, sent all officers a telegram requesting military status desired. Nearly all replied for return to the Survey, and at the same time retain commission in the Engineer Officers Reserve Corps on inactive status. Considerable talk was rife at the time, based on the fact that the Army was supplying a large part of the funds for topographic mapping, that the Topographic Branch would be absorbed into the Army. The officers listed in table 29 chose to remain in the Army.

Maj. A.M. Walker, who had familiarized himself with the work of the French with their Chedan and Arnold relief models, and later had charge of the relief model department at Langres, was loaned to the Corps of Engineers from November 4, 1921, to October 1, 1928, to carry on experiments with relief modeling in order to develop practical methods for Army use. He wrote a training manual on the subject for the Army.

Table 29.   U.S. Geological Survey employees who chose to remain in the Army.

Maj. James W. Bagley	Maj. Leo B. Roberts	Capt. Calvin E. Griffin
Capt. R.M. Herrington	Lt. William C.F. Bastian	Lt. Fred W. Crisp
Lt. John H. Wilson		

Table 28. Roster of Topographic Branch employees that served in the military.

[*, Overseas Service]

Adams, R.E.	Burkland, A.O.*, Capt.	Fairly, J.L.	Hodgeson, H.H.*, Maj.
Aid, Harry	Burt, E.C., Lt.	Fales, J.C.*, 1st Lt.	Holder, C.C., Lt.
Aid, Kenneth	Burton, P.M.	Farney, G.M.*	Holloway, E.V.
Allen, J.B.	Bush-Brown, Junes	Farnham, A.R.	Hopkins, R.W.*
Allen, W.R.*	Cade, A.P.*	Fisher, C.R.*, Capt.	House, James
Alsing, A.A.*	Cahill, S.F.*	Fisher, W.A.*	Hubbard, Bela*
Anderson, C.G., Capt.	Campbell, E.M.	Floore, S.P.*, 1st Lt.	Hudson, H.L.
Anderson, J.L., Lt.	Campbell, R.C.*	Ford, G.R.*	Hughes, F.W., 1st Lt.
Andreas, A.H.L.*	Carson, W.G., Lt.	Fort, F.L.	Hurst, W.P.
Andrews, A.K.	Chaddock, T.H.	Fowler, A.T.*, Capt.	Ingersoll, H.B.
Arthur, B.E.*	Chamberlain, E.D.	Gardner, C.C.*, 1st Lt.	Ireland, E.I.*, Capt.
Ayers, R.E.	Chapman, R.H.	Gehres, W.S.*, Lt.	Jameyson, Bruce*, Lt.
Bagley, F.A.*	Chenault, W.F.*	George, W.O.*	Jenkins, B.A.*, Capt.
Bagley, J.W.*, Maj.	Chinners, C.E.	Gerdine, T.G., Maj.	Jennings, J.H., Maj.
Baldwin, S.T., Jr.*, Lt.	Christiansen, M.C.*	Gertz, C.F.	Joaquin, Hugh*
Ballmer, N.E., Lt.	Clark, H.*, Lt.	Gilchrist, F.G.	Kahlbaum, William
Bandli, E.M.*, Lt.	Clark, M.W.	Gill, J.I.*	Kavanagh, A.J.*, Lt.
Bastian, W.C.F.	Clarke, H.C.O.	Glasgow, L.B.*, Capt.	Keeler, W.W.
Batten, G.L.*	Clement, H.D.	Glaze, A.O.	Keller, J.B.*
Bauer, B.P.	Clement, S.E.	Gott, L.H.*	Kendall, C.B.*, Maj.
Seaman, W.M., Maj.	Clinite, R.G., Lt.	Goudie, C.A.*	Kidd, W.C.
Beames, W.S.	Cohn, H.I.*	Griffin, C.E., Capt.	Kiger, R A.*, 1st Lt.
Bell, F.V.*	Cooke, C.E., Maj.	Griffin, W.H., Maj.	Knox, C.B.
Bell, H.S.*	Crisp, F.W.*, Lt.	Hain, E.L., 1st Lt.	Leavitt, J.B.*, 1st Lt.
Bemis, E.L.*	Cryer, C.T.L.	Hamilton, E.G., Capt.	Lee, L.L.*, Maj.
Berry, R., Capt.	Cunningham, G.M.*	Hammer, B.E.*	Lenovitz, Jacob
Bertenshaw, W.H.	Cunningham, J.W., Capt.	Hannegan, Duncan, Capt.	Leopold, L.S., Capt.
Beyersdorfer, W.M.	Danforth, F.A.*, Lt.	Harlowe, L.S.*	Lindsay, J.C.
Birdseye,C.H.*, Lt. Col.	Davey, C.H.*, 1st Lt.	Harrison, J.P., Capt.	Lipps, J.P.
Birdseye, S.H.*, Lt.	Davey, E.T.	Hartle, W.A.	Lobeck, A.K.
Birkett, D.S., Lt.	Davies, F.A.*	Hartley, J.W.	Lockwood, T.F.*
Blackburn, J.E., 1st Lt.	Dewhirst, H.H.	Hartmann, Chas.*, Lt.	Lonnquest, T.C.
Blount, C.C.	Dickson, D.M.*	Hassan, A.F., Capt.	Lord, I.Q.*
Boston, T.	Dinock, Stuart	Hawkins, G.T., Maj.	Lovett, J.D.*
Bowers, C.M.*	Dod, James	Hayes, S.P.*	MacClintock, Paul*
Bowler, E.W., 1st Lt.	Dolliver, Emerson*	Hayes, W.P.	McCammon, R.L.
Bradshaw, Donald*	Douglas, G.C.*, 1st Lt.	Hearne, C.C.*	McCullough, H.K.
Brislawn, F.L.*	Druhot, G.S.*	Heflin, E.B.	McDonald, H.L., Capt.
Brode, W.R.	Duke, Basil, Capt.	Henderson, W.D.*	McKinley, W.K.
Brooks, E.O.	Eakin, J.R.*, Maj.	Hendrickson, V.J.	McLaughlin, DeWitt, Capt.
Brown, B.W., Lt.	Ebmeyer, G.E., Capt.	Herrington, R.M., Capt.	McLaughlin, Fred, Capt.
Brown, G.W., Jr.	Ecklund, C.A.*, Lt.	Herron, W.H., Maj.	MacManus, Angus
Brown, S.C.*	Ehret, J.N.	Hinds, N.E., Lt.	MacManus, V.D.*
Burchard, L.W., Lt.	Evans, R.T.*, 1st Lt.	Hodges, Guy	McMillen, P.W., Lt.
McNair, E.L., Capt.	Noyes, Donald*	Runyan, R.H.	Sutton, Frank*, Maj.
Mallon, A.E.*	Ogle, A.J.*, 1st Lt.	Russell, R.H.*	Tandy, J.H.*
Maloney, H.A.	Opdycke, A.L.*, Lt.	Rust, Earl	Taylor, H.P.*
Marshall, R.B.	Orrall, S.M.*	Sadler, C.L., Capt.	Taylor, O.G.*, Capt.
Meade, A.P.*, Jr., Capt.	Page, J.H.	Schmidt, R.L.	Thomas, Willis*
Meaders, H.T.*	Parker, S.L., Lt.	Schnurr, Cornelius	Thompson, J.C.

Table 28.   Roster of Topographic Branch employees that served in the military.—Continued

[*, Overseas Service]

Meier, N.C.	Passagaluppi, W.A.*	Schwartz, G.C.*	Timmons, G.E.
Metcalfe, J.B.*, Jr., Lt.	Patrick, J.B.	Searle, A.B., Capt.	Trimble, K.W., 1st Lt.
Miller, R.N.	Patrick, S.L.*	Seitz, R.C.*, 1st Lt.	Triplett, G.A.*
Miller, W.L., Capt.	Peattie, Roderick*	Semper, C.H., 1st Lt.	Tufts, W.O.*, Maj.
Mock, G.A.	Pendleton, T.P., Lt.	Shea, R.I.	Ucker, C.S., Jr.
Monbeck, R.R.*, Lt.	Penick, S.T., 1st Lt.	Shipley, J.J.	Upton, W.B., Jr.*, Lt.
Moncure, T.H.*, 1st Lt.	Perkins, R.G.*	Simmons, H.E., Lt.	Wainwright, D.R.
Montgomery, W.J.	Perkinson, E.V.	Slaughter, T.F., 1st Lt.	Walker, A.M.*, Maj.
Moore, H.A.	Peters, A.C.	Smith, D.B.*	Walker, E.T.*
Moorhead, Bishop*, Lt.	Peters, Gustave*	Smith, G.R.*	Wann, W.C.
Morey, W.H.S., 1st Lt.	Phipps, F.W.	Smith, G.S.*, Lt. Col.	Wardle, C.W.*, 1st Lt.
Morris, Ellison*	Pierce, J.M.	Smith, H.W.	Wasson, Theron*
Mortimore, R.H.*	Pike, Albert*, Capt.	Snell, A.H.*	Watson, D.H.*, Capt.
Morton, R.J.*	Post, W.K.*	Sommers, C.J.*	Wheat, J.H.*, Maj.
Mudd, Kostka*, 1st Lt.	Ranney, T.T.	Southwick, M.L.	Wilcoxon, R.F., Lt.
Muldrow, Robert, Maj.	Rawls, J.M., Lt.	Sparks, P.H.	Williams, A.C.*
Munson, A.T.	Rector, E.R.	Sprague, W.G.*	Wilson, A.V., Lt.
Murlin, D.E.	Reeves, Roscoe*, Capt.	Staack, J.G., Capt.	Wilson, J.H.*, 1st Lt.
Murphy, J.J.	Reineck, R.H.*, 1st Lt.	Stanton, R.D.	Wilson, R.M.*, Lt.
Murphy, T.F.	Rhodes, B.L.	Steele, R.B., Lt.	Woodford, J.F.*
Neff, M.T., 1st Lt.	Riddick, A.L.	Stensland, H.J.*	Woodhouse, E.H.
Nelson, C.L.*, Capt.	Rider, Fred	Stevenson, R.G.*	Woodworth, C.L.
Nelson, O.H.*, 1st Lt.	Roberts, A.C.*, Lt. Col.	Stewart, C.H.*, 1st Lt.	Yarbrough, Arthur*, Lt.
Nessler, C.W.H.*	Roberts, G.H.*	Stoneham, H.L.	Yoakum, B.H.*, Lt.
Newcom, F.C.	Roberts, L.B.*, Maj.	Stonesifer, C.A.*, Lt.	
Newsome, L.O.	Roudabush, M.A.*, Lt.	Stranahan, William, Capt.	
Noble, Mark	Rowell, C.W.*, Lt.	Strange, F.C.*	

Allotments (1907–18)

Funds allotted for topographic mapping for the fiscal years 1907–18, the year preceding Mr. Marshall's regime (table 30).

Cooperation in 1918 was maintained in six states, which contributed the amounts listed in table 31, with the understanding that they were to be used on military surveys. Field work accomplished in the 12 months ending June 30, 1928, is shown in table 32.

Survey Order No. 94

Survey Order No. 94, dated July 1, 1919, is as follows:

"The U.S. Geological Survey is about to supervise extensive topographic mapping in the West Indies. The Republics of Santo Domingo and Haiti have made appropriations sufficient to complete the surveys of their countries, and have requested the Geological Survey to take charge of the work, and to furnish the technical personnel. It is possible that Puerto Rico and the Republic of Cuba will take similar action.

"In order to provide for the administration of this work, the Division of West Indian Surveys is created in the Topographic Branch. The Division is charged with the administration and execution of such topographic surveys in the West Indies and in other Latin American Republics as may be undertaken by the Geological Survey.

"The Chief of the Division of West Indian Surveys will act as the personal representative of the Director and the Chief Geographer, and will handle directly all matters pertaining to that Division, except those which affect the general policy of the Geological Survey or of the Topographic Branch. Although the members of the Survey assigned to this work will be furloughed without pay, the Division is regarded as a part of the Topographic Branch and the Chief of the Division will render to the Chief Geographer monthly reports of the progress of the work and

Table 30.   Topographic mapping allotments for fiscal years 1907–18.

[--, no data]

Fiscal year	Scientific assistants	Topographic surveys	Surveying National Forests	Military surveys and maps	State funds	Total
1907	$9,200	$350,000	$100,000	--	$121,800.00	$581,000.00
1908	9,200	300,000	100,000	--	103,850.00	513,050.00
1909	9,200	300,000	75,000	--	124,410.23	508,610.23
1910	9,200	350,000	75,000	--	180,689.36	614,889.36
1911	9,200	350,000	75,000	--	154,383.04	588,583.04
1912	9,200	350,000	75,000	--	177,853.00	612,053.00
1913	9,200	350,000	75,000	--	161,267.09	595,467.09
1914	9,200	350,000	75,000	--	179,047.96	613,247.96
1915	9,200	350,000	75,000	--	186,800.07	621,000.07
1916	9,200	350,000	75,000	--	166,789.39	600,989.39
1917	9,200	350,000	75,000	$34,808.30	109,440.69	578,448.99
1918	9,200	350,000	75,000	553,328.49	56,000.00	1,043,528.49

copies of all technical data which may be of interest to the Branch. He will keep in close touch with the other work of the Branch in order that the same standards may be maintained.

Table 31.   Allotments contributed by States for military surveys.

State	Allotment
California	$14,000
Illinois	7,000
Maine	5,000
Michigan	15,000
Virginia	4,500
Washington	10,500
Total	$56,000

"Lt. Col. Glenn S. Smith is relieved from his duties in connection with military surveys, and is designated as Topographic Engineer-in-Charge of the Division of West Indian Surveys."

Survey Order No. 97

Survey Order No. 97, dated October 1, 1919, states:

"The resignation of Mr. Robert B. Marshall as Chief Geographer necessitates a change in the organization of the Topographic Branch after a long period during which there has been relatively little change except temporary adjustments required by the war program. Mr. C.H. Birdseye, who has served as one of the division chiefs, has been appointed Chief Topographic Engineer, bringing to his increased responsibility full acquaintance with and appreciation of the technical personnel. The larger opportunity that promises to open for the topographic work of the U.S. Geological Survey can well be met without radical change in either organization or standards of work.

"Mr. Marshall will remain a member of the branch, being enrolled as a topographic engineer on a per diem status, and will thus be available for occasional service."

Table 32.   Field work accomplished in the 12 months ending June 30, 1918.

Activity	Tasks completed
Primary and precise levels extended	7,945 miles; 2,204 permanent bench marks were set.
Total and precise levels extended	273,956 miles since leveling was begun in 1896.
Primary traverse	7,676 miles; 884 permanent marks were set.
Triangulation stations occupied	58; 58 were stations permanently marked.
Square miles mapped	22,431; including 476 square miles of revision.
Total square miles mapped	1,279,890, or 42.2 percent of entire country.

Division of West Indian Surveys

Santo Domingo

Political disturbances in the Greater Antilles of the West Indies, which became a reign of terror in Haiti in July, 1915, led to the intervention by the United States and the occupation of the principal ports by Naval and Marine forces.

Rear Admiral H.S. Knapp, U.S. Navy head of the Military Government of Santo Domingo, wrote the Director of the Geological Survey on April 1, 1917, requesting the estimated cost of necessary basic surveys. The Director replied on May 21, suggesting a topographic survey on the scale of 1:125,000 with a contour interval of 100 feet, the 19,325 square miles at $15 a square mile to cost approximately $300,000, and a reconnaissance geologic survey to cost $41,000. The Survey would undertake the work, if funds were provided, upon conclusion of the war, when topographers, who were now engaged in military surveys, would become available. The Admiral wrote on October 10, 1918, that he had issued an Executive Order No. 207, dated September 12, 1918, setting aside Dominican Government funds, not otherwise appropriated, in the amount of $300,000 to be known as the Dominican Survey fund, that the money probably could not be expended in less than 3 years, and that interest accruals in that period might raise the fund to a total of $341,000, the amount estimated.

A Division of West Indian Surveys was created in the Topographic Branch by Survey Order No. 94, dated July 1, 1919, and Col. Glenn S. Smith was designated Topographic Engineer-in-Charge. The Director had requested that Colonel Smith be detailed to the Survey from the Corps of Engineers, U.S. Army. In April 1919, Colonel Smith and four geologists had conferred in Santo Domingo City and other island capitals with military governors on plans for topographic surveys and preliminary geologic investigations.

The agreement for cooperative topographic surveys provided that the Republic of Santo Domingo would pay the entire expense of the work, that the Geological Survey would furnish the technical personnel and loan the necessary instruments, that the Navy and Marine Corps supplies would be made available for the use of field parties, that subsistence allowance for the Survey members would be arranged on a per diem basis starting at $1.00 per day, that monument caps and bench mark tablets with approved design and lettering would be used, and that members of Survey parties would wear clothing of a uniform nature for identification purposes. In some remote mountainous areas, it was found advisable to have armed Marine guards accompany the field parties on account of bandits.

Topographic Engineer Albert Pike was placed in charge of the project, and he sailed on the U.S.S. *Hancock* from Charleston, S.C., on August 13, 1919, with 18 engineering assistants who were to complete the organization of eight field parties, two triangulation, two primary traverse, one leveling, and three topographic. All had volunteered for the assignment, which carried an increase of 20 percent in compensation. They arrived in Santo Domingo City on August 20. The Santa Domingo personnel are listed in table 33.

Horizontal control was based on the astronomical position of the Santo Domingo lighthouse, as previously determined by the U.S. Hydrographic Office, and upon a baseline 3 kilometers to the northwest. This line had to be cleared through a jungle before it could be measured, and similar heavy tropical growth had to be cleared from nearly all stations made in the extension of triangulation. For vertical control, tidal gauges from which lines of levels could be extended were established at Santo Domingo City, and at Puerto Plata on the north side of the island.

Topographic surveys, by the customary preferred method of planetable sketching and stadia readings, generally were impossible, on account of dense foliage, and had to be executed by tape traverses. There were many trails, but except for periods during the dry season, which extended from early in December to June, these were slippery and full of mud holes. Camp fare, as prepared by native cooks, was none too appetizing.

In September 1921, when the mapping was steadily progressing, instructions were received from the Military Governor, Rear Admiral S.S. Robison, U.S. Navy, that the fieldwork would have to be terminated and the Survey personnel returned to the States, in accordance with the Proclamation of June 14, 1921, relative to withdrawal from the responsibilities assumed in connection with Dominican affairs, as well as the financial condition of the Dominican Republic. A provisional government was organized for the Dominican Republic, and the American Military Governor sailed on October 24, 1922, leaving behind about 1,500 marines to await the establishment of a permanent government. The Republic was evacuated on July 12, 1924.

Topographic mapping was executed on nine 30-minute quadrangles using a scale of 1:80,000 with a contour interval of 100 meters and altogether about 4,500 square miles were completely mapped.

Haiti

The mapping of the Republic of Haiti, which occupied the western one-third of the island, had an area of 10,204 square miles and a boundary line 193 miles long separating it from Santo Domingo, was begun by a Survey party in charge of Eugene L. McNair, who had arrived at Port-au-Prince on January 27, 1920. Plans embodying new methods in mapping that had been suggested by Col. Glenn S. Smith on his visit to the island in April 1919, were followed. By means of aerial photos, the features particularly needed by Cdr. Ernest R. Gayler, U.S. Navy, Director General of Public Works and Engineer in Chief, such as the coast line, cities and towns,

Table 33. Santa Domingo personnel.

Name	Title	Name	Title
Allen, J.B.	Editorial Assistant	Lloyd, W.J.	Topographer Engineer
Baird, W.E.	Editorial Assistant	McArthur, J.O.	Recorder
Bastian, Joe	Rodman	McCook, J.F.	Junior Topographer
Biggins, P.F.	Rodman	MacManus, A.	Editorial Assistant
Britt, I.A.	Recorder	Mock, G.A.	Editorial Assistant
Cannon, E.A.	Rodman	Morrison, Lee	Topographer Engineer
Cash, Elbert	Rodman	Nevitt, J.C.	Clerk
Chambers, G.	Recorder	Ogle, A.J.	Topographer Engineer
Chenault, W.F.	Editorial Assistant	Parnelle, R.E.	Rodman
Chew, R.G.	Recorder	Pike, Albert	Topographer Engineer, in charge
Childs, C.B.	Editorial Assistant	Ramsaur, H.W.	Recorder
Clement, H.D.	Junior Topographer	Rider, Fred	Topographer Aid
Cook, J.E.	Rodman	Robinson, J.W.	Rodman
Davenport, P.E.	Recorder	Runyan, R.H.	Junior Topographer
Davey, T.T., Jr.	Recorder	Seward, V.S.	Editorial Assistant
Dod, J.P.	Recorder	Shalibo, F.L.	Assistant Topographer Engineer
Duck, J.A.	Topographer Engineer	Smith, W.L.	Recorder
Fankhauser, A.	Junior Topographer	Stevenson, R.G.	Junior Topographer
Fisher, C.R.	Assistant Topographer	Swick, J.N.	Rodman
FitzGerald, G.	Editorial Assistant	Teller, L.W.	Editorial Assistant
Guma, G.	Recorder	Terry, H.M.	Recorder
Gayer, J.J.	Editorial Assistant	Thomas, D.J.	Rodman
Grenfell, C.J.	Rodman	Thomas, T.E.	Recorder
Gavken, H.F.	Rodman	Thompson, W.C.	Recorder
Gohlfeld, K.V.	Recorder	Towill, R.M.	Rodman
Gohlfeld, R.L.	Recorder	Tuttle, Basil	Recorder
Hill, A.K.	Recorder	Tyler, W.B.	Recorder
Holder, C.C.	Topographer	Walker, E.T.	Assistant Topographer
Hughes, J.A.	Recorder	Watson, D.H.	Assistant Topographer
Johnson, T.R.	Editorial Assistant	Wender, W.G.	Junior Topographer
Kendall, C.B.	Topographer Engineer	Wiley, C.H.	Recorder
King, J.E.	Junior Topographer	Williford, Paul	Rodman
Latimer, T.M.	Recorder	Wix, O.L.	Recorder
Leavitt, J.B.	Assistant Topographer	Wood, W.J.	Rodman
Lewis, E.C.	Cook	Yoakum, B.H.	Assistant Topographer Engineer

highways, and topography of certain important areas, were to be plotted and drafted. The necessary ground control for the aerial photos was to be supplied by Mr. McNair's field parties. Triangulation was to be extended from stations already established in neighboring Santo Domingo and levels along railroads and highways began at a tidal gauge that was set up at Port-au-Prince.

The agreement for the cooperative Haitian Topographic Survey was similar to that for the Dominican Republic. It provided for an investigation of mineral and botanical resources. The Survey furnished the technical personnel, and loaned the necessary surveying instruments, whereas the Republic of Haiti supplied the money. The first allotment was $10,000; later allotments were behind schedule and never exceeded $25,000, so that field work was limited to triangulation for photo control, and to some leveling. Topographic mapping was not even started.

The first peace time aerial survey as planned by Colonel Smith was a joint enterprise. The photos were taken by a Bagley tri-lens camera, loaned by the U.S. Army, mounted in a DeHaviland hydroplane furnished by the U.S. Marine Corps, which was tendered by a U.S. Navy supply vessel that had quarters for at least six additional officers and space that could be used as a developing and drying room for the photographic film. The photographer assigned by the Survey was Eric Haquinius, a former lieutenant in the Swedish Navy who served with the U.S. Air Force in World War I, became a naturalized citizen of the United States and obtained an international flying license. Mr. Haquinius arrived at Port-au-Prince in April 1920, and conferred with Cdr. Gayler as to the order in which the flights should be made. The first films, when developed, proved to be of poor quality, and later flights were made at lower altitudes in the hope of improving the photographs. Clouds interfered and their prevalence reached such a degree by July 1 that the flights were abandoned until after December 15 when the rainy season should be over.

Triangulation, which had been continued by Rhea B. Steele after Mr. McNair's departure for the States in May, 1921, was discontinued in October when funds were exhausted. Mr. Steele was instructed to mark stations permanently for future use, dispose of property, which included a Ford truck, and return to Washington. He reached Washington in January 1922, and assisted in the office computation of the triangulation. The assistants with the field parties were E.F. Couleur, W.C. Harrison, Sam H. Moyers, L.A. Platt, R.S. Tennyson, and J.S. Ward.

In February, 1921, a Section of Photographic Mapping, with Thomas P. Pendleton, able Photogrammetrist-in Charge, was organized in the Washington office for the purpose of handling the aerial photographs of Haiti, and to investigate the probable usefulness of aerial photographs in making planimetric and topographic maps. The aerial photographs of Haiti possessed all possible defects and presented many perplexities to the Survey photogrammetrists and cartographers. Many of the photos were only fair, were devoid of sharp identifiable natural features, and were of varying scales aggravated by distortion around the edges and by tilt. The ground control was inadequate, triangulation stations were not all recognizable, were not placed so as to appear on the right pictures, and were confined to a comparatively small area; secondary control points such as road forks, junctions of streams, sharp hills, and prominent buildings were generally lacking.

However, by January 1923, most of the material requested from time to time by the military commander for use by the Haitian Department of Public Works was sent as follows:

- Several sets of aerial pictures corrected for scale and distortion;

- several sets of enlargements, the largest corresponding to scales of 1:4,000 and 1:2,000, of 50 towns;

- maps of coast line, scale of 1:400,000;

- maps of 14, 30-minute quadrangles, scale of 1:100,000;

- list of level bench marks;

- list of triangulation stations with geographic coordinates; and

- 5,000 lithographic impressions of the Haitian Coat of Arms, reduced from an oil painting and furnished by a commercial firm.

The American occupation of Haiti terminated August 14, 1934.

Cuba

A geological and topographical survey of the Republic of Cuba was proposed as a cooperative project by Gen. Eugenio Sanchez Agramonte, Secretary of Agriculture, Commerce, and Labor, in February 1919. The Director of the Geological Survey presented a plan for the topographic survey of the island on the scale of 1:63,360 (1 mile equals 1 inch) that would require 8 to 10 years, and suggested an initial appropriation of $100,000. Gen. Mario G. Menocal, President of the Republic, a graduate of Cornell University, understood the value of such a survey, and made it the subject of a special message to the Cuba Congress on June 9. The political situation in Cuba was such that the bill was not considered in that session of the Congress, or in the session that convened on November 3. In August 1920 and May 1921, the Cuban Census Bureau had a large unexpended balance and it was hoped that it might be obtained for surveying the limits of barrios, municipalities, and provinces, and establishing permanent marks thereon. However, the project was finally abandoned when the American forces withdrew from the island in 1922.

Puerto Rico

Puerto Rico was ceded to the United States on August 12, 1898, as a result of the war with Spain, and civil government was inaugurated May 1, 1900. In 1904, the Luquillo National Forest was created, and 3 years later the National Forest Service requested the Geological Survey to mark its boundaries. The work began in March 1908, and was completed in June. E.M. Douglas was in charge of the field party, which surveyed 40 miles of levels and established 98 concrete, stone, or wooden posts. In order to locate the caverns of the reserve, two U.S. Coast and Geodetic Survey triangulation stations were occupied and four new stations were selected, marked, and occupied. In addition to an accurate contour sketch of a narrow strip along the entire boundary, a reconnaissance sketch of 54 square miles of the reserve, not including the summits of the Sierra de Luquillo, was made by C.L. Nelson.

In the winter of 1908 and 1909, in compliance with a request from the Puerto Rican government, the Geological Survey sent a party, consisting of T.M. Bannon, J.P. Harrison, C.H. Semper, and Olinus Smith, to the southern shore of the island to make surveys of the coastal plains in the vicinity of Ponce and Guayama, as a basis for an irrigation system. About 150 square miles were mapped on a scale of 1:12,000, with a 5-foot contour interval.

In January 1912, R.B. Marshall, Chief Geographer, submitted a $50,000 estimate for the mapping of Puerto Rico on the scale of 1:31,680 for about 1,000 square miles of the coastal plains and valleys, contour interval of 25 feet; and $65,000 for about 2,600 square miles of mountainous country, on the scale of 1:48,000 and contour interval of 50 feet.

In 1919, Governor Arthur Yeager and others requested the Legislature of Puerto Rico to appropriate funds for topographic mapping, but it was not until 1921 that the legislature acted favorably. Then funds were set aside for the topographic mapping of a strip 10 miles long by 4 miles wide along the northwest coast between the Guajataca River and the town of Aguadilla, where an irrigation system similar to the one built in Guayama was contemplated. A cooperative agreement was signed by the Director of the Geological Survey and John A. Wilson, Commissioner of the Interior. R.A. Gonzales, Chief Engineer of the Isabela Irrigation Service, requested that the Geological Survey also send an expert geologist to examine the reservoir site of the project, and determine the probabilities of excessive seepage, as the region was of limestone; therefore, George R. Mansfield, geologist, accompanied the topographic party in charge of Dallas H. Watson, which included E.E. Harris, J.O. Kilmartin, J.B. Leavitt, Jacob Lenovitz, J.F. McCook, R.R. Monbeck, W.A. Shumate, H.B. Smith, and B.H. Yoakum. They sailed from New York on January 21, 1922, and arrived in San Juan on the 25th.

It was necessary to measure a baseline and develop a complete system of triangulation, as the stations determined by the U.S. Coast and Geodetic Survey in 1901 were all obliterated; those on hills had been plowed under, and secondary points such as church spires and light houses had been destroyed by an earthquake in 1918. The 40 square miles to be mapped was open country and dotted generally with limestone buttes, an ideal setup for 3-point locations. However, between the buttes were sinks or potholes, presenting a confused or karst topography which, when mapped on the scale of 1:4,000 with a contour interval of 1 meter, proved heavy and slow. Approximately 120 square miles were mapped on a scale of 1:10,000 with a contour interval of 1 meter. Topographic engineers were permitted to show fence lines instead of locating and marking property boundaries, as Mr. Gonzales agreed to accept blueprints of the planimetric base and have his Puerto Rican engineers locate the private property corners from the shown fence lines. When the work was completed, they returned to Washington June 13th, in time for the summer field season, and arrangements were made to defer the inking of the Puerto Rico field sheets until the following winter.

After a lapse of 13 years, the Geological Survey again undertook topographic mapping in Puerto Rico. In 1934, under a grant of funds from the Public Works Administration, B.H. Yoakum was instructed to complete the triangulation of the island, and with the aid of two assistants from the Insular Department of the Interior, accomplished the task in the period February through July. Spirit levels also were extended in the western one-half of the island. With this preliminary preparation, topographic mapping was undertaken the next year by W.E. Chenoweth, who was in charge, J.O. Kilmartin, and F.H. Sargent. They completed an area of 150 square miles, on a scale of 1:25,000, with contour intervals of 1 and 5 meters, in the southwest corner of the island. Here the Lajas Valley was under consideration as an irrigation project.

In 1937, Puerto Rico was declared, by amendments to the Temple Act, in the same status as states with regard to cooperation with the Geological Survey, and an agreement was signed by the Director of the Geological Survey and the Commissioner of the Department of the Interior of Puerto Rico. K.W. Trimble in charge, C.W. Buckey, H.D. Cummings, B.J. Keating, W.D. Leech, R.H. Lyddan, F.H. Sargent, and H.G. Warner were assigned to complete the mapping of the island. Field surveys were executed on a scale of 1:20,000 with 1 and 5 meter contours; publication scale was 1:30,000. A topographer was able to map 4.5 square miles per month on an average. In 1954, cooperation was still in effect and Jay I. Brubaker, resident project engineer at San Juan, was maintaining a systematic revision program.

American Society of Military Engineers

The American Society of Military Engineers was planned in the fall of 1919, with the approval of the Acting Chief of Engineers, U.S. Army, and was organized in the spring of 1920. Its purpose was to perpetuate the friendships, to profit by the experiences of World War I, and to assist in the promotion of preparedness.

Among the charter members of the Society were eight Survey topographers, and among the first officers was Maj. H.R. Richards, who had been troop commander of the 29th Engineers in France, as business manager, and Wm. O. Tufts, former Survey topographer, as treasurer. By the end of the first year, the membership reached 3,500 and, in 1929, it had increased to 8,000. There were 11,000 officers and 285,000 men engaged in military engineering service during World War I.

The magazine of the Society, The Military Engineer, was published bi-monthly until September 1941, and thereafter monthly. It became the official organ of the Federal Board of Surveys and Maps, and contained a section in each issue devoted to mapping performed by the Geological Survey and other mapping agencies and offered a medium for the printing of articles by Survey topographers on special mapping projects. Most of these articles ware prepared at the suggestion of the Chief Topographic Engineer, and are shown in table 34.

Board of Surveys and Maps

The Board of Surveys and Maps of the Federal Government was created by Executive Order No. 3206 on December 30, 1919, "in order to coordinate the activities of the various map making agencies of the Executive Departments of the Government, to standardize results, and to avoid unnecessary duplication of work."

Among the provisions of the Order were:

- The Board to be composed of 1 representative of each of 14 mapping organizations who would serve without additional compensation.

- The Board to establish a central information office in the U.S. Geological Survey for the purpose of collecting, classifying, and furnishing to the public information concerning all map and survey data available in the several Government Departments and other sources.

The Board was organized in January 1920, and by amendatory Executive Orders of March 17, 1920, and January 27, 1921, the Board was enlarged by the addition of four other U.S. Bureaus. The first public meeting was held in the auditorium of the Interior Department on March 9, 1920, to which non-federal agencies were invited, and at the second public meeting on July 12, 1920, representatives of 22 non-federal agencies were named to constitute an Advisory Council to the Board in order to carry out the provisions of the Executive Order of December 30, 1919, touching the relations of the Board with the public.

On January 8, 1924, the Society of American Military Engineers was added as the 23d non-federal agency and the

Table 34. Articles prepared by Survey topographers for The Military Engineer.

Employee	Article	Publication date
Bagley, J.W.	"The Tri-Lens Camera in Aerial Photography and Photographic Mapping. Discussions"	May–June 1920
Bagley, J.W.	"Topographic Surveying from the Air"	Nov.–Dec. 1923
Bagley, J.W.	"Stereophotography in Aerial Mapping"	May–June 1924
Bagley, J.W.	"Study of Searchlight Triangulation"	July–Aug. 1926
Bagley, J.W.	"Surveying with the Five-Lens Camera"	Mar.–Apr. 1932
Birdseye, C.H.	"Surveying the Colorado Canyon"	Jan.–Feb. 1924
Birdseye, Sidney	"Aerial Topography on Tropical Surveys"	May–June 1936
Davey, C.H.	"Interpretation of Aerial Photographs"	Mar.–Apr. 1924
Druhot, G.S.	"Mapping the Hawaiian Islands"	Jan.–Feb. 1932
Gerdine, T.G.	"Topographic in Los Angeles County"	Sept.–Oct. 1926
Gill, W.H.	"Minnesota's Politico Geographical Boundary"	October 1941
Gill, W.H.	"The Mystery of Springs"	November 1941
Gill, W.H.	"Ups and Downs"	December 1941
Gill, W.H.	"Third Dimensional Maps'	January 1942
Gill, W.H.	"A Ghost Boundary That Won't Stay Laid'	February 1942
Gill, W.H.	"How Far Is It?"	April 1942
Gill, W.H.	"These Things We Call Maps"	May 1942
Gill, W.H.	"Key to a Better Understanding of Maps'	Feb.–Mar.–Oct. 1945
Matthes, F.E. and Evans, R.T.	"Map of Grand Canyon National Park'	May–June 1926
Nelson, O.H.	"Mapping the City of Chicago"	Jan.–Feb. 1929
Pendleton, T.P.	"Planimetric Maps of the Tennessee Valley Authority.	Sept.–Oct. 1935
Roberts, L.B.	"Conversion of Hachures to Contours"	Jan.–Dec. 1921
Sargent, R.H.	"Aerial Surveys in Southeastern Alaska"	May–June 1928
Sargent, R.H.	"Photographing Alaska from the Air"	Mar.–Apr. 1930
Smith, G.S.	"A Review of Topographic Mapping"	Mar.–Apr. 1924
Trimble, K.W.	"Some Old Yorktown Maps"	Sept.–Oct. 1931
Wheat, J.H.	"Airplane Photography as an Adjunct to Topographic Mapping"	Jan.–Feb. 1920
Wilson, R.M.	"Triangulation with the Planetable"	May–June 1931
Wilson, R.M.	"A New Photo—Alidade"	Nov.–Dec. 1937
Wilson, R.M.	"Transit Traverse"	June 1943

Society's magazine, The Military Engineer, was made the official organ of the Board.

The Board held public meeting in 5 or 6 months of the year, and executive meetings more often. A chairman was elected for each year. O.C. Merrill, Chief Engineer of the Forest Service, was the first chairman; C.H. Birdseye served as secretary, and J.H. Wheat was in charge of the Map Information Office.

The Board was abolished by Executive Order No. 9094, May 25, 1942, and its functions were transferred to the Bureau of Budget. The Secretary of the Interior designated J.G. Staack, Chief Topographic Engineer, as Departmental representative to keep the Bureau of Budget informed as to Survey operations. The Map Information Office was continued under the Topographic Branch.

Schoolcraft Quadrangle

During World War I, topographers serving with the Army in France made use of airplane photographs in correcting old maps and making new maps along the battle front. In 1919, the Schoolcraft, Mich., Quadrangle was photographed by the Air Corps and single-lens airplane photographs were used in mapping a full quadrangle. The usual horizontal control for the quadrangle was extended with sufficient ties to the land net so as to make a careful adjustment of the net to be used as secondary control. The methods of transferring the data such as roads, houses, swamp outlines, and drainage lines were by crude methods compared to those used at the present time. This photographic data was transferred by pantograph on one-half of a sheet, and on the other one-half the desired data were inked, and the photographs were then bleached so that only the desired information remained on them. These were photographed to scale and transferred by tracing. Even with this tedious method of transferring data to the planetable sheets, the information proved very valuable to the topographer and saved many hours of work, meandering swamps, lakes, etc. He also was able to intersect fence lines and take vertical angles, determining elevations for sketching purposes without sending a rodman into some areas. A report on the benefits of this method was submitted by C.L. Sadler to his Regional Engineer.

The Soil Conservation Service of the Agriculture Department had many square miles in the Missouri Valley section, and the Geological Survey obtained prints of these photographs. Since adjusted section lines could be used as secondary control in these areas, topographers continued to make use of the aerial photograph information, though the transferring by pantograph method was rather tedious. In 1937, Daniel Kennedy and Lee Walker designed a portable projector that could reduce the data to the scale of the map, and the information could be traced directly on the field sheets. This proved so much more valuable than the crude pantograph methods that other topographers in the Missouri Valley section began making portable projectors, each one an improvement on the previously made projector. This method of utilizing photographic data was continued until the planimetric bases were made by Multiplex machines.

Survey Order No. 100

Survey Order No. 100, dated February 26, 1920, reads as follows:

"In order to obviate overlap of the fields of work of the Section of Topographic Map Editing in the Publication Branch, and the Section of Inspection in the Topographic Branch, these two sections will be consolidated into one, to be known as the Section of Inspection and Editing of Topographic Maps, and placed under the administrative and technical control of the Topographic Branch. In addition to its inspection of topographic maps, this section will continue to do the work which the Section of Topographic Map Editing is now doing for the Engraving Division and the other units of the Geological Survey. The Engraving Division will continue to pay the salaries of those who are now on its roll and engaged in such work. Inasmuch as it may be difficult to strike a happy mean between the suitable preparation of original maps, or correcting proofs of maps, and the economical or effective operation of the plant for reproducing them, I expect to be called on frequently to weigh commercial economy against technical desirability.

"Mr. W.M. Beaman is designated as Topographic Engineer-in-Charge of the Section of Inspection and editing of Topographic Maps. Mr. James McCormick, whose long experience in critical work on the Survey's maps has fitted him for highly valuable service in a special field, has been recommended for appointment as the Survey representative on the U.S. Geographic Board and is specifically charged with research in connection with names of localities and features, and with such other duties as may be pre-scribed by the chief of the section."

Vocational Training under the Veterans Bureau

The Survey participated in the Army's Vocational Training Program for veterans of World War I, by assigning topographers to the Missouri School of Mines at Rolla, from 1920 to 1923, to teach the science of surveying and mapping. Charles E. Cooke, who had been furloughed for several weeks in the spring of 1912, 1913, and 1914 to lecture at the University of Chicago, was the senior topographer assigned to Rolla. He took over the work that C.E. Bardsley had started and was later assisted by T.T. Ranney and W.B. Brewer, who instructed the classes in the field. From these classes, the Survey, in 1923–24, gave appointmemts to the veterans wounded in France (table 35).

Table 35. U.S. Geological Survey appointments to veterans wounded in France.

Samuel D. Farmer	Delar Kimble	James L. Sanders
Milton J. Harden	Edward C. Kruse	Jack T. Schultz
Harry P. Jones	Harry S. Milsted	Cecil A. Turner
Daniel Kennedy	Samuel O'Hara	

About one-half of the Meramac Springs, Mo., 15-minute quadrangle was mapped by the vocational training students.

Arkansas–Mississippi Boundary Line

S.S. Gannett was one of three commissioners, appointed by the U.S. Supreme Court, by a decree of March 22, 1920, to survey and mark a portion of the Arkansas–Mississippi State line, commencing about 1 mile below Friars Point, Coahoma County, Miss., on an abandoned bed of the Mississippi River left dry by the avulsion which occurred about 1848. The true boundary had been fixed by the Treaty of Peace concluded between the United States and Great Britain, in 1783, as the middle of the main channel of navigation, and surveys had been made in 1816 and 1833. The Commission of 1920 recognized that the boundary line, in part, followed the deepest water in Horseshoe Lake, (or Old River, or Pecan Lake), a cut off oxbow on the east side of the river, and restored to Arkansas about 5 square miles. Continuous high stage in the river prevented field work before August 1, and the final report of the Commission was submitted on February 28, 1921.[34]

Minnesota–Wisconsin Boundary Line

The state of Minnesota, in 1916, instituted a suit against the state of Wisconsin in order to have that part of the State boundary line from St. Louis Bay up the St. Louis River to the falls near Fond du Lac finally determined. The court handed down an opinion March 8, 1920, and on October 11, 1920, appointed three commissioners to survey and mark the line which "must be ascertained upon a consideration of the situation as existing in 1846 and accurately described by the Meade chart as over water not less than 8 feet deep." At the first meeting of the commissioners, S.S. Gannett was elected chairman.

In January 1921, the survey was commenced on the ice and was completed March 19. The line surveyed was 18.4 miles in length and the instrument used was a transit-theodolite with 6.5-inch circle and reading to 10 seconds of arc. Measurements were made with a 300-foot steel tape. Rectangular coordinates were computed for each angle and suitable reference marks were established on shore. The commissioners report was submitted June 25, 1921, and confirmed by the Court on February 27, 1922.

Survey Order No. 106

Survey Order No. 106, Changes in Administration, Topographic Branch, dated April 3, 1922, was as follows:

"The death of George R. Davis, Topographic Engineer-in-Charge of the Pacific Division, necessitates the rearrangement of the administrative work in the field divisions. The funds available for topographic surveys have been so materially reduced that it is not advisable to appoint a new division chief to take Mr. Davis's place.

"Thomas G. Gerdine is designated Topographic Engineer-in-Charge of the Pacific Division to include the States and Territories of California, Washington, Nevada, Hawaii, and Oregon.

"Glenn S. Smith is designated Topographic Engineer-in-Charge of the Rocky Mountain Division, to include the States of Montana, Idaho, Wyoming, Utah, Arizona, New Mexico, Texas, North Dakota,

South Dakota, Nebraska, Colorado, Kansas, Oklahoma, Louisiana.

"Mr. Smith also will retain supervision over the work in the Division of West Indian Surveys.

"The only change made in the Atlantic and Central Divisions is that Louisiana is transferred from the Atlantic to the Rocky Mountain Division."

Survey Order No. 99 Amendment

On July 22, 1922, Survey Order No. 99, dated January 12, 1920, and amended June 4, 1921, is further amended to read as follows:

"During the absence of the Chief Topographic Engineer, the division and section chiefs will act for him with full authority in the following order:

"Mr. Glenn S. Smith

Mr. Frank Sutton

Mr. W.H. Herron

Mr. T.G. Gerdine

Mr. E.M. Douglas

Mr. W.M. Beaman."

The Brazilian Centennial Exposition (1922)

The Topographic Branch participated in the Brazilian Centennial Exposition held in Rio de Janeiro, in celebration of the independence of Brazil, as proclaimed on September 7, 1822. By authority of the Secretary of the Interior, and the Director, Lt. Col. Glenn S. Smith, was appointed by the Secretary of War to serve with Maj. Wallace W. Kirby, and Capt. Russell M. Herrington (with Maj. Edmund A. Buchanan, Cavalry, substituting for Captain Herrington) Corps of Engineers, U.S. Army, on a committee for the organization of a mapping exhibit at the Exposition. The Committee met on May 11, 1922, and a week later submitted its conclusions and recommendations.

The purpose of the exhibit was stated to be "to present to the Latin American countries such examples of various topographic maps, and of the standards and methods employed thereon, which have been executed by the U.S. Government for the development of the natural resources of the country, as well as for scientific and military purposes, as will demonstrate their equality, if not superiority to, the

---

[34] U.S. Geological Survey Bulletin No. 817, p. 200.

standards, methods, and products accepted by European governments."

That "for the purpose stated, the cooperation of the Topographic Branch, U.S. Geological Survey, Department of the Interior, the Corps of Engineers, and the Military Intelligence Division, is advantageous in that it demonstrates that military mapping in time of peace is a civilian function, of great economical value in the gencral development of the country * * *."

It was believed that an exhibit could be organized as follows:

1. A progressive, connected exhibit of instruments, methods, and products of the Geological Survey, as the maker of the base topographic map of the United States, from its organization to the present time, including various index maps, examples illustrating the use of airplane photography, and examples of the topographic maps made for the Dominican and Haitian Republics and for Puerto Rico.

2. An exhibit, practically continuous with the Survey's, of models and photographic processes with which the Corps of Engineers was experimenting, showing the various stages of map reproduction, modeling, and the development of a "pictorial relief map" from the contoured sheets of the Geological Survey.

3. An exhibit of the general maps produced by the Military Intelligence Division.

4. An exhibit, in operation, of a selected portion of the Mobile Reproduction Train of the Corps of Engineers that was used so effectively during World War I, consisting of a press truck, a process truck, and transfer truck. The three trucks at the Exposition were kept continuously in action during the open hours, printing the Map of the Atlantic for distribution.

Among the printed matter for distribution was 15,500 copies of a souvenir pamphlet turned out by the Survey's Engraving and Printing Division, of 44 pages between cardboard covers, containing 8 pages of engraved quadrangle maps in three colors, 12 pages of photographs and, 20 pages of descriptive text in Portuguese and English. Its title was "Joint Exhibit of Military Maps and Mapping prepared by the War Department and the Department of the Interior."

In order to have information relating to the material, methods, products, and mechanics of the exhibit available at all times, and thus achieve for it the greatest amount of good, the following experts were assigned to the exhibit:

*   Maj. William M. Beaman, Topographic Engineer, (furloughed from August 16, 1922, to May 31, 1923), representing the Department of the Interior.

*   Maj. Wallace W. Kirby, Corps of Engineers, and three employees of he Engineer Reproduction Plant, representing the War Department.

Colorado River Surveys

The canyons of the Colorado River have excited the interest of explorers, scientists, topographers, and engineers since Don Lopez de Cardenas stood on the rim of the Grand Canyon in 1542. Spanish missionaries and soldiers had visited the lower reaches of the river 2 years before.

The Geological Survey has always felt a proprietary interest in the Colorado River. Maj. John Wesley Powell made the first recorded exploration through its mysterious canyons in 1869. Ten years later, he helped to organize the U.S. Geological Survey, and became the second Director, March 18, 1881, to June 30, 1894.

In 1923, C.H. Birdseye, Chief Topographic Engineer, in planning a Colorado River expedition so as to take full advantage of the lessons learned in previous expeditions, and to achieve the maximum results, considered carefully the many necessary preparations, the personal risks involved, and the selection of the personnel. He decided to head the expedition, and selected the following well-seasoned, experienced men:

*   Eugene C. LaRue, hydraulic engineer, who for several years had been studying the problem of utilizing the waters of the Colorado;

*   Raymond C. Moore, State Geologist of Kansas, who for 2 years had been making geological investigations in Utah and Arizona;

*   Roland W. Burchard, topographic engineer, who had already surveyed the lower stretches of the river;

*   Emery C. Kolb, of Grand Canyon, who in 1911, with his brother, made a boat trip for the purpose of taking moving pictures from Green River, Wyo., through the canyons to the Gulf of California;

*   Lewis B. Freeman, of Pasadena, Calif., a well known explorer and boatman;

*   Leigh B. Lint, of Weiser, Idaho, and H.E. Blake, Jr., of Monticello, Utah, two husky youths with 2 years experience in boating the rapids of the Colorado the preceding year with Kelley W. Trimble, topographic engineer, on surveys on the Green River through the Canyon of Lodore to Green River, Utah, and on the San Juan River above its junction with the Colorado;

*   Frank B. Dodge, of Honolulu, a skillful boatman, expert swimmer, and general utility man, capable of filling any position from instrument man to cook; and

*   Frank Word, of Los Angeles, cook.

At Bright Angel, about 90 miles below Lee's Ferry, the party was joined by Herman Stabler, a Hydraulic Engineer of the Conservation Branch of the Survey, who made a trip down the river to Lee's Ferry in 1922, and at Havasu Creek by Felix Koms, who replaced Frank Word as cook.

The expedition left Lee's Ferry on August 1, 1923, in four boats especially made in the design which Nathan Galloway, a Utah trapper, had found best adapted for the canyons that he navigated in 1895 and several times afterwards. Soap Creek Rapids, the second in Marble Canyon and 11 miles downstream from Lee's Ferry, was encountered at the end of the next day, and through a small radio set, the party kept in touch with outside affairs. Thus the party learned, while perched in comparative safety up the slope above high water in the vicinity of Lava Falls, one of the most dangerous on the whole river, during 4 days in September, that the outside world believed that the flood had been disastrous to the expedition. Water rose 21 feet during the flood, and measured in volume increased from 10,000 second feet to 125,000 second feet.

When the river had subsided, the journey was resumed. Diamond Creek, 45 miles distant, was reached at the end of 10 days. Here the party was met by packers with food, letters, and newspapers. Letters were sent out assuring relatives that all was well with the party. Similar deliveries of supplies and mail had been made at Bright Angel and Havasu Creek. This service was in charge of Roger Birdseye, a cousin of Colonel Birdseye, and a resident of Flagstaff, Ariz. His trips to the river were hampered by washed out trails. In Havasu Creek there was no trail, not even a footpath. Below the beautiful walled-in valley, the home of the Havasupai Indians in the "Land of the Sky Blue Water," Havasu Creek plunges through a narrow precipitous gorge, dropping vertically over four scenic falls. On July 8 and 9, a topographic engineer who was mapping remote corners of the Grand Canyon National Park, ran a stadia traverse down Havasu Creek and set a vertical angle bench mark near its junction with the river, as a tie point for the river party.

After four restful days, such as were usually spent at supply points, the party moved from Diamond Creek down the remaining 27 miles to Reference Point Creek, where the survey was tied to previous surveys carried upstream in 1920. A continuous stadia traverse had been carried 251 miles, from an elevation of 3,117 feet to an elevation of 1,122 feet, closing with an error of less than 5 feet. In this stretch were 87 rapids and many small bits of rough water. The survey was made with a planetable and alidade, with a micrometer for the measurement of long distances, if necessary. Such cases, however, were rare and the usual length of sight was not more than 1,500 feet. Wherever possible, differences of elevations were determined by direct level readings, but in a few cases, the use of vertical angles was necessary. All distances and level determinations were double checked. The survey was completed on October 13,

and the remaining 203 miles of river to Needles was covered by October 19, when the party was disbanded.[35]

Survey Order No. 114

Survey Order No. 114, Changes in Administration, Topographic Branch, dated January 2, 1925, was as follows:

"Since the retirement of Frank Sutton, on March 7, 1924, Glenn S. Smith has been in administrative charge of the Atlantic Division, and the Chief Topographic Engineer has been handling the administration of the Rocky Mountain Division. It is desirable to relieve the Chief Topographic Engineer of the immediate supervision of a field division, but the work of the Topographic Branch is in such condition that field administration can be handled by three division chiefs, rather than four. The field administrative work of the Topographic Branch is therefore rearranged as follows:

"Glen S. Smith is designated Division Engineer-in-Charge of the Atlantic Division, to include the States of Maine, West Virginia, New Hampshire, Ohio, Rhode Island, Indiana, Massachusetts, Vermont, Kentucky, Tennessee, Connecticut, North Carolina, New York, South Carolina, Pennsylvania, Georgia, New Jersey, Florida, Maryland, Alabama, Delaware, Mississippi, and Virginia.

"W.H. Herron is designated Division Engineer-in-Charge of the Central Division, to include the States of Michigan, South Dakota, Wisconsin, Nebraska, Illinois, Kansas, Minnesota, Oklahoma, Iowa, Texas, Missouri, Wyoming, Arkansas, Colorado, Louisiana, New Mexico, and North Dakota.

"T.G. Gerdine is designated Division Engineer-in-Charge of the Pacific Division, to include the following States and Territory of Washington, Nevada, Oregon, Utah, Idaho, Arizona, Montana, California, and Territory of Hawaii."

Temple Act

Demands for topographic mapping were far in excess of those that could be accomplished with the funds that were provided for the work. After carefully considering the demands for topographic surveys, and the rate at which they were progressing, the Federal Board of Surveys and Maps proposed a plan for the early completion of the mapping of the United

---

[35] See "Surveying the Colorado Grand Canyon," by C.H. Birdseye, The Military Engineer, January–February 1924.

States. This plan was enacted into law, and was approved by the President on February 27, 1925, and is known as the Temple Act. The Temple Act reads as follows:

> "An Act to provide for the completion of the topographical survey of the United States.
>
> Be it enacted by the Senate and House of Representatives of the United States of America in Congress assembled, That the President be, and hereby is, authorized to complete, within a period of 20 years from the date of the passage of this Act, a general utility topographical survey of the territory of the United States, including adequate horizontal and vertical control, and the securing of such topographic and hydrographic data as may be required for this purpose, and the preparation and publication of the resulting maps and data: Provided, That in carrying out the provisions of this Act, the President is authorized to utilize the services and facilities of such agency or agencies of the Government as now exist, or may hereafter be created, and to allot to them (in addition to and not in substitution for other funds available to such agencies under other appropriations or from other sources) funds from the appropriation herein authorized, or from such appropriation, or appropriations as may hereafter be made for the purpose of this Act.
>
> "Sec. 2. That the agencies which may be engaged in carrying out the provisions of this Act are authorized to enter into cooperative agreements with and to receive funds made available by any state or civic subdivision for the purpose of expediting the completion of the mapping within its borders.
>
> "Sec. 3. The sum of $950,000 is hereby authorized to be appropriated out of any moneys in the Treasury not otherwise appropriated, to be available until the 30th day of June, 1926, for the purpose of carrying out the provisions of this Act, both in the District of Columbia and elsewhere as the President may deem essential and proper.
>
> Approved February 27, 1925."

The act authorized the completion of the mapping in 20 years, and its sponsors contemplated gradually increasing appropriations for both geodetic and topographic work; however, the act only authorized, and did not appropriate, the necessary funds, and has not resulted in expediting the topographic mapping of the country to any great extent.

Training Films

The first training film was a three-reel silent film entitled "The Making of a Topographic Map," prepared from field and office pictures taken in 1924–25, and was ready for distribution in 1926.

This film was sent to the chiefs of field parties with instructions to arrange to show it to local technical or educational groups so they would become more familiar with the work of the Topographic Branch and the need for topographic maps. The following "Story of Topography" also was sent:

> "The early chapters of the 'Story of Topography' as told in the history of the Geological Survey were written more than 40 years ago in the silence and isolation of the rim of the "Great American Desert" on the high summits of the Rocky Mountain ranges, and midst the clamor and excitement of the mining camps of a day long past. Associated with those scenes are the names of Clark, Renshawe, Bodfish, Bien, Nell, Thompson, Wilson, Gannett, and numerous contemporaries who are called to mind whenever mention is made of the early work of the Survey.
>
> "Work in such remote fields of operation necessitated preparations that included provision of shelter and food, and for protection against the dangers of a comparatively unknown and unexplored territory.
>
> "The topographer of that period often times shot with his rifle quite as accurately as with his telescope, was as successful with his fishing lines as with his contour lines, communed as freely with roving Indians as with the voices of nature. In fact, he was a combination of engineer, draftsman, explorer, packer, horseman, mountain climber and diplomat. Those were the days of endurance, romance, and adventure, and they developed men whose memory is held in reverence by the generations of engineers that have followed them.
>
> "This early period of reconnaissance and exploration has been followed by one which has witnessed an unusually rapid development of methods, improvement in instrumental equipment, the establishment of orderly procedure, the adoption of standard scales and contour intervals for different types of country, material additions to the data shown on published maps, and requirements for greater refinement and accuracy in all the field and office processes.
>
> "The engineer's environment and his work of today are in striking contrast with those of the early period. Irrigation has brought to the desert blooms and forage and farm products, with a corresponding agricultural population. Railroads have penetrated to the most remote mountain recesses; and mining companies with smelters and other reduction plants have developed the frontier mining camp to a well organized city.

"The scenes of our story have broadened and grown until they include every part of the United States. Little remains of those experiences to topographic pioneering. Few indeed are the areas in which the topographic engineer can not depend upon a Ford to solve his transportation problems, and although there are still numerous areas, as the pictures will show, where the pack train and camping outfit are necessary, a very large percentage of the topographic work is accomplished with more modern and more prosaic equipment. With a new environment has cone new standards of accuracy for the work, a much wider use of the maps, and a corresponding increase in requests for surveys from many sources.

"The topographic surveys were begun in order to supply the immediate needs of the geologist by providing a base map upon which the results of his geologic investigations could be graphically shown. Today these maps are used by all the departments of the Federal Government, by various state organizations throughout the country, and by engineers, miners, educators, land investors, travelers, and representatives of the general public, whose sincere personal interest in all the activities and plans of the Topographic Branch of the Geological Survey is indicative of their urgent need for the information which they find on these topographic maps.

"During the field season which, which usually begins in April, and ends in November, you may encounter the topographer, with his mounted planetable and telescopic alidade, "running out" the section-line highways of the central western states, working across the sand dunes along the borders of the Great Lakes in Michigan, combing the topographic features in the intricate glacial drift country of Wisconsin, scaling the Appalachian Range, following the tortuous windings of the rivers in Pennsylvania, Tennessee, and Kentucky; penetrating the deep pine woods of Maine, battling with the winds that blow so consistently across the trans-Mississippi plains, floundering through swamps to reach reliable location points, climbing to the very summits of the high mountain ranges of the West, where snow abides throughout the year, or mapping the valley floors of the mighty rivers throughout our land.

"Although the glamour and romance of those olden, golden days of pioneering and adventure are no more, the modern knight of the planetable finds no less of inspiration, and feels no less of enthusiasm in his work than the hardy men of the early times. The character of his dreams may have changed, but they are none the less attractive and realistic, for they enable him to vision the future, its increased activities, more comfortable and more numerous homes,

improved sanitation, and higher standards of living, with a feeling of satisfaction that his own work is making a distinct contribution to the forces which will have accomplished these results.

"In the moving pictures that are to follow, an attempt has been made to show, in logical order, the varied activities of the Topographic Branch of the U.S. Geological Survey, including all office and field operations that contribute to the preparation and publication of the topographic maps."

This film also was used extensively by technical societies and engineering schools, and in 1937, was superseded by a two-reel, silent, black and white film titled "Stereoscopic Mapping from the Air," which showed the use of stereoscopy in the preparation of maps from aerial photographs.

"Reconnaissance Mapping with Trimetrogon Photography," is a sound and color film that was produced in 1944 by the Pan American Institute of Geography and History, Commission on Cartography, in cooperation with the Geological Survey. It gives a detailed portrayal of the methods used in preparing maps by use of photographs taken with trimetrogon cameras. Continuous requests are received for the use of this film in the training programs of colleges and the military services.

"Topographic Mapping by Photogrammetric Methods," is a sound and color film that was produced in 1947 by the Pan American Institute of Geography and History, Commission on Cartography, in cooperation with the Geological Survey. This film shows modern technical procedures employed by the Geological Survey in making standard topographic quadrangles by the Multiplex method of mapping. Colored spectacles are supplied with the film for viewing the three dimensional scenes. This film has been loaned to many technical societies, engineering schools, Federal and State mapping agencies, and others interested in topographic mapping.

Besides the "Topographic Mapping by Photogrammetric Methods" film with sound and color, films are available for limited circulation to specialized groups and others interested in topographic mapping as follows:

- "Leveling for Topographic Mapping"

- "Transit Traverse for Topographic Mapping"

- "Triangulation for Topographic Mapping"

- "Supplemental Control for Topographic Mapping"

- "The Preparation of Topographic Manuscripts for Reproduction"

- "Negative Scribing for Map Reproduction"

- "The Helicopter as an Aid in Alaska Surveys" (Silent-Color)

During a year it is estimated that these training films are shown approximately 80 times before technical and educational societies, colleges and universities, and service groups, and are viewed by about 8,000 persons.

Topographic Instructions

"A Manual of Topographic Instructions" (432 pages) to replace all former instructions relating to topographic work was published in 1928, as U.S. Geological Survey Bulletin No. 788. The manual was issued in several parts, each covering a general subject, and prepared by Branch members best qualified for the work. The subject of each part, author, and date of issuance is shown in table 36.

Table 36. Contributions to U.S. Gological Survey Bulletin No. 788.

Chapter	Author	Date of issuance
Administration	Helen M. Frye	November 15, 1926
Triangulation	E.M. Douglas	December 22, 1926
Transit Traverse	E.M. Douglas	December 18
Leveling	E.M. Douglas	December 13
Topographic Mapping	W.M. Beaman	April 2, 1928
Map Compilation from Aerial Photographs	T.P. Pendleton	February 27, 1928

Late in 1947, the Research and Technical Control Branch of the Topographic Division began preparation of a loose-leaf manual of instructions to replace Bulletin No. 788, now obsolete in part as a result of photogrammetric developments, improved methods, and changing map requirements. However, the first three chapters were not reproduced and distributed until December 1950. The manual probably will contain 144 chapters. A Table of Contents has been compiled to list the proposed chapters; it is, of course, subject to revision, as are all other chapters. More and more chapters are being written and distributed each year. Each chapter will be prepared and issued as a separate unit for operational use, and each part, when complete, also will be issued as a circular in the Survey's numbered series of publications.

Topographic Branch Memo

On June 13, 1927, the following memorandum was sent to the members of the Topographic Branch.

"Mr. Hersey Munroe, Topographic Engineer-in-Charge of the New England Section, will be relieved on June 15 and designated as Inspector and Instructor for the Atlantic Division.

"Mr. Albert Pike, Topographic Engineer-in-Charge of the Middle Atlantic Section, will be relieved on June 15 and designated Topographic Engineer-in-Charge of the New England Section.

"The Middle Atlantic Section will be administrated by the Division Engineer-in-Charge, Atlantic Division until further notice."

Maine–New Hampshire Boundary Line

In 1874, the boundary line between Maine and New Hampshire was resurveyed, and in 1927, the Legislatures of the two States authorized a retracement and remarking of the boundary line from Salmon Falls northward to the Canadian line. A.T. Fowler was given this assignment and he resurveyed and monumented 112 miles of the north-south boundary line between Maine and New Hampshire. This work was completed in 1929.[36]

Oklahoma–Texas Boundary Line

S.S. Gannett was appointed Commissioner under the decree of the U.S. Supreme Court rendered January 3, 1927, to run, locate, and mark the boundary between the States of Oklahoma and Texas, along the true 100th meridian of longitude west from Greenwich, extending north from its intersection with the south bank of the South Fork of Red River, to its intersection with the 36-degree 30-minute parallel the approximate northeast corner of the Panhandle of Texas, according to the surveys of 1858–60. The state of Texas employed an astronomer in 1892 to determine the 100th meridian at its intersection with the Red River.

From 20 triangulation stations, established by the U.S. Coast and Geodetic Survey in 1923 and in 1927–28, with geographical positions on the North American Datum of 1927, offsets were made with a 300-foot steel tape to the true meridian. In the 133.6 miles of the line, the longest offset measurement was 3,620.23 feet, the shortest was 197.78 feet. The meridian, as established, was 4,040 feet east of the old line at the south end, and 880 feet east at the north end, making the strip of land included between the two lines about 44.6 square miles. The number of concrete monuments set to mark the true meridian was 160.

The instrumental work was executed by Eugene L. McNair. Rodmen, chainmen, and other temporary employees were hired locally, an equal number from each state. Mr. Gannett submitted his report on July 15, 1929,[37] and the Court approved the line thus established on March 17, 1930.

---

[36] U.S. Geological Survey Bulletin No. 817, p. 174.

[37] General Land Office Service Bulletin, December 1926, p. 315.

Fiftieth Anniversary

Thursday, March 21, 1929, just 50 years after President Hayes appointed Clarence King Director of the new U.S. Geological Survey, was largely devoted to events in celebration of this momentous day by Survey members and friends.

The first of three events was a noonday reception at the White House. President Hoover had been a geological assistant with the Survey during the seasons of 1893 and 1894 in California. Mrs. Hoover, who had also majored in geology at Stanford University, related several amusing incidents in connection with her search for facts.

In the afternoon, at a meeting in the National Museum, Secretary of the Interior, Ray Lyman Wilbur, and several prominent scientists, outlined the growth and scope of the Survey's activities.

In the evening, more than over 600 people attended a banquet at the Washington Hotel, which was followed by an elaborate entertainment consisting of skits by every Branch, and dancing. The Topographic Branch had prepared a booklet for the anniversary, containing a brief humorous account, in verse, of topographic mapping methods, written by E.I. Ireland, and cleverly cartooned by Gerald FitzGerald. The oldest topographic engineer in point of service was presented with an inscribed loving cup:

1880–1930

Hersey Monroe

From Fellow Engineers

U.S. Geological Survey

Fifty Years of Service

The topographic maps that had been necessary to provide an accurate base upon which to represent the facts ascertained by geologic work had now attained so high a degree of exactness that they were sought for themselves alone, by all classes of people, from engineers to vacation tourists.

Nine states, and the District of Columbia, are now entirely mapped, and the percentage in other states ranges from 8 percent to 88.6 percent. Of the total continental United States, exclusive of Alaska, 43.6 percent had been mapped. Cooperation was continued with the Air Corps, U.S. Army, whereby aerial photographs were furnished for use in topographic mapping. Bulletin No. 788 "Topographic Instructions of the Geological Survey" was published.

New Mexico–Texas Boundary Line

By a decree of the U.S. Supreme Court, dated April 9, 1928, Mr. Gannett was appointed Commissioner to reestablish the boundary line between New Mexico and Texas northwest of El Paso, from parallel 32 degrees southward to parallel 31 degrees, 47 minutes on the international boundary between the United States and Mexico. The legal boundary was the middle of the channel of the Rio Grande, as it existed on September 9, 1850, and to recover it the Salazar-Diaz triangulation and traverse of 1852 was reproduced on the ground as closely as possible from the angles, bearings, and distances given in the records. The river was noted for its habit of swinging back and forth through a 4-mile valley in flood times, and the new survey found it in its old channel for a distance of only 1.5 miles, and away from it as far as 2.5 miles.

The instrumental work of 1929 and 1930 along the disputed boundary was executed by E.L. McNair, assisted by L.B. Leachman. The total length of the boundary line was 25.17 miles; the number of concrete monuments set, each exceeding half a ton in weight, was 105. The new survey caused a net loss of approximately 2,496 acres (3.9 square miles) by Texas to New Mexico. The land was valued at about $300 per acre. Robert H. Reineck, topographer, added the culture to the topographic map of about 60 square miles, which the U.S. Bureau of Reclamation executed on the scale of 1:24,000 with a contour interval of 5 feet. Mr. Gannett submitted his report on July 17, 1930, and it was confirmed on March 23, 1931.

Survey Order No. 128

Survey Order No. 128, dated September 10, 1929, made the following personnel changes in the Topographic Branch:

"Two resignations in the Topographic Branch necessitate a change in administrative personnel. Colonel Smith has asked to be relieved of administrative duty as chief of the Atlantic Division, a duty, which he has fulfilled with great effectiveness. Colonel Birdseye leaves the Government service to fill an important position. As a field topographer, as an engineer officer in France, and as the chief administrative officer of the Topographic Branch, Colonel Birdseye's record has been brilliant. He has added much to the engineering efficiency within the organization as well as to its outside reputation for high standards of work. Best of all, he has won the confidence, respect, and affection of his associates.

"Mr. John G. Staack, after 25 years service in the Topographic Branch, has been appointed Chief Topographic Engineer. His wide experience and close acquaintance with both the work and the personnel of the Branch warrant confidence that he will take full advantage of the larger opportunity afforded through the Secretary's policy of expansion.

"Mr. Albert Pike, who has served so successfully in the administration of the Atlantic Division, will succeed his retiring chief, and his designation as

Division Chief means the continued efficiency of the topographic work in the East.

"Colonel Smith will not sever his connection with the Survey but will be available for occasional duty as the Director may wish, to make use of his wide experience, or the Secretary of the Interior may desire to continue him as his representative on the Southern Appalachian National Park Commission.

"Until Mr. Staack returns from the field, Colonel Smith will serve as Acting Chief Topographic Engineer."

The appointment of John G. Staack as Chief Topographic Engineer carried with it the assurance that the Secretary of the Interior supported a policy of expanding the mapping program of the Survey. This meant, for Mr. Staack, the preparation of many plans and accompanying detailed estimates for new mapping projects, though appropriations for many of them were not forthcoming. The application of the new science of photogrammetry to topographic mapping called for constant study and frequent analyses of costs. The reputation of the Geological Survey for producing the finest topographic maps was founded on searching the ground with the planetable and alidade, and upon reproduction by copper plate engraving. However, it was early recognized that aerial photography could expedite the ground surveys and furnish the location of hundreds of minor points, thus increasing the accuracy of the map, and copper plate engraving was being challenged on cost by photolithography.

Appropriations and Personnel (1929)

In 1929, the personnel of the Topographic Branch consisted of 256 employees; 169 topographic, geodetic, and cartographic engineers of various grades; 71 engineering field aides and draftsmen of various grades; and 16 clerks in the clerical force.

In 1929, Field Divisions numbered three: Atlantic, Central, and Pacific. A summary of mapping for 1929 is show in table 37 with appropriatioms shown in table 38. Appropriatioms by state for topographic surveys are listed in table 39, and the expended funds for 1929 are listed in table 40.

Map of Roseau River Valley, Minnesota—Staack and Sadler (1929–30)

In 1929, at the request of the International Boundary Commission, the mapping of the Roseau River Valley was undertaken on a scale of 1:31,680 and contour interval of 2 feet. The total area of this project was approximately 600 square miles. Approximately 200 square miles of this area was mapped in

Table 37. Summary of mapping for 1929.

[--, no data]

Publication scale	New (square mile)	Resurvey (square mile)
1:12,000	57	86
1:20,000	266	--
1:24,000	--	321
1:31,680	878	800
1:62,500 Hawaii[a]	9,498	1,449
1:125,000	2,257	1,393
Total		
New mapping	12,956	--
Resurveys	--	4,049
Revision	--	328

[a] Advance sheets with 10-foot contour interval published at 1:62,500.

cooperation with the Department of Drainage and Water, state of Minnesota. The reason for this survey was complaints from local residents on the Canadian side that the extensive ditching of this valley area had increased flood conditions and was causing great damage on the Canadian side. Also, complaints were made by the United States residents that a dam that was contemplated on the Canadian side would back water onto the American side and cause considerable damage. A very large percentage of this valley area was covered with peat bog swamps, which furnished very good grazing land when properly drained. Consequently, the state of Minnesota had placed ditches along each section line and, after leveling off, had constructed county roads on top of the spoil banks.

Table 38. Appropriations for 1929.

Appropriation	Amount
Topographic surveys	$780,288
Transfers and repayments for work performed for other Federal units	133,002
State cooperative funds (27 states)	476,527
Total	1,389,817

The topography on the project was by Daniel Kennedy, F.L. Whaley, S.B. O'Hara, M.J. Harden, F.K. Van Zandt, E.J. Fennell, and C.L. Sadler, who was in charge.

Two hundred and twenty-two miles of transit traverse and 280 miles of levels were run and 54 bench marks were set to control this project. Since a large percentage of the permanent bench marks had to be set in the swamp areas, which comprised peat moss to a depth of sometimes more than 6 feet, and the freezing during severe winters had a tendency to push telephone poles and monuments out of the ground,

Table 39. Appropriations by state for topographic surveys during fiscal year 1929.

[--, no data]

State or project	Appropriation for topographic surveys	Repayments for work performed for other Federal units	Total Federal funds	State cooperative funds	Total funds
Alabama	$12,091.46	--	$12,091.46	$10,000.00	$22,091.46
Arizona	8,278.05	--	8,278.05	--	8,278.05
Arkansas base map	1,565.76	--	1,565.76	1,500.00	3,065.76
California	37,485.01	$1,432.31	38,917.32	56,911.99	95,829.31
Colorado	7,194.95	--	7,194.95	2,000.00	9,134.95
Connecticut	3,468.97	--	3,468.97	4,988.27	8,457.24
Hawaii	27,562.63	--	27,562.63	24,887.55	52,450.18
Idaho	28,009.71	--	28,009.71	--	28,009.71
Illinois	46,378.05	27,199.88	73,577.93	40,015.77	113,593.70
Iowa	1,133.04	--	1,133.04	1,997.48	3,130.52
Kentucky	93,486.85	8,525.09	102,011.94	66,145.32	168,157.26
Maine	29,823.77	13,498.65	43,322.42	33,934.56	77,256.98
Michigan	14,629.53	--	14,629.53	15,053.54	29,683.07
Mississippi		48,962.70	48,962.70	--	48,962.70
Missouri	7,486.62	--	7,486.62	6,288.32	13,774.94
Montana	2,769.02	--	2,769.02	15,626.92	18,395.94
Nevada	32.05	4,797.42	4,829.47	--	4,829.47
New Hampshire	14,736.03	--	14,736.03	33,677.90	48,413.93
New Mexico	13,868.86	--	13,868.86	12,726.85	26,595.71
New York	6,098.81	--	6,098.81	10,424.50	16,523.31
North Dakota	8,694.39	--	8,694.39	12,965.09	21,659.48
Oklahoma	16,690.31	--	16,690.31	15,566.73	32,257.04
Oregon	11,345.82	--	11,345.82	1,735.17	13,080.99
Pennsylvania	15,211.13	--	15,211.13	22,456.46	37,667.59
Tennessee	72,920.44	--	72,920.44	14,000.00	86,920.44
Texas	36,418.51	4,531.63	40,950.14	33,867.41	74,817.55
Utah	7,421.02	--	7,421.02	329.44	7,750.46
Vermont	13,925.03	4,029.51	17,954.54	8,184.73	26,139.27
Virginia	68,795.74	3,012.03	71,807.77	12,020.04	83,827.81
Washington	8,618.73	--	8,618.73	3,367.98	11,986.71
West Virginia	203.78	--	203.78	--	203.78
Wisconsin	13,431.98	6,329.47	19,761.45	15,855.32	35,616.77
Wyoming	708.68	--		708.68	708.68
Books for library	157.39	--	157.39	--	157.39
Computing	[a]4,207.53	--		4,207.53	4,207.53
Contingent	11,176.74	--		11,176.74	11,176.74
Field district offices	600.00	--	600.00	--	600.00
Instruments (field)	[a]8,137.10	--	8,137.10	--	8,137.10
Stationery (field)	[a]225.00	--	225.00	--	225.00
Inspection and editing	[a]8,887.42	--	8,887.42	--	8,887.42

Table 39. Appropriations by state for topographic surveys during fiscal year 1929.—Continued

[--, no data]

State or project	Appropriation for topographic surveys	Repayments for work performed for other Federal units	Total Federal funds	State cooperative funds	Total funds
Map information	3,496.36	--	3,496.36	--	3,496.36
One-millionth maps	7,747.64	--	7,747.64	--	7,747.64
Miscellaneous repay	--	10,683.44	10,683.44	--	10,683.44
Office salaries	[a] 8,742.64	--	8,742.64	--	8,742.64
Photographic mapping	[a] 3,939.81	--	3,939.81	--	3,939.81
Relief maps	3.85	--	3.85	--	3.85
Total expenditures	[b] 687,806.21	133,002.13	820,808.34	476,527.34	297,335.68
Unexpended balance	[c] 92,482.17	--	92,482.17	--	92,482.17
Total	780,288.38	133,002.13	913,290.51	476,527.34	1,389,817.85

[a] Presents 37.5 percent of total cost; balance of 62.5 percent included in charges for cooperation.

[b] $452,541.39 expended on State cooperation.

[c] Includes $57,945.84 Great Smoky and $33,924.78 Shenandoah Park funds available during fiscal year 1930.

it was imperative that some method of setting bench marks be evolved that would ensure their permanent location. It was decided to make the bench mark posts with reinforced concrete, 6 inches in diameter and 6 feet long. The forms for the posts were set on a plank with small holes bored in them so that the reinforcement wires could extend out at least 16 inches. A bench mark tablet was placed in the upper end of the monument. Holes 2 feet in diameter were then dug for placing these monuments, the extra reinforcement wires being bent at right angles in all directions, and then a couple of buckets of concrete were poured into the holes, making a collar at the bottom of each monument. A majority of the posts reached to hardpan clay, but it is believed that those set in the peat formation will remain permanently located.

Because of a small contour interval, innumerable elevations were determined on each square mile and indicated on the planetable sheets to the correct tenth, the decimal point being the location of elevation.

Table 40. Expended appropriations for 1929.

Appropriation	Amount
Topographic surveys	$726,870.62
Transfers and repayments for work performed for other Federal units	127,368.76
State cooperative funds	441,851.95
Total	1,296,091.33

An examination of the base map showed that a rise of the river considerably below flood stage caused the flow of water in these ditches to reverse itself and the water was spilled into the Red River Valley country. These section line ditches were beneficial rather than detrimental in the flood control of the Roseau River. Also, a dam on the Canadian side that would raise the river as much as 3 feet at the boundary would cause it to spill over through a low gap within one-half mile of the boundary, and the water would again return to the Canadian side.

Aerial Photography

In 1930, aerial photographs were taken by commercial contractors, of a strip along the upper Columbia River in the state of Washington, of Zion National Park, and Bryce Canyon National Park in southern Utah. About one-half of the maps of the two national parks were produced by stereo-photogrammetric processes with the Hugershoff aerocartograph.

In 1931, aerial photographs of Lakeport Quadrangle, Calif., were taken by a commercial contractor. The quadrangle was then revised and partially remapped by stereo-photogrammetry.

Orders were received for the forced retirement of husband or wife, if both worked for the government, and of certain other employees, in April 1933. Government salaries were reduced 15 percent and a 3-month furlough without pay was suggested. The Democratic Administration (The New Deal) that was elected on a platform that promised to bring the country out of the depression and effect a 25 percent reduction in governmental expenditures proposed to effect the latter by reducing personnel. Complying with the Secretary's orders was a distasteful duty for Mr. Staack, as most of the Topographic Branch members who were separated were his friends. In several months, The New Deal found that reduction of governmental personnel merely added to the ranks of the unemployed. Emergency relief measures were necessary and the Public Works Administration was one of the agencies created for the purpose of increased employment, opportunities for work, and stimulation of business.

Public Works Administration Projects

The Survey received $1,000,000 from the Public Works Administration (PWA) in October 1933. Under PWA allotments, the Topographic Branch was able to accomplish some planned mapping sooner than expected and to enjoy some definite advantages. Field parties were equipped with new trucks. They were permitted to hire the extra pack horse and the extra rodman, cook, or packer that they had always needed. The new engineers required careful training, and the larger field parties made additional work for the party chiefs. They also were bothered by new rules. Rodmen, chainmen, recorders, cooks, and packers were not to work, more than 40 hours a week, or 8 hours a day for 5 days and they were allowed annual and sick leave, the same as permanent employees with Civil Service status. In Washington, the added burden of the expanded program fell heavily upon the Chief Topographic Engineer and his assistant, E.L. Hain, for they were unable to employ additional clerical help.

Public Works Administration funds were available until the end of 1940. The Survey kept wage standards high, $2,000 per annum for instrument men and, with the first allotment, achieved the highest record among Government bureaus; 739 man hours for each $1,000 expended. With the second allotment, the man hours average was 705 because equipment and trucks were purchased. Indirect employment not counted, survey productivity was highly rated.

Public Works Administration funds were prorated among the states and Puerto Rico. However, some of the states had CWA, ERA, FERA, WPA, or PWA funds[38] and asked the Topographic Branch for the loan of supervisors, instruments, and automobiles. In the Atlantic Division, South Carolina had such funds, and in the Central Division, Arkansas, Colorado, Kansas, and Missouri also requested aid.

In South Carolina, from September 1933 to September 1935, the Hagood, Camden, and Killian Quadrangles were completely mapped; supervision was made of horizontal and vertical control, 6,000 square miles were photographed and planimetric maps of this area were begun. K.W. Trimble was first in charge of this work, followed by C.B. Kendall and A.T. Fowler.

From December 1935 to October 1941, the mapping of the city of Denver, Colo., was undertaken as a Works Progress Administration Project No. 674. Under WPA regulations, the

Topographic Branch could share only administrative costs. Work had to be in or near the city, as men could be employed only 3 days in the week at less than $1,000 annual wage. Fred Graff, Jr., topographic engineer, was selected to take charge of the project, and four topographic engineers H.S. Senseney, S.P. Floore, S.G. Lunde, and M.J. Gleissner were assigned as heads of the sections of control, planimetry, topography, and editing. About 1,200 persons on relief were trained in all phases of topographic mapping. The 60 square miles were covered by 92 maps, for which the field surveys were on a scale of 200 feet equals 1 inch with contour interval of 2 feet.

The city of St. Louis map was among the many relief projects undertaken in Missouri from 1933 to 1940. In 1933, the City of St. Louis, having curtailed construction operations, had surplus engineers whom they desired to keep employed on a relief basis. The Director of Public Works contacted the State cooperating official and a program was worked out, providing that the city would supply engineers to extend the necessary horizontal and vertical control, and draftsmen to ink all maps prepared. The mapping to be executed on the scale of 200 feet equals 1 inch, with a contour interval of 5 feet. The former St. Louis map, made in 1889, on a scale of 200 feet to the inch, with a contour interval of 3 feet, was prepared in sheets covering 1 minute in latitude and 2 minutes and 10 seconds in longitude. The same subdivisions were used on the new mapping program. F.W. Hughes was in charge of the project. All city, county, town, railroad, and public service, etc. maps were assembled for use in this compilation. The City of St. Louis received a topographic map on the scale of 1 inch equals 200 feet, and the Geological Survey took the original manuscripts, inked them in three colors, and then reduced and assembled them into quadrangles, on the scale of 1:24000, by photography and photolithography.

The PWA program was huge success. Everyone worked hard but there were no complaints. All felt that they were makers of men, as well as maps, and were gratified with the results.

A clipping from a small town newspaper in Arkansas furnished an example:

"Manna from Heaven"

"Two men came to town today. They said they were topographic engineers with the U.S. Geological Survey and would like to make their headquarters here while mapping surrounding quadrangles. Next week there would be about 32 in the group, including wives and children. They could use unemployed boys and would make all purchases at the local stores. They would stay a year or more.

"The hotel will reopen, including its dining room. All discharged men and women will be reemployed. Stores will open their doors and this town will live again.

"Truly manna from heaven."

---

[38] Civil Works Administration; Emergency Relief Administration; Federal Emergency Relief Administration; Works Project Administration; or Public Works Administration.

Louisiana–Mississippi Boundary Line

S.S. Gannett served for the third time as sole commissioner on a boundary dispute when the U.S. Supreme Court, by decree rendered April 13, 1931, appointed him to run, locate, plat, and mark the boundary between the state of Louisiana and the state of Mississippi, between latitude 32 degrees 39 minutes on the north, and the division line between Issaquena and Warren Counties, Miss., (as extended westward) on the south. This was a case of relocating the boundary as it existed prior to the avulsion of 1912–13, the action of the river cutting off one of its sweeping oxbows and isolating a portion of Carroll Parish, La. The boundary was the center of the navigable channel of the river. Mr. McNair executed the instrumental work in October and November 1932, and found 2.5 miles of the northern boundary filled with silt, which afforded a normal survey, and base for five concrete monuments and 12 creosoted wooden posts at intermediate angle points. The total length of the line was about 10 miles. Mr. Gannett's report was submitted on May 17, 1933.

Personnel Notes

In January 1932, a committee was appointed to prepare news items twice a year for distribution to Topographic Branch personnel. The items were brief and became even more brief, limited as to location and date of field assignments. After 2 years, the "Personnel Notes" were discontinued in order to save time and paper.

Members of the Branch were always interested in the assignments of their colleagues, and the Topocraft Club, composed of Masons of the Branch, at its annual meeting in January 1935, selected a committee to prepare news notes. C.H. Davey, W.H. Griffin, J.P. Harrison, H.H. Hodgeson, and J.M. Whitman were members of the committee, and as a private enterprise, with official sanction, the notes were not restricted to a bare statement of facts. One hundred mimeographed copies of "Topocraft Newsnotes" (34 pages) were issued to club members in April 1935, with the suggestion that they be circulated among non-members. So many expressions of appreciation were received that "Topocraft Newsnotes No. 2" (75 pages) were issued in July 1936, after the men were located on their field assignments. and "Newsnotes No. 3" (74 pages) was presented in August 1938. The tragedy of World War II discouraged further issues.

Mr. Hodgeson's retirement on November 30, 1942, permitted him to do what he had wanted to do for some time; write news notes for the Topographic Branch. The mimeograph machine in the Sacramento office was his private property, and could be used by him after office hours and on weekends. The enterprise was again a private one, with official sanction, and was supported by subscribers who paid enough to cover the cost of paper and stamps. Subscribers increased from 200 to more than 300 in the later years.

The "Personnel Notes, Pacific Division," made its first monthly appearance on January 31, 1943. At first only three pages, it grew in size as items were contributed by a volunteer correspondent in each Division and the Washington office. Its growth was stopped at 10 pages, which was reached in July 1944, because that was the maximum weight that could be mailed for a 3-cent stamp.

At the end of 4 years, the notes were discontinued by Mr. Hodgeson, and R.T. Evans decided to continue them in Washington as the "Topographic Branch Notes." They enjoyed sustained popularity, but after 5 years, Mr. Evans discontinued their preparation and v. IX. no. 12, December 1, 1951, was the last issue.

Vermont–New Hampshire Boundary Line

The U.S. Supreme Court on May 29, 1933, sustained the findings of the Special Master as to the location of the true boundary line between Vermont and New Hampshire, and decreed that the boundary line, in dispute for more than 150 years, should follow the western side of the Connecticut River at low water mark, "as the same is or would be if unaffected by improvements on the river." Ten dams across the river for the generation of hydroelectric power, and the lakes about them, had obliterated the original low water mark for much of the 200 miles of river involved.

The Court's decree meant, for Vermont, the acquisition of a new island or so, and for New Hampshire, the privilege of maintaining most of the bridges. S.S. Gannett, notwithstanding the fact that he had retired from active service on June 30, 1932, at 71 years of age, was appointed by the Court on January 8, 1934, as Special Commissioner to locate the boundary line and to mark certain specified points. From the large "Mud Turtle" monument, built in 1895 and 1897, common to boundary lines of Massachusetts, Vermont, and New Hampshire, a careful third order transit traverse reference line was extended along the railroads and highways nearest the Connecticut River, tied to seven triangulation stations in the course of the 206 miles, and terminated at the Canadian boundary monument No. 519, which marks also the northeast corner of Vermont.

Monuments, weighing approximately 1,200 pounds each, and markers together, numbered 103. Except the few set on dams and bridges, the monuments and markers were erected on the west bank of the river, out of reach of extreme high water. The bronze tablet imbedded in the top of each marker was stamped with an arrow and the number of feet from its center to the State line. Check lines of third order spirit leveling were extended from nearby bench marks to all markers. The geographic positions of the bench marks also were determined, adding more than 70 additional reference points near the boundary line. Elevations: Turtle Monument (low-water level), 177 feet; river at northeast corner of Vermont, 1,102 feet.

Fieldwork started in May 1934, was temporarily discontinued on September 28, 1934, upon the sudden death of the chief of the field party, E.L. McNair; resumed May 1936, by Clyde B. Kendall, and completed in August 1936. Mr. Kendall also was in immediate charge of the office computations of geodetic positions of reference monuments and state line points from January 1936, to the middle of September 1936. The Connecticut River drained a generally wooded country, but large floods occurred in 1927, and in March 1936, the latter destroying a number of bridges. Mr. Gannett submitted his bound, book-size report to the Court on November 16, 1936.

Mapping of the Tennessee River Basin

On August 31, 1933, an act of Congress created the Tennessee Valley Authority for the development of a unified water control plan for the entire Tennessee River Basin, an area of 40,600 square miles, including parts of seven states. Closely, and inevitably linked with consideration of water control and use are such phases of land planning as relocation of highways and railroads, distribution of electric power, erosion control, forestation, industrial development, and shifts of population. Planning of this scope would require complete and accurate base maps, and a cooperative agreement was made with the Geological Survey in the following December. As the need for maps was urgent, their production by photogrammetric methods was necessary and the Survey's leading photogrammetrist, Thomas P. Pendleton, was placed in charge of the work.

The agreement stipulated that the quietest process of covering the valley was the preparation of planimetric maps, without contours. These maps were made by office compilation, using the radial line method, with three-lens aerial photographs. Correct geographic position was secured by means of secondary triangulation or third order traverse, the latter spaced about 5 or 6 miles apart, and closed on U.S. Coast and Geodetic Survey control stations. Details of road alignments were obtained by automobile traverses, wherein distances were taken from the speedometer and alignments from compass bearings. After the office compilation, the maps were given an overall check in the field by planetable. The size adopted for the individual maps was the 7.5-minute quadrangle on the scale of 1:24,000, and it would require about 766 maps. More than 200 men, with headquarters at Chattanooga, worked on the project, and it was completed by the end of 1936.

The temporary planimetric maps were to be replaced with topographic maps with contours, and studies were made in the Washington office of two new stereo photogrammetric plotting machines of German manufacture, which seemed to possess advantages over the Survey's one Hugershoff aerocartograph. The Zeiss Stereoplanigraph, which was set up for study, was as complicated and costly as the aerocartograph. The Multiplex Projector, which Heinz Gruner and E.O. Hester exhibited and demonstrated on May 1, 1934, was small, comparatively simple, and not expensive, and the Survey ordered several units. They arrived from Germany, three at a time, late in 1935.

From the use of the stereoplanigraph in mapping a part of the Bushkill Quadrangle, Pa., and the use of the Multiplex in mapping the Oswald Dome Quadrangle, Tenn., it was found by analytical studies that the accuracy of the former, accommodating 4-lens photographs, possessed about three times the accuracy of the latter in respect to vertical measurements, could compete with the planetable for cost and accuracy on contour intervals greater than 5 feet, and was the best plotting machine tried out by the Survey. However, the Survey never had money enough to buy it. The Multiplex was comparatively inexpensive.

The mapping of the Oswald Done Quadrangle disclosed distortions in the projected models and other defects in the Multiplex instrument. Correspondence with the Zeiss Company revealed that the Multiplex was intended only for reconnaissance mapping, and that if finer work was desired the Survey could redesign any parts of the instrument and have the Bausch & Lomb Optical Company of Rochester, N.Y., manufacture new parts, or even new instruments.

Accordingly, the Survey assigned Russell K. Bean to develop and redesign the Multiplex instrument. Beginning with the recalibration of the projectors, changes, as they were believed desirable, were progressively made, such as the conversion of the normal projectors to wide angle projectors; later their increase in size in order to accommodate 64 by 64 millimeter diapositive plates rather than the 45 by 60 millimeter plates; enlargement of the reduction printer, which was the special projection camera used in making the small diapositive plates that were employed in the projectors in the form of reduced positive copies of the original aerial negatives; later its redesign so as to make it as universal as possible to accommodate film from a variety of precision mapping cameras. This latter involved consideration of the focal length of such cameras, their angular field of view, and the distortion characteristics and range of light transmission of the lens employed.

Before the completion of the planimetric program, the Tennessee Valley Authority offered to set aside and furnish space for photogrammetric equipment if the Survey would furnish Multiplex instruments and personnel to operate them. This was done, and as Mr. Pendleton already was in charge of the planimetric program, the stereophotogrammetric unit also was placed under his supervision, with Mr. Bean in charge of the unit.

The first bar, with nine projectors, was used in April 1936. For this experimental work, two topographers from each division volunteered. They were Paul Blake and William S. Higginson, Atlantic Division; Robert O. Davis and Milton J. Harden, Central Division; and Emmett J. Coon and George A. Fischer, Pacific Division.

To these, known as the "Original Six," were added two more topographers in July 1937, and an additional eight by the following January. Then operations were still on a three-shift basis, but with an increase of equipment, it was planned to gradually expand the force to a number that would permit continuous two-shift production.

In August 1941, Mr. Pendleton was transferred to the Washington office and Joc K. Bailey was left in charge of the Chattanooga office. In April 1948, Mr. Bailey was transferred to the Atlantic Division headquarters at Arlington, Va., and John G. Groninger was left in Chattanooga to close out that office. Multiplex units and operators were distributed to all divisions and Mr. Groninger reported to the Rocky Mountain Division on July 25, 1949.

Science Advisory Board

In order to make more effective the advice of scientists toward the solution of governmental problems, the Science Advisory Board was established by Executive Order No. 6238, dated July 31, 1933, and was continued to December 1, 1935. The Board was under the jurisdiction of the National Academy of Science and the National Research Council; the latter created at the request of President Wilson in 1916, and perpetuated by Executive Order No. 2859, dated May 11, 1918, and was vested with authority to appoint committees to deal with specific problems in the various departments.

Composed at first of nine members, to which six were later added, the Board constituted 18 subcommittees. Of these, the Committee on Mapping Services of the Federal Government was constituted in response to a request by the Bureau of the Budget for an investigation of the 28 Federal agencies, which were reported as engaged in surveying and mapping activities to a greater or lesser extent, with a view of determining what, if any, regrouping or consolidation was desirable in the interest of greater efficiency and economy. The members of this committee are shown in table 41.

Table 41.   Science Advisory Board Committee members.

Name	Affiliation
Douglas Johnson	Professor of Physiography, Colombia University, New York City, Chairman.
Isaiah Bowman	Director, American Geographical Society.
W.L.G. Joerg	Research Editor, American Geographical Society, New York City.
C.K. Leith	Professor and Head of Geology Department, University of Wisconsin.
Robert E. Randall	President and Chief Engineer, R.H. Randall & Co., Toledo, Ohio.
Frank W. DeWolf	Professor and Head of Geology Department, University of Illinois.
C.C. Williams	Dean, College of Engineering, University of Iowa.

The result of intensive studies by the committee was a comprehensive report containing the recommendation that the Coast and Geodetic Survey, the Topographic Branch of the Geological Survey, together with the Division of Engraving and Printing, which reproduces its maps, the Lake Survey (War Department), and the International Boundary Commission, United States, Alaska, and Canada (State Department), be consolidated under one central agency.[39] It was recommended further that the central agency could function best as an independent organization, reporting directly to the President. Failing this disposition, for reasons of expediency, the second suggestion placed the agency in the Department of the Interior. The committee's third choice was the Department of Commerce.

It was fully recognized that many Federal agencies needed subsidiary surveying services under their control, and that other regrouping and consolidation was not considered as being altogether impracticable. The consolidation would unite, under efficient direction, all major Federal land and water surveys, except cadastral surveys of the General Land Office. All changes contemplated, except that involving the International Boundary Commission, had repeatedly been urged following previous studies of this problem. It was mentioned that to the proposed central agency would gravitate, in time, many of the other mapping activities; however, the plan did not become effective. Some of its provisions were worthy of note.

1. The Geological Survey would retain, under its exclusive control, a limited number of skilled topographic engineers, estimated at 15 to 18, and draftsmen, 4 to 5, to constitute a subsidiary surveying and mapping unit sufficiently large to serve its needs.

2. The system of commissioning properly qualified personnel, already established by law in the Coast and Geodetic Survey, which was to serve as the nucleus of the consolidated mapping service, would be extended to appropriate personnel of the agencies, or parts of agencies, transferred for the purpose of ensuring a unified service with high morale, and of securing and retaining men of superior quality for high grade technical work.

3. More effective utilization of field topographers of the Topographic Branch, in a large central agency, would do away with the uneconomical employment of these men in office work.

4. In the Division of Engraving and Printing, development of glass engraving should result in material savings over engraving on copper with improvement rather than deterioration in product.

5. Very large savings could be expected to result from extensive stereoscopic methods in planimetric surveys.

---

[39] Science Advisory Board Second Report, 1934–35, p. 226–241.

6. If the central mapping agency were enabled to prosecute an enlarged mapping program, one that called for expenditure of $15,000,000 annually, the savings due to consolidation could conservatively be estimated at between $2,000,000 and $3,000,000.

7. The committee emphasized the fact that executive agencies of the Federal Government were handicapped by lack of adequate topographic maps, and endorsed the long established plan of Geological Survey to complete, as soon as possible, the topographic maps of the country on the scale 1:62,500 (1:125,000 if necessary to complete mountainous areas far advanced on that scale.)

8. It recommended further that the central mapping agency, if and when established, should review the whole policy of state, county, and other local cooperation in Federal mapping, for this cooperation could retard the progress of the overall plan, as it sometimes involved mapping on larger scales, which required more time.

Moreover, the larger scales are not multiples of the scales previously considered as standard. About 1908, the Survey began making some maps on the scale of 1:31,680, and about 1923, some on the still larger scale of 1:24,000, both as cooperative enterprises with state or other local agencies. At first regarded as special scales, they were later accepted as "standard," and since 1925, several state cooperative programs have been undertaken for production of these larger maps on the 50-50 basis. Supposedly, standard maps are thus based on three incompatible scales; one representing aliquot parts of a million, one so many inches to the mile, and one so many feet to the inch. The trend probably would continue toward yet other and larger "standard scales" unless a definite policy in respect thereto was adopted and adhered to.

The Coast and Geodetic Survey, the oldest scientific Bureau in the Federal Government, possessed in its inland geodetic unit a smaller mapping agency than the Topographic Branch of the Geological Survey. It seemed logical to consider transferring the smaller unit to the larger one. To offset this idea, the Coast and Geodetic Survey seemed to establish a policy of publicizing its work in frequent popularly written articles in engineering and scientific magazines, in which attention was called to its mapping operations, although it was well known to scientists and engineers that geodetic and vertical control work constituted only a small part of topographic mapping.

The Coast and Geodetic Survey was able to set aside funds for research work in photogrammetric instruments, designing and building, among other things, a nine-lens camera for aerial photography. During World War II, it was assigned an area for urgent military mapping. Afterwards, with the photogrammetric equipment and trained personnel, it aspired to execute topographic mapping; to assume a portion of the mapping program, in addition to the primary control work. It was encouraged early in 1946, by the Bureau of the Budget, to undertake the mapping of Alaska, as it had already charted the long indented coastline. Secretary Ickes objected to this arrangement, stating that it was an encroachment upon the established and organized function of the Geological Survey.

In March, 1947, an understanding was reached by the Directors of the two Surveys, recognizing the existing responsibilities of each organization, whereby the Coast and Geodetic Survey would complete the topographic mapping of 7.5-minute quadrangles touching coastal waters. In order to avoid overlapping and duplication of surveying operations, and to ensure integration of efforts, the two Surveys, through their designated liaison officers, would coordinate their plans in time to present estimates for the ensuing year to the Bureau of the Budget.

American Society of Photogrammetry

The American Society of Photogrammetry, a nonprofit professional and technical society was incorporated in 1934, under the laws of the District of Columbia. The object and business of the Society is to advance knowledge in the science of photogrammetry, and to act as a clearing house for the dissemination and exchange of information pertaining to the practice of photogrammetry; to foster a spirit of understanding and cooperation among interested individuals and organizations; to provide a medium for interchange of ideas; to hold meetings for the presentation and discussion of papers; and to publish and distribute books, periodicals, treatises, circulars, and papers related to the uses and application of photogrammetry.

"Photogrammetric Engineering," the quarterly journal of the Society, is the outstanding publication in its field, and is the only one in this country devoted exclusively to photogrammetry and closely associated practices.

In 1944, the Society published a reference text entitled "Manual of Photogrammetry." All copies of this popular edition have been sold, and a revised edition is now available.

By means of technical committees, working without compensation, the Society has a number of notable achievements to its credit. Some of the more important are: (1) the development of standard specifications for aerial photography, which were adopted as standard government specifications by the Treasury Department Procurement Division; (2) the development of precision mapping camera specifications that were adopted by most governmental agencies; (3) the compilation of an extensive Bibliography of photogrammetry; (4) the development of specifications for map accuracy; (5) the compilation and publication of an authoritative list of definitions and terms used in photogrammetric surveying and mapping; and (6) the development of "job descriptions" covering the various grades of position classification of photogrammetric engineers and photogrammetrists.

Income of the Society is derived primarily from membership dues. Sustaining membership is composed of individuals and commercial organizations who desire to contribute in a financial way toward support of the Society.

The Society is International in scope, having members in practically every country of the world, as well as taking an active part in promoting the International Society of Photogrammetry.[40]

Topographic Division personnel have taken an active part in the organization and maintenance of the American Society of Photogrammetry as shown in table 42.

Chapters in the "Manual of Photogrammetry" written by members of the Topographic Division are shown in table 43.

---

[40] Photogrammetric Engineering Journal, v. XVI, no. 2, April 1950.

Table 42. Topographic Division employees who served with the American Society of Photogrammetry.

Presidents of the Society	
1935, C.H. Birdseye	1946, Gerald FitzGerald
1942, T.P. Pendleton	1952, G.D. Whitmore

First Vice Presidents	
1941, T.P. Pendleton	1951, G.D. Whitmore
1945, Gerald FitzGerald	

Second Vice Presidents	
1946, R.M. Wilson	1953, J.I. Davidson
1947, R.K. Bean	

Directors	
R.E. Altenhofen	W.S. Higginson
C.H. Birdseye	David Landen
W.E. Harman, Jr.	R.M. Wilson
Daniel Kennedy	R.K. Bean
G.D. Whitmore	C.H. Davey
R.E. Ask	Gerald FitzGerald
J.L. Buckmaster	H.T. Kelsh
R.O. Davis	M.K. Linck

Secretary-Treasurers	
1940, J.H. Wheat	1944, E.J. Schlatter
1943, H.M. Towsend	1946, M.K. Linck

Honorary Membership Award	
1947, Thomas P. Pendleton	1954, George D. Whitmore

Sherman Fairchild Photogrammetric Award
1944, Col. Gerald FitzGerald, Development of the Trimetrogon Method
1948, Harry T. Kelsh, Design and construction of Kelsh Plotter
1951, Russell K. Bean, Development of the Twinplex, etc.

Talbert Abrams Award
1949, William E. Harman, Jr. "Negative Quality Required for Stereoplotting Equipment"
1951, Morris M. Thompson, "A New Approach to Flight Planning"

Table 43. Articles by Topographic Division employees published in "Manual of Photogrammetry."

In the 1944 Preliminary Edition, the following chapters have been written by Topographic Branch personnel:
Chapter I, "Principles of Surveying Rectangular Coordinates and Standard Horizontal Datum," R.M. Wilson.
Chapter VI, "Practical Tilt Correction for Single-Lens Aerial Photographs," C.P. Van Camp
Chapter XI, "Multiplex Instrument and Its Use," T.P. Pendleton
Chapter XIII, "Oblique Photographs for the Surveyor," R.M. Wilson
Chapter XIV, "Completion Surveys," T.P. Pendleton

The authors in the 1952 Second Edition were:
Chapter I, "The Development of Photogrammetry,", G.D. Whitmore
Chapter II, "Elements of Photogrammetric Optics,", R.E. Ask
Chapter IV, "Aerial Photography," W.E. Harman, Jr.
Chapter VIII, "Radial Triangulation," H.T. Kelsh
Chapter IX, "Rectification," R.E. Altenhofen
Chapter XV, "The Operation of the Multiplex," E.I. Loud, Jr.
Chapter XX, "Ready Reference List," David Landen

Topographic Mapping Policy Committee

In August 1937, the new policy of selling three-color advance sheets made it necessary that the several photo-lithographed drawings be more carefully drafted as to registry and weight of lines. Besides, other problems crowded upon the Section of Inspection and Editing, and to aid in their solution, the Director, early in 1939, appointed a Topographic Mapping Policy Committee of eight section chiefs, with Joseph Hyde Pratt as chairman, and R.M. Wilson as secretary, who would make a study of the details of map preparation practices and changing procedures involved in multicolored photo-lithographic editions of maps, and make recommendations. Meetings were held once or twice a month, and often visiting division engineers and other Survey officials were invited to participate in the discussions. Some of the problems considered by the committee and sub-committees were:

1. Rigid rules for defining the perimeter of urban areas in which only landmark buildings are shown; a colored overprint to mark the urban area.

2. Specifications of map accuracy and field tests.

3. Map scales; maps of the 1:31,680 category to be engraved on the 1:24,000 scale by special request of the cooperator.

4. The contour interval for each area to be mapped to be determined by a responsible field engineer, so as

to avoid congestion of contour lines, except where they indicate cliffs; the use of two contour intervals on a single quadrangle map to be avoided; the contour intervals of 5, 25, 50, or 100 feet to be used as standard west of longitude 103 degrees, as far as possible, and contour intervals of 5, 10, 20, 40, or 100 feet to be used east of longitude 103 degrees.

5.   Omission of projection lines within the map in the interest of clarity; on some maps preference is for grid lines or rectangular coordinates.

6.   Quadrangle names and numerical indexing system.

7.   A new text on the back of maps.

8.   Proper width of roads; only outside tracks in railroad yards.

9.   Use of intermittent stream symbols, in east as well as west.

10.   Proper placing of ocean shorelines, river shorelines and contour crossing.

11.   Symbol for sewer disposal plant.

Recommendations, if approved, were embodied in Topographic Branch technical memoranda and were communicated to the Federal Board of Surveys and Maps, to other governmental mapping agencies, and to cooperating states, cities, or counties.

Special Committee

On February 2, 1940, the Chief Topographic Engineer sent the following memorandum to members of the Topographic Branch:

"A Committee has been created in the Topographic Branch, Geological Survey, which has as its object the following:

"(1) To notify members of the branch in advance of the date of retirement of employees, and to offer an opportunity to those who wish to contribute (voluntarily) toward a gift honoring them on their retirement.

"(2) To attend to the purchasing and sending of flowers upon receipt of notification of the death of a member of the immediate family of a member of the branch; this to be restricted to the wife, husband, or children of a member of the branch. (This is not meant to preclude the sending of flowers to other than immediate relatives, but collections for such contributions should originate in the section most interested.)

"(3) To arrange for the sending of Christmas remembrances to retired employees , shut-ins, and families of deceased members.

"(4) To arrange for interment in Arlington Cemetery of those entitled to this service.

"Too often in the past have the above services been left to the initiative of some special friend of the person concerned, with the result that there has been unintentional discrimination, some being remembered on retirement or in the time of trouble, while others have been ignored. The establishment of this Committee is an effort to systemize and unify our efforts. The membership is as follows: R.H. Reineck, chairman; J.M. Whitman; R.A. Kiger; Delar Kimble; F.W. Borden.

"It is desired to stress again the fact that all contributions are to be voluntary, the opportunity being given to participate in the above services if it is your desire."

This committee is still functioning in 1954. As members are retired, new members are chosen. Delar Kimble has always been Treasurer. R.T. Evans and A.J. Ogle have been chairmen. The last chairman was O.H. Nelson, with Helen M. Frye, Delar Kimble, and Wm. B. Overstreet as members.

American Congress on Surveying and Mapping

In June 1941, the American Congress on Surveying and Mapping, a non-profit association, was organized to advance the sciences of surveying and mapping in their several branches, in furtherance of the public welfare, and in the interests of both those who use maps and surveys, and those who make them. Its aims were to establish a central source of reference and union for its members, to contribute to public education in the use of surveys and maps, and to encourage the prosecution of basic surveying and mapping programs, especially those programs which are paid for, in whole or in part, with public funds.

Many members of the Topographic Branch have joined and have taken an active part in this organization and have contributed scientific articles for publication in its official journal, "Surveying and Mapping," which is published quarterly. Topographic Branch personnel who have served as officers and members of committees are shown in table 44.

World War II

Japan made a surprise attack on Pearl Harbor, Hawaii, on December 7, 1941, forcing war on the United States on

Table 44. Topographic Division employees who served on The American Congress on Surveying and Mapping.

President	
George D. Whitmore in 1944, while with TVA	Gerald FitzGerald in 1951

Treasurer and Acting Executive Secretary.	
R.T. Evans (In June 1950, he received Citation Certificate for Outstanding Professional Services.)	

Board of Directors	
Gerald FitzGerald	C.F. Fuechsel
R.M. Wilson	R.H. Lyddan

Topographic Mapping Division	
J.G. Staack, Chairman	C.F. Fuechsel

Division of Control Survey	
R.M. Wilson, Chairman	J.L. Speert

Research and Development Committee	
T.P. Pendleton, Chairman	C.F. Fuechsel

Technical Division of Cartography	
C.H. Davey	L.E. Marsden
O.H. Nelson	C.W. Buckey
R.L. Moravetz	W.B. Sears
A.F. Striker, Chairman	Roscoe Reeves

Technical Division of Topographic Surveying and Mapping	
C.S. Maltby, Chairman	E.J. Fennell

Publication Committee	
C.F. Fuechsel	A.F. Striker
J.O. Kilmartin	M.M. Thompson
G.D. Whitmore, Chairman	C.S. Maltby

Editor-in-Chief	
J.I. Davidson	R.C. Eller

Associate Editors	
R.M. Wilson	J.P. Burns
A.F. Striker	

Membership Committee	
A.C. Stiefel	C.R. Graham

Program Committee	
B.P. Taylor	

Public Relations Committee	
A.C. Stiefel, Chairman	

Professional Status Committee	
R.H. Lyddan	

Representative to National Research Council	
C.H. Davey	

Northern California Section	
H.H. Hodgeson, Vice President	R.H. Moore, Secretary-Treasurer
C.A. Biever, Executive Board	

December 8, 1941. Need for topographic maps was sharply emphasized and the Geological Survey no longer possessed the capacity to produce them fast enough, so all mapping agencies, governmental and commercial, were called upon. Photogrammetric methods, producing more maps in less time, had to be used. Mr. Staack resisted efforts to have the Topographic Branch taken over by the Army, but he assisted 189 members of the Branch into the Armed Forces, approved the transfer of some members to other mapping agencies, and retained many in the Survey organization by justifying their exemption from the draft on the basis of their indispensability to the mapping effort. During World War II, the former members of the Branch who died with the Armed Services are listed in table 45.

Table 45. Members of the Topographic Branch who died with the Armed Services.

March 1, 1945 Ensign Desmond M. Yetter, in accident on shipboard.

March 10, Lt. Paul B. Kovacs, in accident on Iwo Jima.

April 1, Pvt. Thomas F. Roberts, by enemy fire in Germany.

World War II Record

Contour and Relief Mapping [41]

During the war period from December 1941, to September 1945, the Topographic Branch was engaged in essential war work as follows:

1. Conducted the necessary field surveys of areas in the strategic war zone of the United States as defined by the War Department, and prepared the resulting topographic maps for reproduction.

2. Compiled and prepared aeronautical approach charts of the continental United States, and target charts of Yugoslavia for the Army Air Forces.

3. Revised from aerial photographs the data on existing local topographic maps of Holland, France, and Italy, and redrafted the revised maps according to specifications for the Army Map Service.

4. By stereophotogrammetric methods, compiled topographic maps from aerial photographs for the Army Map Service of areas in France, Italy, Yugoslavia, Luzon, and the Sulu Archipelago of the Philippines, Austria, Belgium, China, Formosa, Ryukyu Rhetta, and Japanese City Plains.

---

[41] See file 404, and Central Region Engineer Departmental War Records, C.L. Sadler, June 1, 1944.

5.   Prepared relief shaded maps of Japan, China, Manchuria, Korea, Formosa, Hainan, and South Central Europe for the Army Air Forces, and the General Staff, War Department.

Personnel

On December 1, 1941, Secretarial appointees of the Topographic Branch numbered 485. The peak of employment of Secretarial appointees reached 745 on September 30, 1942, which included 59 on military furlough, leaving a net working force of 686. On June 30, 1944, Secretarial appointees numbered 668, which included 119 on military furlough, leaving a net working force of 549; and on June 30, 1945, Secretarial appointees numbered 635, which included 141 on military furlough, leaving a net working force of 494. Also there were 48 Secretarial appointees whose services were terminated when they entered on active duty with the Armed Forces, who retained their reemployment rights with the Branch and are eligible for reinstatement when discharged from the Armed Services. Actually, the Branch had 189 Secretarial appointees in the Armed Services during World War II.

Photo Mapping Section

The Photo Mapping Section at Arlington, Va., completed all of its foreign mapping assignments given to it by the Army Map Service. As of September 1945, it furnished stereophotogrammetric map compilation from aerial photographs as shown in table 46.

Table 46. Stereophotogrammetric map compilation from aerial photographs, as of September 1945.

Country	Square miles of coverage
France	9,192
Italy	780
Yugoslavia	16,152
Luzon	3,060
Sulu Archipelago	145
France and Belgium	2,173
Austria	4,529
China	19,800
Ryukyu Rhetta	466
Formosa	3,126
Japan City Plains	208
Total	59,551

In addition, the Chattanooga, Tenn., Stereophotogrammetric office furnished topographic map compilations from aerial photographs of the following coverage for the Army Map Service.

China	20,673

Relief Shaded Maps

Relief shaded maps of Hainan Island, off the coast of China, Hokaido Japan, Formosa, Southern and Northern Japan proper, and of occupied China, Manchuria, and Korea were prepared and completed on time in all cases. The last manuscript, that for the China-Manchuria-Korea area, was transmitted for reproduction on January 20, 1946. The map for South Central Europe was completed and delivered in April 1944. In all, 11 relief shaded maps were prepared and delivered to the War Department.

Owing to the Branch's prominence and leadership in the application of stereophotogrammetric mapping methods in the field of topographic mapping, its stereophotogrammetric training facilities at Arlington, Va., were utilized for training officers and civilians of foreign countries, and personnel in the Army Map Service, War Department. Six officers from the Turkish Army, 2 civilians from San Salvador, and 20 persons from the Army Map Service received training in the operations of the Multiplex stereo projectors.

Survey Order No. 140 Modification

The Modification of Survey Order No. 140, dated February 23, 1942, reads as follows:

"The last two names ("Mr. Wheat or Mr. Birdseye") in Survey Order No. 140, dated January 26, 1939, are hereby replaced by "Mr. Pendleton," so as to make the order read as follows:

"In the interest of coordination and efficiency, it is hereby directed that all outgoing requests for aerial photographic data, whether in the form of letters, requisitions, or specifications, shall be submitted to the Topographic Branch and, before being presented for signature, shall bear the initials of Mr. Staack, or Mr. Pendleton."

Appropriations and Personnel (1943)

On March 18, 1943, upon recommendation of Director W.C. Mendenhall, prior to his retirement, Mr. Thomas P. Pendleton, the Survey's leading photogrammetrist, was appointed Chief Topographic Engineer and Mr. Staack, Assistant Chief Topographic Engineer. Recognition was made of the steadily growing burden upon the administrative officers of the Branch and these appointments were designed to furnish relief by spreading the duties. The Branch had expanded in personnel to 2.5 times its size when Mr. Staack became its Chief in 1929, and emphasis had shifted to photogrammetric methods of mapping with changed technical and business problems continually increasing in volume, diversity, and complexity. Charles H. Davey became Acting Chief of the Photo Mapping Section.

In 1943, the Topographic Branch consisted of 602 employees, departmental and field, and appropriations expended are shown in table 47.

Table 47. Appropriations to be expended, 1943.

[These amounts were 2.5 times the amount expended in 1929]

Activity	Funds
Topographic Surveys	$689,030
Work performed for the War Department	2,199,124
Transfers by other Federal Units	193,180
State cooperative funds	337,279
Total	3,418,613

Pan American Institute of Geography and History

The Pan American Institute of Geography and History (PAIGH) is an organization created by Governments to render them service in the fields of geography, history, and the related sciences. It is financed by a quota system on the basis of population categories.

The purposes of the Institute, in the language of its organic statutes, are to "encourage, coordinate, and publicize geographic, historical, and related scientific activities; and to initiate and execute any studies or research assignments of this character requested by the Member States. In addition, it is charged with promoting cooperation with American organizations interested in these fields of activity, including specifically cooperation among geographic and historical institutes of the Member States. No work of political or sectarian nature may be undertaken."

It is more a coordinating, stimulating and planning organization than one engaged in the execution of special projects. It renders assistance to the scientific, technical, and academic institutions of its member governments and will undertake to execute specific operations, if they are of an international nature and/or if there is no National agency organized to do so. What is accomplished under its guidance is emphatically an accomplishment of the member state itself, or a joint accomplishment of several, as, for instance, the extension of a uniform triangulation net through all of Central America. The Americas have the largest connected arc of triangulation in the world, from the Bering Strait to Southernmost Chile.

The Institute consists of the General Assembly, the Governing Board, the Executive Committee, the National Sections, the Commissions, and their Committees, and the General Secretariat. Each Commission is composed of a representative from each country, and the three members of Commission in each country form the National Section of that country.

A large consulting library is maintained at the headquarters of the Institute, and the Commission on Cartography has a small film library of technical films, produced by the Institute, and available with both English and Spanish sound tracks, for loan to institutions of the Member States.

An important factor of the Consultations on Cartography are the National reports on progress, presented in uniform fashion on forms provided by the Commission.

Committees of the Commission on Cartography:

- Geodesy

- Gravity and Geomagnetism

- Seismology

- Topographic Maps and Aerophotogrammetry (Dr. Wrather, Director, was Survey member) (followed by George D. Whitmore, Topographic Division)

- Aeronautical Charts

- Hydrography

- Tides

- Urban Area Surveying, Committee on Special Maps

The first Pan American Consultation on Cartography was initiated at Washington, D.C., and Mexico City in 1943.

The second Consultation was held at Rio de Janeiro from August 14 to September 2, 1944. Thomas P. Pendleton was the Geological Survey delegate and Col. Gerald FitzGerald represented the Army Air Force.

The third Consultation met at Caracas from August 22 to September 1, 1946. Milton F. Denault of the Topographic Division and Wm. D. Johnston, Jr. of the Geologic Division, were Geological Survey representatives.

The fourth Consultation was held at Buenos Aires from October 15 to November 14, 1948. Harry T. Kelsh, Topographic Division, was sent as Geological Survey Advisor to act for Dr. Wrather as vice chairman of the Committee for Topographic Mapping and Aerophotogrammetry.

The fifth Consultation met at Santiago in October, 1950. George D. Whitmore, Topographic Division, was the Geological Survey representative.

The sixth Consultation was held at Cuidad Trujillo from October 12 to 24, 1952. Gerald FitzGerald, Chief Topographic Engineer, was the Geological Survey delegate.[42] Appropriate Geological Survey exhibits were assembled and sent to these Consultations.

---

[42] From article by Dr. Andre C. Simonpietri, published in The Canadian Surveyor, April 1955.

Technical Memorandum No. 51

Topographic Branch Technical Memorandum No. 51, dated November 3, 1944, reads:

"The Director has approved the omission from all maps of the names of engineers responsible as authors for their preparation. This action has become necessary, as the number of engineers and photogrammetrists entitled to such credit has become so large that it is no longer possible to continue the former practice. Henceforth, the credit line on the published map will read "Mapped by the Geological Survey," together with the year of survey.

"A complete record of those individuals whose names formerly would appear in a credit line of the published maps will be retained in the files of the Topographic Branch. The following rules will cover the application of these new procedures:

"1. For maps contoured by planetable method:

"a. Report names of all topographic engineers engaged in instrumental work.

"b. The actual year in which fieldwork is accomplished shall be considered the date of survey.

"2. For maps contoured by photogrammetric methods:

"a. Report names of all engineers responsible for original photogrammetric mapping and those responsible for field completion or photointerpretation surveys, indicating in each case the type of work performed.

"b. The date of mapping shall be considered the date of the field completion survey, or the photointerpretation work, regardless of the fact that the date of the photointerpretation work may precede that of the actual photogrammetric operation."

Noteworthy Events (1945)

The following events of 1945 are listed as having special concern and poignant impact on the Survey group in Washington, D.C. and Clarendon, Va.

January.—The distribution of the "Manual of Photogrammetry", an 819 page standard reference work published by the American Society of Photogrammetry, and written by various members of the Society. Among the authors were several Survey topographic engineers.

February.—The American Society of Photogrammetry, in compliance with the rules promulgated by the Office of Defense Transportation, limited its annual meeting to the evening of February 15 in the Department of Agriculture Auditorium, when about 300 local members and guests attended. The American Congress on Surveying and Mapping held a regional meeting in the same Auditorium, all day on June 21, instead of an annual meeting, and about 75 attended.

May 8.—VE Day, when radios announced the surrender of German armies.

August 14.—VJ Day, Proclaimed as Sunday, September 2, by President Truman, who also proclaimed August 15 and 16 victory holidays. Twenty holidays had been missed, beginning with May 30, 1942. The new 40-hour week was observed by not working on Saturday 1. Labor Day, September 3, was the first restored holiday. On Sunday, September 30, Eastern War Time, in effect since February 9, 1942, was changed back to Eastern Standard Time.

Memorandum for Field Offices

Memorandum for field offices and field men, dated August 28, 1945, reads:

"The 40-hour week will be established in the field offices of the Topographic Branch (Rolla, Sacramento, Chattanooga, and Clarendon) and in the field effective with the week commencing September 9. The 44-hour week will be in effect in the field through September 8.

"Owing to the relatively short field season and the frequency of climatic conditions unfavorable for fieldwork, it is hoped that field personnel will make every effort to render 40 hours of field work each week, even though this may require Saturday work.

"Effective commencing September 3, 1945, the legal holidays mentioned in Paragraph 35 of the Topographic Branch Administrative instructions will be observed."

Survey Order No. 148

Survey Order No. 148, Appointment of Staff Topographers, dated September 13, 1945, was as follows:

"The greatly expanded mapping activities of the Topographic Branch of the Geological Survey have created a need for additional staff to advise and assist the Chief Topographic Engineer and myself in the planning and implementation, both of the present program and of the long range program that we anticipate is before us. Two new positions as Staff Topographic Engineer have been established in the Branch, and I have requested the appointments of George D. Whitmore and Col. Gerald FitzGerald to fill them.

"Both Mr. Whitmore and Colonel FitzGerald are well known in the Geological Survey, and in other governmental and private agencies engaged in mapping and related activities.

"Mr. Whitmore comes to the Survey from the Tennessee Valley Authority with a background of 30 years in engineering and mapping, a capable administrator and an able organizer, he has held numerous important positions in the field of mapping, and is widely recognized as an outstanding authority on the subject. He is a member of the American Society of Civil Engineers, the American Society of Photogrammetry, and a past President of the American Congress on Surveying and Mapping.

"Colonel FitzGerald as been a member of the Geological Survey for more than 25 years. Although specializing in reconnaissance mapping in Alaska, he has had a wide variety of Survey assignments, including field and office work on control surveys and topographic mapping projects, on both detail and reconnaissance scales. During his 3 years' service in the Army Air Forces, he organized and commanded the Aeronautical Chart Service. He is a member of the American Congress on Surveying and Mapping, and First Vice President of the American Society of Photogrammetry.

"Mr. Whitmore will report to the Survey for duty on September 17, and Colonel FitzGerald at a later date upon his release from active service with the Army Air Forces."

Memorandum for Division and Section Chiefs

Memorandum for Division and Section Chiefs, Topographic Branch, dated December 26, 1945, reads:

I. Pursuant to recent discussions in the Washington office, and clarification of Survey Order No. 148, dated September 13, 1945, issued by the Director, it is desired to establish, as rapidly as possible, two new staff divisions within the Topographic Branch. Effective January 2, 1946, Mr. Gerald FitzGerald is assigned to head the Staff Division on Plans and Coordination, and Mr. George D. Whitmore is assigned to head the Staff Division on Research and Technical Control.

II. Pending formal approval of the revised organization of the Topographic Branch, the duties and responsibilities of the two Staff Divisions are outlined below.

A. Plans and Coordination

1. Map Information Office—Maintains complete files of topographic maps and indexes of control data and aerial photography. Evaluates topographic maps published or distributed by the Geological Survey and furnishes specific map information to the various federal agencies and general public.

2. Estimates and Plans—Prepares budget estimates, general plans for mapping programs, estimates of personnel and space requirements, and keeps status of funds current.

3. Production Control—Maintains production schedules and priorities, compiles and distributes status reports, prepares work load studies and cost analyses.

4. Liaison and Coordination—Maintains liaison with State and Federal officials on mapping matters, and attends conferences and meetings for this purpose. Formulates procedure for staff and operational coordinations within the Topographic Branch.

B. Research and Technical Control

Will consist of four sections as follows: Geodesy and Control Surveys, Topographic Surveys, Photogrammetry, and Cartography and Map Editing. This Staff Division will be responsible for:

1. Research and development of procedures, techniques, and equipment.

2. Establishing specifications, standardization of procedures, preparing interim technical orders, and final Instruction Manuals.

3. Inspection of operations, to ensure standard results.

4. In each appropriate section, responsible also for such as:

(a) Plan new control nets, and instruct on utilization of existing control.

(b) Determine general mapping methods and contour intervals.

(c) Instruct on utilization of existing air photos, and prepare specifications and award contracts for new photos.

(d) Instruct on use of source maps and other source data.

III. Immediate and urgent Staff functions will be carried on by the temporary assignment of personnel to the various sections of the two Staff Divisions. Effective January 2, the following will report to and assist the Staff Engineers:

Plans and Coordination: C.F. Fuechsel, E.L. Hain, H.A. Bean, R.L. Moravetz, R.H. Lyddan, and Jackson Schlesinger

Research Technical Control: H.M. Wilson, C.H. Davey, R.K. Bean, R.H. Sargent, and R.F. Wilcoxon

IV. Effective immediately, all memoranda, instructions, orders, outside correspondence, and similar matters, which in any way concern or affect the Staff functions outlined above, either shall be cleared through Mr. FitzGerald's or Mr. Whitmore's office before transmittal, or when this is not feasible for any reason, information copies shall be forwarded immediately to the proper Staff office.

/s/ J.G. Staack
for T.P. Pendleton
Chief Topographic Engineer

Approved:
/s/ W.E. Wrather
Director

Dallas H. Watson, Atlantic Division Engineer

MEMORANDUM:

The retirement of Mr. Albert Pike on February 28 after 56 years of government service, and more than 16 years of which as Division Engineer, created a vacancy in the Topographic Branch in the position of Division Engineer of the Atlantic Division.

Effective April 17, Mr. Dallas H. Watson has been appointed Division Engineer of the Atlantic Division to succeed Mr. Pike. Mr. Watson has served for more than 33 years, except for 1 year and 9 months as an officer during the First World War, as field engineer, supervisory officer, and principal Assistant to the Division Engineer. He enters upon his new duties with demonstrated topographic engineering and administrative capacity, and thorough familiarity with the operations of the Division. He possesses the full confidence of its personnel which gives promise of a successful and effective Administration.

/s/ T.P. Pendleton
Chief Topographic Engineer

Survey Order No. 150

Survey Order No. 150, dated May 17, 1946, is as follows:

"Recent studies of the Survey's organization and facilities for topographic mapping indicate the desirability of now combining all such activities within the Topographic Branch. The resulting standardization of photogrammetric and map compilation procedures, and the equitable deployment of trained personnel and equipment, will not only greatly facilitate preparations for meeting our extensive mapping commitments for the immediate future, but is necessary if we are to carry out administration, research, and field operations with maximum coordination and economy.

"Accordingly, on July 1, 1946, the topographic mapping facilities and commitments now vested in the Alaskan Branch, together with such funds allotted or transferred for the purpose will become the responsibility of the Topographic Branch. The Chief Alaskan Geologist will act as coordinating official, integrating Alaskan topographic mapping activities with other Survey functions in Alaska, and will assist in determining and appraising other Alaska map needs. Therefore, close liaison will be maintained with the Chief Alaskan Geologist in establishing schedules and priorities for both reconnaissance and detailed surveys in Alaska.

"The Topographic and Alaskan Branches are hereby directed immediately to make all necessary plans and arrangements for the activation of this order."

Survey Order No. 151

Survey Order No. 151, dated July 1, 1946, contained the following information:

In recognition of the imperative need to expand the facilities of the Topographic Branch and, at the same time, provide an operational unit strategically located in respect to the several Rocky Mountain States, where much new mapping is required, there is hereby re-established, effective July 1, 1946, a division of the Topographic Branch known as the Rocky Mountain Division. The Division shall be comprised of two states from the Pacific Division (Montana and Wyoming), two states from the Central Division (Colorado and New Mexico), and Alaska.

Supervision of actual field work now being carried on in the four states of the new Division will remain under the direction of the Pacific and Central Division Engineers until such time as the Chief Topographic Engineer determines that the responsibility can be transferred without interfering with mapping programs in operation.

Robert O. Davis, Topographic Engineer, will be designated as Acting Division Engineer with full authority to carry out the Division Engineer's responsibility.

It is directed that necessary plans for the activation of the new Division be undertaken at once by the Topographic Branch.

/s/ W.E. Wrather
Director

Survey Order No. 152

Survey Order No. 152, dated July 1, 1946, contained the following information:

"In order to expedite the re-organization of the Topographic Branch and to consolidate operational functions under central control, the following activities of the Topographic Branch are hereby transferred to the Atlantic Division effective July 1, 1946:

"(a) The Drafting Unit, located in the Medical Building, Clarendon, Va., and now operating under the supervision of the Chief, Section of Inspection and Editing.

"(b) The Multiplex Unit, located in the Old Dominion Building, Clarendon, Va., and now operating under the supervision of the Chief, Section of Photo Mapping.

"The Section of Photo Mapping is hereby deactivated, and its remaining functions will be supervised by the Chief of the Photogrammetry Section of the Technical Staff, until such time as these functions can be absorbed by other appropriate units of the Topographic Branch."

Survey Order No. 157

Survey Order No. 157, dated October 1, 1946, contained the following information:

Mr. Robert O. Davis is hereby designated as Division Engineer, Rocky Mountain Division, Topographic Branch, effective October 1.

Mr. Davis joined the Survey in 1927, and was appointed a junior topographic engineer on June 1, 1929. During his early years in the Survey, Mr. Davis' experience included topographic mapping and control surveys in the Rocky Mountains and the central and eastern United States. He was assigned to the Chattanooga office of the

Survey in April 1936, and was one of the first field topographers to become proficient in the field of photogrammetry.

In June 1942, Mr. Davis was placed on military furlough and was commissioned in the Army Air Force, where he served until April 1946, and was discharged as a lieutenant colonel. During his Army service, he served in various capacities in connection with aerial photography and reconnaissance mapping. On his final tour of duty with the Army, he was Commanding Officer of the Sixth Compilation Squadron located at Jefferson Barracks. Since his return to the Geological Survey in April 1946, Mr. Davis has been Acting Division Engineer of the Rocky Mountain Division.

/s/ W.E. Wrather
Director

Survey Order No. 160

In view of the greatly expanded topographic mapping program and the more widespread use of topographic maps among government agencies and the public, for an increasing number of purposes, both civilian and military, it is necessary to promulgate and adopt certain basic policies with respect to overall specifications for standard topographic maps. The following specifications, treatments, and procedures shall therefore henceforth be adopted as standard within the Geological Survey for the area of the continental United States. Alaska and certain other possessions will be made the subject of special orders after studies as to map needs are completed.

1. <u>Publication scale for 7.5-minute sheets</u>

The preferred publication scale for 7.5-minute sheets for general distribution shall be 1:24000. The 1:31,680 scale shall be used only where local circumstances, or status of previously published maps, are such that the 1:24,000 scale would be clearly inadvisable. In future map revision programs, 1:31,680 scale sheets shall be converted to and republished as 1:24,000-scale sheets as rapidly as circumstances at the time will permit. This policy shall not be interpreted as prohibiting the publication of special editions of any map sheet on other than the 1:24,000 scale.

2. <u>Determination of quadrangle sizes for topographic maps</u>

The present policy of publishing map sheets in 7.5-minute units of the more developed areas, and of areas capable of development within a reasonable period of time, and publishing sheets of the sparsely settled and undeveloped regions in 15-minute units, shall be continued. Where there is doubt as to which sheet size is desirable for any region, project, or area, the 7.5-minute unit shall usually be selected. The contour interval for each sheet, or area, shall be determined after the publication scale has been determined, and the interval shall be appropriate for the selected publication scale. In all cases, whether the published maps are in 15-minute or 7.5-minute units, the horizontal-position accuracy shall be sufficient to meet the 1:24,000-scale specification, in order that the military requirement for accurate-position fire control maps will be met without the necessity of remapping those sheets originally published in 15-minute units. Thus the horizontal accuracy requirement practically compels preparation of all planimetric, and other types of base sheets, to be by modern photogrammetric methods, at scales large enough to meet the horizontal position specification outlined in Item 6 of this Order.

3. <u>Complete 15-minute sheet coverage in the United States</u>

It shall be the objective of the Geological Survey ultimately to supply a complete atlas of 15-minute topographic map sheets of the entire U.S. area. Therefore, where a region, or area, is originally mapped in 7.5-minute, or smaller, units, these shall later be reduced, recompiled, and published as 15-minute quadrangle maps, with appropriate contour intervals, as rapidly as funds and facilities permit.

4. <u>Contour interval in western United States</u>

The 5, 10, 20, and 40-foot contour intervals shall henceforth be the preferred and basic intervals for the entire continental United States. Within those states located generally west of 103 degrees longitude, the 25- and 50-foot contours shall henceforth be used only within those local areas where the status of completed or in-process maps is such that to introduce different contour intervals obviously would be detrimental to efficient use of the maps within any such local area. The 25-foot interval in particular shall be discontinued as rapidly as possible, and the use of the 50-foot interval shall be limited to those groups of 15-minute sheets that cover mountainous terrain, and are within or adjacent to areas where the 50-foot interval has already been used extensively on completed or in-process maps.

5. <u>Half-interval and supplementary contours</u>

On those map sheets having a contour internal of 10, 20, 40, or 50 feet, half-interval supplementary contours may, and in certain types of terrain shall be shown in the flatter areas of the sheet, where necessary to bring out properly the topographic detail in such flat areas. It is not expected that supplementary contours will be shown on all map sheets, but only on those where the terrain is predominately hilly or mountainous, and which also include important flat areas. Under this treatment, the regular interval can always be that which clearly and properly portrays the hilly or mountainous terrain, hence original compilation should be more economical, the maps will be more clear and readable, and the map accuracy specifications will be practical.

On sheets having 25-foot interval and covering mainly steep, mountainous terrain, supplementary 5-foot contours may be shown in the extremely flat areas. As noted in Item 5 preceding, however, the use of 25 feet as a basic or regular contour interval on new maps is expected to be discontinued as rapidly as practical.

All half-interval, and other supplementary contours, shall be differentiated on the compilation manuscript from regular or standard contours by showing the supplementary contours as dashed lines; and shall be differentiated on the published map in a suitable manner to be determined by the Technical Staff of the Topographic Branch.

All supplementary contours shall be of the same relative accuracy as the regular contours. For example, where the supplementary contours are of 10-foot interval, they shall be accurate (90 percent) within 5 feet, or one half of the supplementary interval. In those unusual cases where the use of half-interval supplementary contours in extensive flat areas does not indicate sufficient topographic detail, spot elevations, accurate within one-quarter of the supplementary interval shall be shown at critical points.

6. <u>Map accuracy specifications</u>

In 1940 and 1941, the Bureau of the Budget sponsored a series of conferences, attended by representatives of the principal Federal map-using and map-making agencies, from which were developed accuracy specifications for planimetric and topographic maps. These are, in part, as quoted below:

"STANDARD OF ACCURACY FOR A NATIONAL MAP PRODUCTION PROGRAM"

With a view to the utmost economy and expedition in producing maps which fulfill not only the broad needs for standard or principal maps, but also the reasonable particular needs of individual agencies, standards of accuracy for published maps are defined as follows:

1. <u>Horizontal accuracy</u>. For maps on publication scales larger than 1:20,000, not more than 10 percent of the points tested shall be in error by more than 1/30 inch, measured on the publication scale; for maps on publication scales of 1:20,000 or smaller, 1/50 inch. These limits of accuracy shall apply in all cases to positions of well defined points only. 'Well defined' points are those that are easily visible, or recoverable on the ground, such as the following: monuments or markers, such as bench marks property boundary monuments; intersections of roads, railroads, etc.; corners of large buildings or structures (or center points of small buildings); etc. In general, what is 'well defined' will also be determined by what is plotable on the scale of the map within 1/100 inch. Thus, while the intersection of two roads or property lines meeting at right angles would come within a sensible interpretation, identification of the intersection of such lines meeting at an acute angle would obviously not be practicable within 1/100 inch. Similarly, features not identifiable upon the ground within close limits are not considered as test points within the limits quoted, even though their positions may be scaled closely upon the map. In this class would come timber line, soil boundaries, etc.

2. Vertical accuracy as applied to contour maps on all publication scales, shall be such that not more than 10 percent of the elevations tested shall be in error more than one-half the contour interval. In checking elevations taken from the map, the apparent vertical error may be decreased by assuming a horizontal displacement within the permissible horizontal error for a map of that scale.

3. The accuracy of any map may be tested by comparing the positions of points whose locations, or elevations, are shown upon it with corresponding positions as determined by surveys of a higher accuracy. Tests shall be made by the producing agency, which shall also determine which of its maps are to be tested, and the extent of such testing.

4. Published maps meeting these accuracy requirements shall note this fact in their legends, as follows:

This map complies with the national standard map accuracy requirements.'

5. Published maps whose errors exceed those afore stated shall omit from their legends all mention of standard accuracy.

6. When a published map is a considerable enlargement of a map drawing ('manuscript') or of a published map, that fact shall be stated in the legend. For example, 'This map is an enlargement of a 1:20,000-scale map drawing;' or 'This map is an enlargement of a 1:24,000- scale published map.'"

Since the Geological Survey was a party to the development and adoption of these specifications, as both a map-using and a map-making agency, it is the policy of the Survey to prepare its planimetric and topographic maps to comply with these specifications, wherever this is practical from a procedure standpoint, and when the mapping costs are not unreasonably increased thereby. For difficult terrain covered by dense timber, and where there are no project developments in prospect, it will be permissible to publish maps of slightly less-than-standard contour accuracy, subject to approval of the Chief of the Topographic Branch, but in no case should the horizontal-position accuracy be decreased. It is expected that such sub-standard map sheets will be a small minority of the total sheets published for the United States. Sub-standard maps shall omit the accuracy compliance note in paragraph 4 of the specifications quoted above.

In order to make this policy effective, each manuscript map sheet believed to comply with these accuracy specifications shall carry a certification by a responsible field engineer to the effect that it is his considered opinion that the endorsed map does meet the accuracy requirements. The minimum field operations considered necessary to justify such certifications shall include: (a) thorough field inspection of all sheets or portions of sheets which have been produced by means of stereoscopic plotting instruments; and (b) test surveys, for sample testing as to accuracy compliance, on a reasonable proportion of the map sheets for each project area, regardless of the procedure used in preparing the manuscript maps. Field inspection is defined as an on-the-ground visual comparison of the map sheet with the ground features by competent topographers, and shall also include occasional checks and/or revisions by planetable surveys where the photogrammetric mapping work appears questionable. On completion of the field inspection and planetable revisions, certain of the map sheets in each project area shall be designated as test sheets and on these there shall be executed special surveys which are solely for the purpose of accuracy testing. The selection of the number and type of map sheets to be so tested in any project shall be such as will reasonably ensure that all sheets of each project will comply with the national map accuracy requirements. The procedures adopted for the special test surveys shall be such that the adjusted coordinates of each station of the test survey will have a horizontal-position error not greater than the least plottable distance on the map at publication scale, and the elevation error of any vertical test point will be not greater than 1/10 of the contour interval for the map being tested. Complete records of all test surveys for each test sheet shall be filed at each field division office, subject to later examination by any interested parties.

/s/ W.E. Wrather
Director

activities in the continental United States, Alaska, and the West Indies since that time, except for his military service in World War II. During most of his military service from 1942 to 1945 he was Commanding Officer of the Aeronautical Chart Service and was responsible for the compilation, reproduction, and distribution of all aeronautical charts and related publications required by the Army Air forces. As a part of that program, he directed the photogrammetric compilation of maps of more than 15,000,000 square miles in various part of the world, utilizing the "trimetrogon" method, which was largely developed by him in the Geological Survey. He was awarded the Legion of Merit on his discharge from the Army in December 1945, and since that time has been Staff Engineer and Chief of the Plans and Estimates Division of the Topographic Branch. During Mr. Pendleton's recent absence, Mr. FitzGerald has served as Acting Chief of the Branch.

/s/ W.E. Wrather
Director

Survey Order No. 162

May 26, 1947

T.P. Pendleton, Chief Topographic Engineer, at his own request retired from active service on April 1, following a prolonged illness. Mr. Pendleton joined the Geological Survey in 1905, and conducted topographic surveys chiefly in the western United States until 1918, when he was commissioned in the Corps of Engineers, U.S. Army. Following World War I, he continued mapping assignments in Turkey, Palestine, and the Balkans, as well as in the United States. After a period of private practice, Mr. Pendleton was reinstated in the Survey in 1934. He was largely responsible for the adaptation and development of the photogrammetric process now widely known as the "Multiplex" method to topographic mapping, first of the Tennessee Valley area, and subsequently, of the whole country. An outstanding authority on photogrammetry, surveying, and mapping, Mr. Pendleton's retirement is a loss to the Geological Survey and to the Nation.

Gerald FitzGerald, on my recommendation, has been appointed by the Secretary of the Interior to succeed Mr. Pendleton, effective May 22. Mr. FitzGerald was appointed to the Survey in 1917, and has been engaged in surveying and mapping

Aid for Downed AAF Plane

Karlton B. Miller and Kenneth Carter, Topographic Engineers in the Rocky Mountain Division, were climbing the side of Pikes Peak, Colo., one morning in October of 1947, seeking a spot for a 3-point location, when they met a sergeant who was walking away from a before-dawn plane crash.

According to a letter of high commendation from Brigadier Gen. Leon W. Johnson, commanding the 15th Air Force, as transmitted by R.O. Davis, Region Engineer:

"Despite intense cold and difficult terrain, Mr. Miller escorted the sergeant to safety and then notified the 15th Air Force Headquarters. Mr. Carter followed the injured man's tracks for more than 3 miles up the steep slope of the peak, and upon arrival at the scene of the accident, assisted with the first-aid treatment and the removal of the injured personnel to ambulances. Due to the prompt action of the Survey engineers all men were saved."

Regional Conferences

The interest in river-basin studies has been so intensified in recent years that it has become essential that all the Federal agencies operating within a basin coordinate their programs to assure the accomplishment of our common goal, which is the full development of our natural resources.

An important step toward the accomplishment of this purpose was taken at the Central Division Topographic Branch Conference held at Rolla, Mo., on November 10–12, 1947, by representatives of State and Federal agencies interested primarily in the development of the Missouri River Basin. This conference afforded to officials of the Federal and State agencies an opportunity to discuss their mutual problems; also, for the first time in the history of the Topographic Branch, it gave to the field and office personnel a full chance to exchange helpful views on the technical problems involved in the real objective, down-to-earth mapping. It generally was agreed that the conference was beneficial and that similar conferences should be held at regular intervals, not only in Rolla, but at other division headquarters, and at the Washington office.

While we know and appreciate the sterling qualities of our topographic engineers, we are always grateful when our cooperating officials show their appreciation. Many interesting speeches were given by Central Division cooperating officials and we are quoting from one on "Forty Years of Cooperation with the U.S. Geological Survey" by Dr. M.M. Leighton, Chief, State Geological Survey, Urbana, Ill.

" * * * In this matter of cooperation of the Illinois State Geological Survey with the U.S. Geological Survey in topographic mapping, the word cooperation has meant to me just what it says through all these years. And I have sensed among the officials of the U.S. Geological Survey during my 24 years of State connection that same sense of realization that cooperation is a two-way relationship. Nothing less could have brought the success that has marked our efforts. First of all, there has been the fine attitude of Directors George Otis Smith, W.C. Mendenhall, and Wm. E. Wrather, three fine distinguished men who recognized the magnitude of the task in all of the States, and who possessed the kind of spirit that was necessary to achieve it. Without exception, they have given their full support to the cooperative program.

"The chief topographic engineers were after the same pattern. First we called them geographers, and then chief topographers. The chief geographers were R.M. Wilson, R.B. Marshall, and W.H. Herron. The chief topographic engineers were C.H. Birdseye, J.G. Staack, T.P. Pendleton, and now Gerald FitzGerald. It has been my experience that they never failed to show anything but the greatest of interest in topographic mapping in Illinois. My faith in them continues down to the present time and includes the Chief, Gerald FitzGerald.

"My most frequent contacts have, however, been with the chiefs of the Central Division. There was that good old soul, Major Herron, who represented the U.S. Geological Survey in our office at Urbana through all the field seasons from 1905, until his death in 1930, with the exception of 1 year (1916–17), when he was away during the first World War. Everybody on our staff loved Major Herron. He had a way about him that made you feel that cooperation was something more than the following of regulations, or the terms of the agreement. He was most personable. Almost invariably, when he came to the office each morning and had looked over his mail he had a humorous story to tell us that started the day off refreshingly. We were pleased that he fell in with us as humanly.

"I believe that I am safe in saying that Major Herron and I planned the first large-scale maps for metropolitan areas. At any rate, it was fairly early, and it involved something like the equivalent of twenty-four 7.5-minute quadrangles in the Chicago area, and several in the East St. Louis area. When the large-scale maps of the Chicago area were finished, which was done under the immediate direction of O.H. Nelson, the Western Society of Engineers held a meeting, (December 12, 1927), of Chicago engineers to commemorate this, and Director Smith, Colonel Birdseye, Major Herron, O.H. Nelson, and several State officials were all there to participate.[43] That, by the way, is one of the best methods to get the support of the people. Everyone was enthusiastic about all of the new features of the maps, and their subsequent support could not be questioned.

"Following Major Herron in office was H.H. Hodgeson, who came from the Pacific Coast. Our relations with him were very fine, but rather brief, 1930–31. When he returned to the Coast he was followed by Glenn S. Smith, a man who showed great persuasiveness in presenting the need for topographic maps, and in arranging for cooperative funds. In his wide contacts, he made many friends for the U.S. Geological Survey, and for us.

"In 1940, Colonel Smith retired and I was greatly relieved when it was announced that C.L. Sadler would succeed to the position of Chief of the Central Division. He had done mapping a number of years before in Illinois, but most of it had been in Missouri * * *. We welcomed Sadler as Division Chief because we knew first of all that he was a man of integrity; second, that he was an able topographer; and third, that his public relations would be of the finest.

"Following his appointment, the Office of Division Chief was transferred from Urbana to Rolla. I might have objected, but Sadler's work had been in Missouri for many years, Buehler had always shown

---

[43] See Western Society of Engineers Journal, January 1928.

a great interest in topographic mapping, and it was only fair, after so many years, to yield the distinction to another state. Now that the headquarters of the Central office is so well established here at Rolla, and the personnel have become so much a part of this community, it seems only natural to keep the headquarters here. The ties are so strong that they have the qualities of permanency.

"Of our section chiefs I have spoken of O.H. Nelson and his fine work in the Chicago area. Gentlemen, I want to tell you that he put in hours, day and night, on the work that far out measured union standards. I am thankful that we still have men who are so deeply interested in the work they are doing that they bury their lives in it. Following Nelson, who was Chief of the Great Lakes Section, came F.W. Hughes. Not only is he a gentleman in every sense of the word, but he has ideals in the making of maps and a high conscience in the amount of production that a man should turn out in a day. Let us all hold firm to such ideals. It is production that has made this country great, together with the character that develops from the pride and satisfaction of work well done. Someone referred yesterday to the creative work performed by a topographic engineer, the turning out of a map of an area that shows the character of its surface, defines its drainage, exhibits the culture, and portrays the forest resources. That is the kind of work that a man can put his heart and soul into.

"I want to pay my tribute to all of the topographers and chiefs of parties that have worked in Illinois these 40 years. In the time that I have for these remarks, I cannot possibly name them all, but among them are Captain Sadler, Captain Staack, W.J. Lloyd, J.H. Renshawe, O.H. Nelson, J.R. Ellis, E.L. Hain, J.A. Duck, F.W. Hughes, S.R. Archer, W.S. Gehres, E.W. Gouchenour, H.S. Senseney, F.A. Danforth, A.T. Munson, W.S. Beames, H.E. Simmons, J.A. Shumate, J.T. Schutz, W.E. Baird, W.A. Tingey, A.W. Plushnick, J.C. Hilliard, H.S. Milsted, C.E. Weishapple, and others.

"In the 40 years of the cooperation between the U.S. Geological Survey and the State of Illinois, there have been certain cardinal principles recognized devotedly by both parties. First, there has been the cardinal principle that we are going to do good job of topographic mapping. Each year we have executed a cooperative agreement that served as a record of understanding and commitments. Toward this agreement, we have maintained a spiritual point of view. I cannot remember a single incident when we got out the agreement to settle any controversial point. We realized that we had a high mission and we put our shoulders together to perform it * * *.

"The second cardinal principle is that the Illinois Geological Survey has consistently recognized the topographic branch of the Geological Survey as the central mapping agency. They have the staff, they have the equipment, and they know how good maps are made. We match our dollars and they direct the execution of the work * * *.

"The third cardinal principle has been that the maps to be made should meet both local and general needs, needs with respect to the contour interval, the scale of the map, the amount of detail to be shown, and other features of map making. The Illinois Survey has made its contribution by giving study to the local needs and by conferring with and soliciting suggestions from engineers and engineering groups, soil survey people, geologists, various State departments, and other users of maps * * *.

"The fourth cardinal principle pertains to finances, namely, that both organizations should match their funds dollar for dollar. This principle has invariably been followed.

"Years pass and the world of human affairs continues with changes in personnel. We dislike to see good men reach the age of retirement. We in Illinois especially dislike to anticipate the retirement of C.L. Sadler as Chief of the Central Division next July. He and I respect each other and the interests of each other's organizations. He has done an excellent job in administering our topographic mapping. We shall await with interest the announcement as to who his successor will be, but we know that the right man will be chosen. The U.S. Geological Survey has made no mistakes in their selections * * *."

Map Exhibits

On December 9, 1947, Mr. FitzGerald wrote the following memorandum for the Staff Division and Section Chiefs:

"The Topographic Branch is often called upon to prepare map exhibits in connection with engineering and scientific meetings. These occasions afford excellent opportunities for explaining our mapping operations and presenting the need for expediting our mapping operations and presenting the need for expediting our 20-year program.

"At the National Highway Research Board annual meeting last week, we displayed three panels which were rather effective. They are now in Rooms 6233 and 6221, and will be there for several days so that you may inspect them.

"I will appreciate your comments and suggestions regarding these exhibits and how we can most effectively tell our story and put it over; first, to the various map users; second, the engineering students who may be interested in topographic mapping as a career; and third, the many State and municipal agencies who are all potential cooperators in our mapping program.

"During fiscal year 1947, exhibits of maps, aerial photographs, and related material pertaining to mapping procedures were prepared for the Pan American Institute of Cartography and History at Caracas, Venezuela, the American Mining and Metal Congress at Denver, Colo., Middlebury College, Vt., the American Society of Civil Engineers at St. Paul, Minn., the American Society of Photogrammetry at Washington, D.C., the Utah Centennial at Salt Lake City, Utah, and the National Research Council at Washington, D.C. Many requests were received during the succeeding years.

"In 1950, requests were received for map exhibits for the annual meeting of the American Society of Civil Engineers held in Washington, and the Map Information Office prepared a series of display panels shoving present-day mapping methods and the various types of published maps, together with samples of advance map copy, such as blue-lines, composite prints, and other photostatic reproductions available to engineers. An exhibit showing the latest photogrammetric mapping techniques was also prepared for the annual meeting of the American Society of Photogrammetry in Washington. An exhibit of special interest to petroleum engineers was supplied for the annual meeting of the American Association of Petroleum Geologists in Chicago. Assistance was also given in selecting maps, aerial photographs, and other illustrative material for exhibits sponsored by state, county, university, and other local groups. Similar requests have been received through fiscal year 1954."

Foreign Visitors

Recognition of the Geological Survey's preeminence in the field of mapping is further evidenced by the number and type of visitors during fiscal year 1947, and succeeding years.

In addition to those from our own country, civil and military representatives of the following foreign governments have spent considerable time studying and observing field and office methods used by the Topographic Branch in preparing maps: Canada, Mexico, Cuba, Brazil, Chile, Columbia, Peru, Venezuela, Belgium, Denmark, France, Great Britain, Greece, Netherlands, Norway, Sweden, Egypt, South Africa, China, India, Palestine, Australia, and the Philippine Islands. In practically all of these countries, the development of new mapping methods was considerably curtailed during World War II. Letters of appreciation received from these foreign visitors during the year indicate that the technical assistance and advice given by members of the Survey is contributing not only to the progress of mapping throughout the world, but to a mutual understanding of good will and a friendly spirit of cooperation.

May 21, 1948

Memorandum

To: All personnel concerned

From: Chief, Topographic Branch

Subject: Map appraisal and classification

In accordance with the Memorandum of Understanding, dated October 21, 1946, between the War Department and the Geological Survey, and subsequent informal agreements with other agencies, the Geological Survey is committed to maintain distribution stocks of topographic maps produced by these agencies. Standards and procedures for appraisal and classification of these maps are hereby established. All such topographic maps, prior to their reprinting by the Geological Survey in accordance with the policy outlined in the Director's memorandum dated May 18, 1948, shall be appraised and classified as to quality. The classification of each map shall be in accordance with the degree of its compliance with current requirements for a topographic map of the best quality, on the same scale as the particular one under consideration. The classification shall be by numbers 1–5 inclusive, with number one (1) indicating a map of the quality required to comply with current standards. Succeeding numbers will indicate maps with progressively greater deviations from those standards.

This classification does not furnish criteria to be used in estimating or judging current public requirements for more detailed map information in areas covered by existing topographic maps. It is thought that the number of requests for the resurvey of an area, with weighted consideration of their sources, is the best basis for evaluating the serviceability of existing maps for current general purposes.

Under this classification of existing maps, it will be possible to have maps in the 7.5-minute, 15-minute,

30-minute, and 1-degree series covering all or portions of the same area. Each of these maps may require appraisal and classification. All of the overlapping maps may have the same or different classifications, depending upon their relative deficiencies, when compared with current requirements for a class 1 map in the series.

A topographic map will normally retain its appraised classification of quality for a period of about 10 years, or until a reprint edition becomes necessary, when it should be re-examined with a view of its reclassification, if appropriate.

Topographic maps of special areas, or on special scales, which are not published in quadrangle form, are not recommended for classification. However, they may be classified if it is thought necessary or desirable.

Geological Survey Maps

An increasing public demand for accuracy and for fidelity of expression in topographic maps has resulted in the adoption by the Geological Survey of specifications for horizontal and vertical accuracy that conform to the National map accuracy standards. Test surveys are necessary to evaluate the degree of compliance with the accuracy standards; however, accuracy in location and elevation does not assure fidelity of expression of the physiographic features that the contours should portray. Field and office procedures and practices have been developed in an effort to assure the high standard of quality that is desired for the general purpose maps of quadrangle areas. Thorough field and office inspections are necessary to assure that fidelity of expression has been attained.

All maps that comply with the current standards for horizontal and vertical accuracy, and which represent physiographic features with fidelity, are of excellent quality and shall be classified no. 1 maps. Other maps shall be classified according to their degree of compliance with the adopted standards of accuracy and fidelity of expression as set forth above. Field inspections, test surveys, and office examinations of Geological Survey maps, currently being made, indicate that most of them are no. 1 maps (of excellent quality) and that the remainder of them are no. 2 maps (good quality), since they have been found to be deficient only to a minor degree. Earlier maps of the Geological Survey, which were made in compliance with less rigid specifications and standards, may merit any one of the five quality classifications. Some of the deficiencies which may contribute to substandard qualities are:

- Township and section boundaries omitted

- Woodland areas not indicated

- Use of an inappropriate contour interval

- Weak or poor portrayal of physiographic forms

- Substandard accuracy of locations, or elevations, or both

- Revision from aerial photographs or stereo-compilations not followed by a field check

- Drafting of a substandard quality.

Maps by other Federal Agencies

Other Federal organizations are engaged in topographic surveys, primarily to produce maps to meet particular requirements, and as a preliminary phase of other activities for which they are responsible. However, many of the resulting maps portray topographic information which is, or may be, of considerable interest and value to the general public. In order that such useful information may not be withheld, and that its distribution may be centralized, other Federal organizations, including the Department of the Army, by agreement are furnishing the Geological Survey with their maps of quadrangle areas for reproduction and distribution. These maps are subject to the same factors of strength or weakness that appear to be inherent in all maps. They may be of excellent quality in all elements, or may be deficient, in varying degrees, in one or more elements that are considered essential components of first class maps, according to current standards.

All of the topographic maps that are distributed by the Geological Survey should be appraised, and classified according to uniform standards of quality. Manifestly, those standards should be the same as have been adopted and specified for Geological Survey maps. Moreover, it is recommended that the classification of any particular map be made known to a map user, or purchaser, upon request, or when the classification indicates that a map is substandard for an expressed purpose or for a known or intended use.

Classification

In order that the classification of all topographic maps of quadrangle areas may proceed in a uniform and impartial manner, the standards by which they shall be appraised and classified are set forth as follows:

1. Topographic maps that are to be included in this class must fulfill the following requirements:

(a) The map must comply with the national standards of map accuracy substantially as set forth in Geological Survey Order No. 160.

(b) It must be of standard content, which includes the representation of all pertinent details of cultural, topographic, and hydrographic features, within the limitations of the scale of the map, by means of approved symbols and treatments. The approved map symbols are shown on map symbol sheets, and treatments are specified in technical instructions, which are issued by the Geological Survey.

(c) Its contours must adequately delineate and express the physiographic forms on the ground with such fidelity that their character is clearly evident on the map.

(d) The map must employ a contour interval that is appropriate for showing all significant topographic details of the area, on the scale employed, without over burdening the map with unnecessary contours. Attention is called to the fact that current mapping instructions authorize the use of supplementary contours to assist in portraying significant details that would escape contours drawn at the regular interval.

(e) The map must be reasonably complete in its representation of current cultural detail as it exists upon the ground at the time of classification. Comments by map users indicate that a map may be expected to remain up to date for an average period of about 10 years. However, maps of sparsely settled areas may remain up to date for a longer period and areas subject to intensive development may become out of date in a much shorter period. First quality maps will normally be expected to retain their classification as such for approximately 10 years after completion of field work in original production or revision, unless criticism from map users indicates shorter period. It should be noted that, to be accurate and complete, the compilation, or the revision of the cultural features of a map from aerial photographs must always be supplemented by field interpretation, field checking, or both.

2. Topographic maps that fail to meet the requirements of no. 1 maps in <u>not more than one</u> of the listed requirements for that category shall be included in this classification. Most of the topographic maps made by the Geological Survey between 1920 and 1938 were rigidly controlled; many had planimetric bases resulting from photogrammetric compilations; and the others had culture and drainage located by detailed and extensive plan-etable surveys in conjunction with the mapping of the topography. Such maps may be of less accuracy than that specified for first class maps. The period of mapping, however, is not conclusive evidence of the

quality classification of a map. Some maps produced prior to 1920 may be of a quality to be included in this class, and some produced after that date may be sufficiently deficient to warrant their exclusion.

Specifically, a no. 2 map may be deficient in <u>one</u> of the following elements.

(a) It may be below standard accuracy if large parts of its area are densely forested. Such maps, if compiled by modern photogrammetric methods, will usually meet the horizontal accuracy specifications, but are likely to be deficient in vertical accuracy. It would not be economical to improve the vertical accuracy of such maps sufficiently to qualify them for the no. 1 classification.

(b) Its content may be less than that required of a no. 1 map. This may consist of the omissions of township and section lines of the public land surveys, drainage details, trails, houses, or the departure from common practice in the classification of buildings or in their treatment in the urban areas. Minor revision of such maps could improve them sufficiently to permit their classification as no. 1 maps.

(c) The map may be deficient in topographic expression. The term "topographic expression" or "expression" is difficult to define clearly. Nevertheless, it covers one of the more important elements that distinguish a first class topographic map. In general, it refers to that appearance of the contouring that denotes, at a glance, the character, or type, of the country that has been mapped, or the fidelity with which the land forms have been portrayed. Weakness in topographic expression may be, and often is, evidenced by omission of small but characteristic details in the topography; by the erratic spacing of contours, indicating slopes that are not natural in the area; by forms or shapes that are obviously foreign to the area; or by various other deficiencies such as drainage head or ridge top contours improperly drawn or spaced.

(d) The map may employ a contour interval that is inappropriate for its scale and the relief encountered in the area. If such maps are first class in other respects, they could be made to qualify as no. 1 maps by recompilation, or field surveys to add supplementary contours in some cases, and in others to change the basic interval to one that is appropriate for portraying the relief encountered, in the required detail, without overburdening the map.

(e) Cultural developments within an area, after completion of field surveys, may make a map deficient in its representation of these details at any later date. As has been stated, the culture will usually will be considered to be out of date in about 10 years after the date of survey, and the map then will be placed in class no. 2 unless, or until, the culture has been revised.

3. Topographic maps shall be included in this classification if they are sufficiently accurate and complete to be generally useful, but which have too many known deficiencies to qualify for classification as no. 2 maps. Most of the maps produced by the Geological Survey in the period from 1900 to 1920 will fall in this category, and may be classed as no. 3 maps. There was a marked improvement in the quality of maps in this period over those that were produced in prior years. This improvement was in response to an increasing awareness, on the part of the general public, of the usefulness of map information and related efforts to furnish better maps by increasing their accuracy and the amount of detail shown. It was during this period that spirit level and transit traverse control came into general use, although both were initiated prior to 1900; the Beaman stadia arc was invented and planetable and stadia became the accepted method of mapping quadrangle areas of moderate to low relief, and increasing use was made of the larger standard map scales in order to permit mapping in greater detail than had previously been thought necessary or desirable.

During this period, when economic considerations limited the amounts of both basic control and instrumental surveys of areas, personal skill on the part of the topographer was a very important factor in the interpretation of land forms and the delineation of their character, and in the estimation of differences in elevation, and of distance in the location and portrayal of minor features. Therefore, authorship sometimes must be considered in determining whether a map should be classed as no. 3 (fair) or no. 4 (poor). Modern mapping methods, together with required procedures and practices, have greatly reduced the effect of the personal element on the quality of. topographic maps.

It should be noted that many maps in this classification (no. 3) could be converted into class no. 2, or even class no. 1 maps by revision surveys or limited resurveys where the character of the deficiencies, and their extent, would make that treatment economically advantageous.

4. Topographic maps that obviously are defective in the major elements of accuracy, content, and expression, to an extent that disqualifies them as no. 3 maps, but which are the best or only maps available, should be classified in this category of no. 4 maps. Included in no. 4 will be those maps listed in earlier classifications as "rapid reconnaissance,"

and others of similar grades. These maps are of such substandard quality that their revision would not be warranted. They are placed in this class for the period during which they will be sold and distributed, pending resurvey of their areas and publication of maps of better quality.

5. Topographic maps that would otherwise fall in no. 4, but whose areas are covered by other maps of better quality at the same scale, and maps in the 30-minute and 1-degree series whose areas are completely covered by maps in the 7.5-minute or 15-minute series, should be classified as no. 5 maps. They will neither be shown on state index maps, nor reprinted in a sales edition, but file copies and manuscript material will be retained so that specific requests for copies of these maps may be filled.

/s/ Gerald FitzGerald
Chief, Topographic Branch

Approved:
/s/ W.E. Wrather
   Director

Survey Order No. 173

Survey Order No. 173, Change in Organization Structure Nomenclature, dated December 15, 1948, is as follows:

As of January 1, 1949, the term Division shall be used in lieu of the present term Branch, and the term Branch shall be used in lieu of the present term Division, with two exceptions:

(a) The present Atlantic, Central, Rocky Mountain, and Pacific Divisions of the Topographic Branch shall become Regions of the Topographic Division.

(b) The present Divisions of the Geologic Branch shall be abolished and their constituent Sections shall become Branches of the Geologic Division.

/s/ W.E. Wrather
Director

December 27, 1948

Memorandum

To: All Topographic Branch Employees

From: Chief, Topographic Branch

Subject: Change in Organizational Nomenclature

The Geological Survey under date of December 15, 1948, issued Order No. 173, making changes in the terms used to designate its principal organizational subdivisions, and changing the titles of the present field divisions of the Topographic Branch. This action was taken in accordance with a request of the Secretary of the Interior that all bureaus of the Department conform to the system of nomenclature recommended to the departments and agencies by a congressional committee.

In accordance with Survey Order No. 173, the organizational designation of the Branches of the Survey has been changed to Divisions, effective January 1, 1949. There are now four divisions in the Geological Survey as follows:

Conservation Division
Geologic Division
Topographic Division
Water Resources Division

The four principal field offices of the Topographic Division will be designated by the term Region. The two staff Divisions now become Branches.

In order to comply fully with the suggested nomenclature, changes in the designation of subordinate units also are made: The term "District" shall be used instead of "Area;" "Project" shall be used instead of "Sub-Area." The term "Project Office" shall be used to designate subdivision of a region in a semi-permanent status, such as the Chattanooga office.

The organizational titles of officials in the Topographic Division as of January 1, 1949, shall be as shown in table 48.

Table 48. Organizational titles in the Topographic Division, as of January 1, 1944.

Name	Title
Gerald FitzGerald	Chief Topographic Engineer
Robert L. Moravetz	Assistant to Chief Topographic Engineer
George D. Whitmore	Chief, Research and Technical Control Branch
Robert H. Lyddan	Chief, Plans Coordination Branch
Dallas H. Watson	Atlantic Region Engineer
Ralph F. Wilcoxon	Assistant Atlantic Region Engineer
Daniel Kennedy	Central Region Engineer
Frank W. Hughes	Assistant Central Region Engineer
C.A. Ecklund	Pacific Region Engineer
C.P. Van Camp	Assistant Pacific Region Engineer
Robert O. Davis	Rocky Mountain Region Engineer
T.V. Cummins	Assistant Rocky Mountain Region Engineer

The title of Area Chief is changed to District Engineer and the title sub-Area Chief is changed to Project Engineer.

The titles of the Trimetrogon Section and the Special Map Projects Section, both of which are special operational sections in the Washington office, are unchanged.

Because of the increase in the scope of the mapping activities of the present Topographic Branch, and the resultant increase in complexity of organization and number of personnel, many employees do not have current information as to the organization structure of the Branch. The following is a brief outline of the organization as it will exist after January 1, 1949.

The Washington office is composed of the Chief Topographic Engineer, the Assistant to the Chief Topographic Engineer, and number of administrative and clerical employees; two staff branches covering planning and technical activities, and known as the Plans and Coordination Branch, and the Research and Technical Control Branch; and two special operational sections, the Trimetrogon Section (John I. Davidson, Chief) and the Special Map Projects Section (O.H. Nelson, Chief). All other operational activities are performed in the field regional offices. There are four regions with headquarters and areas as follows:

<u>Atlantic Region</u>

Headquarters: 1109 North Highland Street, Arlington, Va.

Regional Engineer: Dallas H. Watson

Geographical Area: Maine, New Hampshire, Vermont, Massachusetts, Connecticut, Rhode Island, New York, New Jersey, Pennsylvania, Delaware, Indiana, Ohio, Kentucky, Virginia, West Virginia, Maryland, Tennessee, North Carolina, South Carolina, Mississippi, Alabama, Georgia and Florida.

<u>Central Region</u>

Headquarters: Rolla, Mo. (Box 133)

Regional Engineer: Daniel Kennedy

Geographical Area: North Dakota, South Dakota, Minnesota, Wisconsin, Michigan, Nebraska, Iowa, Illinois, Kansas, Missouri, Oklahoma, Arkansas, Texas, and Louisiana.

<u>Rocky Mountain Region</u>

Headquarters: Denver Federal Center (Box 2858 Lakewood Branch) Denver, Colo.

Regional Engineer: Robert O. Davis

Geographical Area: Montana, Wyoming, Colorado, New Mexico, and Alaska.

Pacific Region

Headquarters: Sacramento, Calif. (Box 346)

Regional Engineer: C.A. Ecklund

Geographical Area: Washington, Idaho, Oregon, California, Nevada, Utah, and Arizona.

In addition, Multiplex work carried on in cooperation with the TVA is conducted at the Chattanooga Project Office (John G. Groninger, Project Engineer).

/s/ Gerald FitzGerald
Chief, Topographic Branch

Approved:
December 27, 1948
(Date)

/s/ W.E. Wrather
Director

Survey Order No. 186

Survey Order No. 186, dated October 14, 1949, is as follows:

In view of the increasing interest in natural resources on the continental shelf, and the requirements of the military map users for coastal data, topographic maps of coastal and navigable-water areas shall henceforth show underwater contours, as well as certain obstructions to navigation, and features of landmark value to map users, whenever these data can be obtained from hydrographic charts or from surveys made for their preparation.

1. Underwater contours

The water areas of topographic maps along the coast, and the inland waters navigable from the coast, shall show the depth of the water by means of underwater contours of an appropriate interval. The units of measure and contour interval shall be correlated to the similar information shown on overlapping and adjacent hydrographic charts; however, no underwater contours shall be shown below the 100-fathom line.

Inasmuch as the datum of the topographic map is based upon mean sea level, the shoreline shown is mean high water, and the datum of hydrographic charts is based upon either mean low water or mean lower low water, explanatory notes showing the difference in datums for that area shall be added on the margin of the map.

2. Obstructions to navigation

In order to present more nearly complete information on the map concerning the coastal areas, certain types of obstructions to navigation shall be shown. In general, the policy shall be to show all offshore features, compatible with the scale of the map, that are of a reasonably permanent nature and that can be seen from the shore at low tide.

/s/ W.E. Wrather
Director

Reprinted August 1951.

Tours of Duty at Division Headquarters

On October 28, 1949, Mr. FitzGerald sent the following memorandum to the Atlantic, Central, Pacific, and Rocky Mountain Region Engineers:

"In accordance with decisions reached at the Regional Engineers Conference held last May, plans should be made to send a representative of each region to Washington for a 1-month period of staff duty this winter.

"One of the principal advantages to be gained by such assignments is through the opportunity they afford for the representatives to get to know the men here and to observe and discuss the operations that are carried on in this office. We also plan to take advantage of the information, criticism, and suggestions on technical instructions that can be offered by these men who are on, or close to, actual operations. To a lesser extent any of the representatives might be given a specific project while in this office—a project which he may be specially qualified to handle, or which may give the individual an opportunity to demonstrate some special talent.

"The dates of the various tours may be generally arranged to suit your local circumstances, although this office would prefer that men from two of the regions be here at the same time. It would also be well if the tours were arranged sometime during the months December to February, inclusive.

"It is preferred that selections for these assignments be made from your P–4, P–5, or P–6 personnel, although especially talented or promising men in the P–3 grade are not excluded. Also, the individual selected should be either (1) engaged in regional headquarters work; (2) considered as potential for future regional or main office permanent assignment; or (3) located in some part of your organization where general knowledge of staff activities would be especially helpful.

"Please consider this matter carefully and forward to this office your recommendations, including any alternate suggestions, sometime early in November."

Letter from Chief, Topographic Division

On January 2, 1950, Mr. FitzGerald sent the following letter to all Topographic Division employees:

"I am sure you realize that in terms of accomplishment, the year 1949 has been a notable one for the Topographic Division. We have made real progress in the production of hundreds of new standard quadrangles covering widely separated areas in the United States and Alaska, and we have also continued to make progress in technical research of our field and office methods in an intensive effort to improve the quality, as well as the quantity, of our work. This is a good record.

"Because of the very heavy work load we have all been carrying to meet production schedules, it has been difficult these last few years to maintain the close bond of working relationships that has been a cherished tradition of our group for more than 50 years. Our present method of work and the size of our organization these days tend to keep all of us separated for most of the year and oftentimes in rather restricted fields of operation. Under such conditions, I know it is sometimes difficult to maintain the "esprit de corps" and team work which have long been characteristic of the Topographic Division, and which we all know is essential if we are to remain one of the great mapping organizations of the world.

"The next year promises at the outset its full quota of problems, which from past experience I am confident we can take in stride. Every individual member of the Washington office and the four Regions of the Division working together can make 1950 a memorable year in the history of topographic mapping.

"Your own personal contribution to the work of the Topographic Division is important because our kind of a job depends on team work. I'm proud to be a member of the Topographic Division team with you and take this opportunity of sending greeting for the New Year."

Survey Order No. 192

Survey Order No. 192, dated February 6, 1950, is as follows:

In order to simplify the preparation of Alaskan maps at scales of 1:250,000 and larger, and to facilitate matching with sheets of the National Canadian topographic map series, the Geological Survey will henceforth use for these maps the Universal Transverse Mercator (UTM) Projection instead of the polyconic projection used heretofore. Quadrangle sheets will be bounded by the same meridians and parallels as would be used with the polyconic projection, and there will be no noticeable change in the shape, size, format, or appearance of the maps. The words "Polyconic Projection" that have hitherto been included in the legend at the lower left corner of the map sheet will be replaced by "Universal Transverse Mercator Projection, Zone --." Quadrangle maps that are currently in process of preparation may be completed and printed on the polyconic projection if the work has progressed to the stage where a change of projection does not appear justified.

This order does not apply to Alaskan maps and charts prepared for publication at scales smaller than 1:250,000.

> /s/ W.E. Wrather
> Director

Survey Order No. 193

Survey Order No. 160, dated January 2, 1947, promulgated basic specifications for the national topographic map series of the continental United States, exclusive of territories and island possessions. Similar specifications for the national map series of Alaska are outlined herein.

1. <u>Quarter-million-scale series</u>

In accordance with approved programs, the entire area of Alaska shall be covered as rapidly as feasible with a new series of 1:250,000-scale topographic maps. The original sheets of the 1:250,000-scale series shall be prepared in an expeditious manner, from whatever control, photographs, and other source materials are immediately available. The content and accuracy of the first edition shall be of as high an order as is consistent with available source data and an expedited compilation program. Those sheets for which the original compilation is necessarily of sub-standard content or accuracy shall be regarded as provisional, and a revised edition shall be prepared whenever the availability of additional control and/or better source materials would result in substantial improvement over the original maps.

Contour interval usually shall be 200 feet, except (1) the interval of 100 feet may be utilized for sheets covering flat terrain, provided that the source data available will yield standard-accuracy 100-foot contours; (2) on extensive mountain slopes, the intermediate contours should be "feathered out" in accordance with current instructions of the Topographic Division for steep-slope treatments, but with the 1,000-foot index contours being shown clearly and completely; and (3) in areas where vertical control and/or suitable aerial photographs are lacking at the time, the first-edition provisional sheets are compiled, the contour interval may be either 500 feet or 1,000 feet.

2. Mile-per-inch series

The publication scale for this series of Alaska sheets shall be 1:63,360, or exactly 1 inch equals 1 mile. Maps at this scale will be prepared, as rapidly as funds permit, for those areas of Alaska for which there is specific and well justified need. The contour interval shall be 200 feet, 100 feet, 50 feet, or 25 feet, depending on the type of terrain. Supplementary contours, including 12.5-foot contours, may be utilized in the usual manner to show details in the flatter areas of rugged-terrain regions, provided that means are available for plotting the supplementary contours with the customary accuracy.

The procedures and source materials used in the preparation of the 1:63,360 sheets shall be such that there will be reasonable assurance that the maps will generally comply with the Federal standards of map accuracy, as outlined in Survey Order No. 160, but routine accuracy testing and field certifications of these maps will not be required.

3. Larger-scale special maps

Occasionally, the needs of engineering investigations, geologic and mineral studies, construction works, or other projects of restricted area having similar requirements, will justify the preparation of topographic maps at relatively larger scales and with relatively smaller contour intervals. When practical, such maps shall be prepared for publication at a scale of 1:24,000, and with contour interval from either the 50-25-foot series, or from the 40-, 20-, 10-, 5-foot series, depending on type of terrain and other circumstances. Half-interval or other supplementary contours may be utilized on the large-scale sheets where necessary and appropriate for topographic detail in the flatter areas of regions of rugged terrain. Whenever practical, such maps shall be

of standard accuracy as defined in Survey Order No. 160, shall be field tested for accuracy in the usual manner, and the field sheets shall carry notation as to accuracy by an appropriate field engineer.

/s/ W.E. Wrather
Director

Memorandum to Trimetrogon Section

The following memorandum, dated May 15, 1950, to Geological Survey employees, Trimetrogon Section, from the Chief Topographic Engineer states:

"The Commanding Officer of the Aeronautical Chart Service, Department of the Air Force, has notified the Geological Survey of plans to consolidate all charting operations within the A.C.S. and to discontinue, insofar as possible, contract work with other government agencies. After a long period of discussion between the two agencies on the subject, an agreement has been concluded which provides for a gradual curtailment of work assignments to the Survey by the Chart Service. Under the terms of this agreement, an effort will be made to complete all assigned work within the next 2 years.

"No material change in the personnel strength of the Survey's Trimetrogon Section is contemplated at this time. It is planned, during the transition period, that all available production capacity will be shifted to the domestic mapping program, which is a primary responsibility of the Geological Survey, and is now understaffed.

"The termination of active participation by the Geological Survey in the aeronautical charting program will mark the end of a significant period. Although the Geological Survey pioneered the compilation of aeronautical charts 25 years ago, the accomplishments of the present Trimetrogon Section during the past 9 years have been outstanding in the history of cartography. Methods and techniques were developed by this group in the Survey to carry out one of the largest and most important war mapping programs undertaken during World War II. "Trimetrogon mapping" has been used, and is being used, by civil and military organizations in many countries throughout the world. The Geological Survey is proud of this record of achievement and, when possible, plans for the full use of this trained group in carrying out the expanded national topographic mapping program."

Topographic Division Bulletin

On March 12, 1951, Mr. FitzGerald sent the following memorandum to the Atlantic, Central Pacific, and Rocky Mountain Region Engineers:

"The first issue of the Topographic Division Bulletin will probably be distributed about April 1, 1951. The work of assembling the material and preparing copy for the initial edition is largely the result of the special assignment of George S. Druhot, District Engineer from the Pacific Region. Articles in the first copy have been selected from available papers and no attempt has been made to canvass the Regions for material. For this reason, the initial edition may appear to be somewhat out of balance as being representative of the entire Division.

"With the completion of this special assignment to the Pacific Region, Mr. Druhot returns to the Pacific Region and the work of preparing future editions will be delegated to the Regional offices, the Trimetrogon Section, as well as the Washington office. The success of this undertaking will depend on the amount of effort and cooperation put into it by members of the Division. Unless a continuous flow of articles from all field offices can be maintained, interest will not be general and the Bulletin will be unable to maintain its purpose, which is the dissemination of general information regarding new ideas, methods, equipment, administrative and technical practices and procedures as they affect both the office and the field.

"An Editor-in-Chief has been designated in the Washington office, and will have general responsibility for the assembling of material, and the publication of the Bulletin. He will also be charged with obtaining papers and discussion material from Division headquarters. Supporting the Editorial Chief will be five Assistant Editors, one from each Region, and one from the Trimetrogon Section. It will be the duty of the Assistant Editors to collect, edit, and transmit, through the Region Engineer, or the head of the Trimetrogon Section, at least one long article every 3 months. These articles should be related to the activities, organization, history, or ideals of the Topographic Division, and they should be designed to instruct the reader in an interesting way. A length of 2,400 words, 8 pages of double-spaced typing, is suggested as most desirable, but articles may vary in length from 1,500 words to 7,500 words. In addition to a long article each quarter, several brief items of interest should be sent from each Region. Discussions of statements or articles in previous issues, questions, and suggestions are particularly desired.

Short reviews of other published papers and reports on technical meetings will also be items of interest.

"After selected articles are edited by the Assistant Editors, and reviewed by the Region Engineers or the Editor of the Trimetrogon Section, they will be neatly typed, double spaced, and two copies each will be forwarded to Washington. A short, but suitable, title should head each article and the author's name should appear as a byline under it. The number of words should be indicated in pencil in the upper right corner of the first page. The date should be shown on the last page. If good photographs are available, prints or negatives, about two for each article should be attached. These should be completely described and marked. It is planned to publish the Bulletin quarterly, four issues in a volume. Your personal interest and cooperation in this venture is necessary to ensure its success."

Training Course in Topographic Mapping Methods

A training course for new field engineers of the Topographic Division was operated in the vicinity of Clemson College, Clemson, S.C., from June to September, 1951. The course was conducted by the Atlantic Region office and was started in order to train employees added to the staff in response to the Division's new workload connected with defense mapping requirements.

One week of each trainee's time was spent on training for transit traverse, the first half day being devoted to the use of the transit and allied equipment, including the principles involved in making azimuth observations on the Sun and Polaris. The remainder of the week was spent in actually running a traverse line; each trainee rotating through a simulated field party so that each man received at least a half day of experience in each position.

Another week was devoted to training in leveling methods, including work in writing descriptions, setting bench marks, accomplishing the administrative paper work, etc. The system used called for two instruments and two recorders with only one pair of rodmen. Here too, all men changed jobs after a half day. In this way, checked elevations were obtained at each turning point.

The training parties operated in an area that is to be mapped by planetable methods. (Townville and Seneca Quadrangles, S.C.) The data that they produced would later be used in the mapping phase. No permanent bench marks were set, but otherwise all procedures were the same as would be followed in actual productive work.

Diplomas were awarded to 36 college-trained engineers from 12 East Coast institutions. Several of the men who completed this training course will go directly into the field, and some will be entrusted immediately with the supervision of field parties. A number of very promising men will return to college for another year, but have already demonstrated ability that will make them desirable for future employment. Credit for the success of this training course is due to W.K. McKinley, director of the course, E.A. Krahmer, F.B. Neary, G.B. Grunwell, E.E. Van Brunt, E.I. Loud, Jr., and A.C. Stiefel, instructors.

Survey Order No. 213

Survey Order No. 213, Change of Regional Boundaries—Topographic Division, dated December 27, 1951, is as follows:

> In order to provide for a more equitable distribution of work load and to expedite the mapping of high priority areas for the Department of Defense, a realignment of Regional boundaries of the Topographic Division is approved. Florida, Alabama, and Mississippi will become a part of the Central Region and Texas a part of the Rocky Mountain Region.
>
> There will be a transition period during which time certain projects will be completed in these areas by regional personnel who have been previously operating in the state concerned. At the same time, additional, or new, projects will be undertaken by personnel of the region to which the state has recently been added.
>
> /s/ W.E. Wrather
> Director

Table 49.   Revised listing of the states and territories in each region.

Atlantic	Central	Rocky Mountain	Pacific
Georgia	Florida	Texas	Arizona
Tennessee	Alabama	New Mexico	Utah
Kentucky	Mississippi	Colorado	Idaho
Indiana	Louisiana	Wyoming	Washington
Ohio	Arkansas	Montana	Oregon
West Virginia	Oklahoma	Alaska	Nevada
Virginia	Kansas		California
North Carolina	Missouri		Hawaii
South Carolina	Illinois		
Maryland	Michigan		
Delaware	Wisconsin		
Pennsylvania	Iowa		
New Jersey	Nebraska		
New York	South Dakota		
Connecticut	North Dakota		
Rhode Island	Minnesota		
Massachusetts			
Vermont			
New Hampshire			
Maine			
Puerto Rico			

Topography—The Lay of the Land

We learn much by reading the contours carved by nature. In fact, our first objective in the process of developing facts concerning our resource base is that of covering the land with adequate topographic maps, and in order to properly appreciate and evaluate the major changes that have taken place in this activity from 1945 to 1952, it is necessary to look back into the preceding years.

Development of photogrammetric mapping required many years; in fact, it is still going on. Much effort has been devoted to developing new methods, and to investigating and testing photogrammetric instruments, which include the stereoautograph in 1921, the aerocartograph in 1927, and in 1935, the Multiplex aeroprojector.

In 1936, engineers of the Tennessee Valley Authority saw the value of the Multiplex and hailed it as the best solution for their map problems. Cooperation with the Geological Survey to provide planimetric maps of the region was continued in a plan to map the entire area topographically with the Multiplex. In this respect, the Tennessee Valley served as a proving ground for these instruments and a training school for operators. Many of these former trainees are now supervising photogrammetric activities in Survey field offices, or have done their part in designing new equipment and developing new procedures.

Expansion of mapping activities during the past decade also must be viewed in light of military demands, because the Nation, insofar as mapping is concerned, has been under a war economy almost continually since December 7, 1941. The Survey's military mapping program is greater in 1952 than at any time during the 11 years that have elapsed since attack on Pearl Harbor. More than 50 percent of the mapping capacity of the Survey is being used to map areas selected by the Department of Defense. Indications point to a continuance of such a program for several years.

Military needs for maps were supplied in large part by other mapping organizations such as Army Map Service, Forest Service, Coast and Geodetic Survey, and private companies. The map-making facilities of these Federal organizations were so much augmented by the effort that they are still considered important factors in map production. Their standard quadrangle maps of domestic areas are published and distributed by the Geological Survey.

As of June 30, 1945, it was estimated that about 48 percent of the United States had been covered with topographic maps produced by the Geological Survey, but less than half of these were considered adequate according to modern standards. Work was continued on eight sheets of the international map of the world on a scale of 1:1,000,000 and on preparation of the United States transportation map for the Public Roads Administration.

Topography Streamlined and Modernized

In 1946, the Geological Survey made an intensive study of administrative requirements, because it was evident that expansion had outstripped existing facilities and that reorganization would be necessary.

A staff was set up to assist the chief topographic engineer in coping with the manifest growing problems. Four regional offices were organized to fit in with the staff plan. The Rocky Mountain Region was reactivated with offices at Denver, Colo. The others are Atlantic Region, Arlington, Va.; Central Region, Rolla, Mo.; and the Pacific Region, Sacramento, Calif. Each region was converted into a complete mapping unit with all the facilities required to produce maps except for reproduction. Map reproduction is centralized in the District of Columbia.

Under this administrative setup, the Topographic Division in the course of these 8 years has increased its annual man-years of effort from 532 to 2,279. This has been possible because funds available for topographic mapping have in the same period been increased from about $2,000,000 to about $16,000,000. Production has jumped from about 200 quadrangles a year in 1945 to more than 1,000 in 1952. In square miles, this means an increase from 14,000 to 70,000.

During the past few years, mapping in Alaska also has been accelerated. All of the 1:250,000-scale maps are either published or ready for publication. A remapping program has been started in the Hawaiian Islands.

With New Instruments a More Exact Science

Among the instruments developed to facilitate mapping operations, is the Polastrodial, originally conceived in 1942, which was further perfected in 1947, to determine mechanically the azimuth of the north star, Polaris, at any time and at any latitude in the United States.

The next year an electrical survey net adjuster was completed. With this instrument, complicated networks of transit traverse or levels can be adjusted in a fraction of the time required for least square computations. Also, electromechanical elevation meters were used for the first time to measure elevations continuously along roads at speeds of 15 to 20 miles an hour, and a vacuum chamber was obtained in which precision altimeters in sets of six are calibrated.

Helicopters were first used for transportation in 1948 (fig. 22), in Colorado and later in Alaska, Washington, and California. Supplemented by walkie-talkies for communication, they have revolutionized the old methods of extending control in the mountains. Stations high in mountainous terrain, which formerly required several days of pack train travel, can now be reached from base camps in a matter of hours, or

minutes, so that full advantage can be taken of brief lulls in sieges of inclement weather.

Electronic methods for obtaining ground control and for positioning aerial photography were employed in Alaska during 1949 and succeeding years, when Shoran was used by the U.S. Air Forces to control thousands of square miles of photography for the Geological Survey.

Also in Alaska during this period, the Geological Survey's first attempt to obtain ground elevations for topographic mapping by airborne electronic methods with the Airborne Profile Recorder was attempted. An area of some 78,000 square miles in central Alaska was surveyed from the air with this instrument during a single summer season.

Also, in 1951, distance measurements by means of an electronic method similar to Shoran was used in California in connection with ordinary horizontal angle measurements to determine the elevation of points for supplemental control. Portable radio communication made it possible to synchronize operations between the base station and several widely separated parties traveling over the area to be controlled.

A major accomplishment has been the appraisal and classification of all topographic maps prepared by other agencies, and sold or distributed by the Survey. This work involved the examination of nearly 12,000 maps, in order to evaluate their general usefulness and application to the needs of the expanding economy. This work was completed in 1950, and from the results of this appraisal it was estimated that only about 25 percent of the country was adequately mapped to supply the current general needs.

Progress in the development of new instruments for mapping by photogrammetric methods has been striking. Within the last several years, the methods for making topographic maps have undergone a revolutionary change from basic procedures carried out primarily by engineers working with surveying instruments in the field, to basic procedures carried on primarily in air conditioned office buildings by technicians working with aerial photographs and precise plotting instruments.

In addition to testing and using equipment resulting from its own research, the Geological Survey has adapted techniques and instruments developed elsewhere. Several kinds of stereoplotting instruments of the universal type, manufactured in Europe, have been installed for certain specialized photogrammetric projects.

Abandonment of copper-plate engraving in favor of color-separation drafting on metal-mounted paper or other scale-stable media has made it possible to complete the preparation of maps for reproduction in the region offices. This has done much to make possible the ten-fold increase in map production over pre-war years, most of which has been accomplished since 1945.

In 1948, after several conferences with other map-making agencies, a mutual understanding was reached in regard to map content and map symbols. This resulted in making feasible the interchange of maps between agencies requiring different symbolization and content without the necessity for extensive redrafting.

Map Information Office

The Map Information Office was reorganized in 1945 to continue one of the functions of an earlier organization; the Federal Board of Surveys and Maps, which was abolished by executive order in 1942.

One of its primary duties is to determine, on request, whether there is any mapping data available such as vertical and horizontal control, topography, and aerial photography for areas proposed for mapping. It also has the duty of supplying information regarding these three subjects to the general public. To some extent it distributes control data, sells local maps over the counter, and takes care of purchase orders for aerial photographs.

For reference purpose, it maintains, as far as possible, a complete set of all Geological Survey maps, including advance copies of current work, as well as various domestic maps and charts made by other organizations. One of its objectives is to extend knowledge of the availability and value of topographic maps.

Systematic Appraisal of Mapping Needs

Needs of other federal agencies for maps received constant attention. Recently an effort has been made to more systematically determine needs of State map-using groups through the formation of State mapping advisory committees. This information is necessary for determining priorities so as to prepare broad general mapping plans and schedules covering the entire country. Within these plans successive annual programs are formulated with due consideration to budgetary and capacity limitations. Production control has been established and operates to determine rates of progress, costs, and other factual data about production. This is useful for sound management.

An Assist to the Air Force

During the early years of the war, a new group was established primarily to compile small-scale reconnaissance charts of areas throughout the world for the U.S. Air Force, utilizing trimetrogon photography. This method of compiling charts was an entirely new development by the Survey for use in Alaska, and was devised for the purpose of rapidly mapping huge areas necessitated by the demands of World War II.

The responsibilities and duties originally assigned to this group have been greatly increased during the past 7 years. The quality, accuracy, and overall standards of mapping have increased considerably, and the group is now compiling, editing, drafting, and preparing negatives for maps and charts up to the final stage prior to printing.

Activities include the compilation of large-scale planimetric and topographic maps of standard accuracy by Kelsh plotter and Multiplex, the compilation of maps from radar scope photography, the preparation of radar prediction charts, the extension of vertical and horizontal control to remote areas by means of the photo alidade, and the performance of operational research to determine new methods and procedures, besides various drafting and photographic activities, including the preparation of shaded relief plates.

The volume of production since 1945 by trimetrogon methods has reached a total of more than 10 million square miles of small-scale maps and charts, largely of foreign areas. This was done at a cost of about $5,500,000.

Foreign Activities of the Topographic Division

The Topographic Division has been engaged in foreign mapping activities of various kinds for many years. This includes the preparation of military maps and charts during war periods, the assignment of topographic mapping experts to advise and counsel foreign countries on techniques and methods of topographic mapping, and the continuing work of the Special Maps Section on special Air Force projects. Other than military mapping, the recent foreign work of the Division has been connected with, and related to, natural resources development programs under the congressionally approved foreign aid programs.

With the advent of Point Four and the Economic Cooperation Administration, additional requests for specifications and advice on contract terms for aerial photography required considerable time and effort by the Survey specialists.

Mapping specialists have been sent to foreign areas for two main reasons, (1) to inspect photography and map compilation in order to make sure that specifications are being met and, (2) to advise, direct, and instruct the foreign agencies in modern mapping methods.

An effective effort in the foreign field is the Brazilian mapping project, which initially called on the Geological Survey to produce topographic maps of areas containing strategic minerals and carry on, at the same time, training of native engineers in the science of topographic mapping by planetable method. This initial effort, undertaken by Charles W. Buckey, in 1941 to 1942, was subsequently continued by Edward A. Krahmer, Milton F. Denault, John G. Collins, and Edward F. Barner. This experience proved that the effort and expense were well justified in providing high quality bases on which to pursue investigations and record information necessary to the development and production of mineral deposit areas.

Because the primary responsibility of the Topographic Division is domestic mapping, much of the foreign work has been carried on at the request of agencies of the Department of Defense. The exception to this procedure has been, as stated above, mapping projects or consultant services coordinated with foreign aid activities. Advice on specifications for aerial photography and topographic maps is offered, and training in photogrammetry and topographic mapping is a primary responsibility of the Division. An agreement was made between the Geological Survey, the Coast and Geodetic Survey, and the Inter-American Geodetic Survey whereby certain training responsibilities of South and Central American technicians would be undertaken by each organization.

The Geological Survey, with more than 250 operating stereo-photogrammetric units, and almost 3,000 skilled employees with 75 years of experience in the field of topographic mapping, must necessarily furnish leadership and accept responsibility for the topographic and photogrammetric phases of this training, not only for the South and Central Americans, but for trainees from other areas of the free world as well. Implicit in the agreement between the Survey, C&GS, and the IAGS is the requirement that the foreign trainee, coming to the Survey, be able to speak English fluently, have a degree in engineering or science, have had experience in the mapping field, and, if possible, have a connection with a map-making agency in his home country to which he will return.

In addition to the contractual mapping in Brazil, the Topographic Division writes specifications, and has negotiated a contract, for aerial photography, topographic map compilation, and mosaics for a considerable area in Saudi Arabia, where water resource investigations were the primary objective under a Point IV Project. Specifications for aerial photography and advice on contract terms have been furnished the Mutual Security Agency (MSA) for areas in Angola and Mozambique, Africa. Four geodetic engineers were recruited by the Survey under a project cosponsored by the Economic Cooperation Administration and the Geological Survey, for operations in Kenya, Uganda, and Tanganyika, Africa, in 1950, directed by British Colonial Surveys, and the assignment was completed in 1952. Preliminary advice was furnished to MSA on a detailed large-scale topographic mapping project in the Philippines, which was later transferred to the Corps of Engineers for supervision of compilation by a commercial contractor.

The type of foreign activities carried on by the Topographic Division do not indicate assignments of personnel to foreign countries, nor mapping in those countries. Nevertheless, the Survey feels a strong responsibility in the promulgation of knowledge and appreciation of the uses and benefits of topographic mapping and takes every opportunity to advance this theme in advising on problems of development in foreign fields.[44]

---

[44] From J.L. Watkins' article published in the Topographic Division Bulletin, June 1952.

Summary of Progress

In summing up the accomplishment of the Topographic Division from 1945 to 1952, it can be said that the period has been one of expansion, refinement as to accuracy, and standardization as to methods and results.

Progress has been made toward more complete cooperation between Federal agencies to prevent overlapping of activities in the field of domestic topographic mapping, and to utilize each other's efforts. Some disruption of the efforts to meet civilian requirements has been caused by the priority of military needs, but this was balanced by the fact that the military agencies contributed extra funds toward mapping. Of the almost $70,000,000 available for mapping since 1946, approximately 13 percent or $9,000,000 was advanced by the Department of Defense.

Cooperating funds allotted during these 8 years by State agencies amount to more than $7,000,000, as compared with about $17,000,000 covering the entire period of cooperation, which was first initiated by Massachusetts in 1885.[45]

Inventions

U.S. patents issued to Geological Survey photogrammetrists are listed in table 50:

---

[45] From Department of the Interior Booklet, "Years of Progress 1945–1952."

Table 50. U.S. patents issued to U.S. Geological Survey photogrammetrists.

[Other patents are pending]

Name	Patent number and date issued
R.M. Wilson	No. 2,261,201, Photoalidade, November 4, 1941.
H.T. Kelsh	No. 2,341,031, Mapmaking and Projection Instrument, October 12, 1948.
	No. 2,492,870, Stereoscopic Projection Mapmaking Instrument, December 27, 1949.
	No. 2,552,975, Photogrammetric Projection Machine, May 15, 1951.
J.G. Lewis	No. 2,321,033, Rectoblique Plotter, June 8, 1943.
	No. 2,364,082, Mapmaking from Aerial photographs, (Photoangulator, December 5, 1944).
	No. 2,561,386, Stereoblique Plotter, July 24, 1951.
J.L. Buckmaster	No. 2,542,640, Optical Transfer Instrument, February 29, 1944.
	No. 2,352,641, Transfer Methods and Instruments
	No. 2,370,143, Camera Lucida Instrument and Prismatic Units.
David Landen	No. 2,487,814, Mapping Instrument (Topo-Angulator), November 15, 1949.
Mary E. Lower and Charles C. Shockey	No. 2,680,909, Tilt-Setting Means for Rectoblique Apparatus, June, 1954.

Suggestions Awards

In November 1943, the Department of the Interior inaugurated a suggestions system to muster the best ideas of all its far flung staffs to solve its problems and to improve the Government service in war. The system was laid out, and it now patrols a clear channel along which an ever increasing and welcomed stream of new ideas from the field and office flow to market on their merit. Alongside this stream, is a return flow of recognition and awards to suggestors and inventors whose ideas are adopted and plowed back into the benefit of Interior operations (table 51).

Award of Excellence

Ronald M. Wilson, Chief, Computing Section, Topographic Branch, received the Award of Excellence of the Department of the Interior and $1,000 for devising a method of exercising mapping operations control in the United States and Puerto Rico.

The Wilson Alidade, designed by Mr. Wilson in 1932, and subsequently modified and improved, was patented on November 4, 1941. This alidade makes possible in the office most of the operations possible in the field with an ordinary plane table and telescopic alidade. It can be used with oblique aerial photographs taken with the panoramic camera. The instrument consists of an adjustable print holder, a telescope for reading horizontal and vertical angles, a centering microscope for placing the station point on the map sheet exactly in the vertical axis of the viewing telescope, and a straightedge connected with the vertical axis of the telescope so that directions from the station point can be plotted on the map sheet. Topographic features may be located, and their attitude determined, provided oblique photographs of the same terrain have been taken from two or more different points of view. Thus the sketching of contours may be controlled, just as is done through using the intersection method with a planetable. This instrument was an important factor in the success of the Trimetrogon method of mapping.[46]

---

[46] "History of Photogrammetry in the United States," by David Landen, Photogrammetric Engineering Journal, v. XVIII, no. 5, December 1952.

Table 51.   Selected members of the Topographic Branch who won Suggestions Awards through 1954.

1944	1948	1951—Continued	1954
William A. Fisher	Thomas V. Cummins	Jay M. Hatfield	Arnold H. Appleby, Jr.
Edward P. Hodge	Irwin Gottschall	Hugh B. Loving	Ralph L. Beitle
William C. Thompson	Sol M. Greisman	George M. O'Meally	Charles W. Birdseye
Edward W. Tibbott	Charles E. Grossman	Houston T. Walsh	Ashley R. Corbin
Dallas H. Watson	**1949**	**1952**	John J. Currey
Ralph F. Wilcoxon	Lester W. Ganahl	Edward T. Addi	Charles O. Fiola
1945	Anslow Harshaw	Thomas Cumerford (2)	Alfred M. Fraser
William A. Allen	Kenneth B. Keegan	William J. Jones	Thomas A. Hughes
Walter S. Beames	Pituman Ratchford	Russell M. Laird	Josephine C. Koblenz
Glenn S. Harper	John I. Roney	Charles I. Miller	Joe T. Long
Annie L.C. Kennedy	**1950**	Janet Raleigh	Hugh B. Loving
Edward I. Loud, Jr. (2)	Roy E. Carey	Jack Shinagawa	Sidney Meizell
M. Edith Lower	Max M. Chesy	**1953**	Jesse E. Mundine
Anton Navratil	Ramiro G. Cintron	Robert E. Altenhofen	Douglas A. Pryor
Louis S. Soreide	Ralph W. Evans	Elmer Elshire	John E. Ray
1946	Charles E. Grossman (2)	Wilbur D. Graham	Richard A. Skeen
Dale E. Crangle	Carl L. Huffaker	Edward I. Loud, Jr. (2)	F. Neil Wick (3)
William A. Fisher	Reynold E. Isto	Charles O. Fiola	Richard B. Wong
John R.P. Wilson	Kenneth B. Keegan	William A. Fisher	
1947	Guy W. Strohm	Eugene G. Gomolka	
Dale E. Crangle	Elwood J. Thacher	Charles W. McCaw	
Charles E. Grossman (2)	**1951**	Joseph F.W. Smith	
Lauretta H. Hawk	Frank Chiovitti	Glen L. Staley, Jr.	
Archie T. Munson	Georgia H. Gass		
Herbert J. Vinton	Sol M. Greisman (2)		

Superior Accomplishment or Merit Promotion Awards

Interior Department Order No. 2144 from Secretary Ickes, dated December 17, 1945, Rewards for Superior Accomplishment, states:

"Attached is a copy of 'The Dept of the Interior Program for the Granting of additional Within-grade Salary Advancements as Rewards for Superior Accomplishment' which has my approval, and the approval of the Civil Service Commission, and a memorandum on procedures.

"I believe that the fair and equitable application of this plan will be a valuable incentive to employees and a contribution to good administration.

"The present Departmental Committee on Especially Meritorious Promotions will continue as the Departmental Committee on Rewards for Superior Accomplishments."

Members of the Topographic Branch who have received Superior Accomplishment or Merit Promotion rewards from February 1944 to June 1954 are listed in table 52.

Table 52. Members of the Topographic Branch who have received Superior Accomplishment or Merit Promotion rewards from February 1944 to June 1954.

Edward T. Addi	Jerry R. Harris	Clarence W. Nottage
C. C. Barnekov	Robert B. Hart	Thomas P. Pendleton
Ralph W. Berry	Wm. V. Kennedy	Delmar W. Pinkston
Wayne R. Broaddus	Frank Larner (2)	Thaddeus T. Ranney
Florence E. Butler	Chester R. Lloyd	Marvin B. Scher
Dewitt R. Cabaniss	John H. Lycett	Lucy W. Sollenberger
Henry C. Collins	Harold J. McMillen	Louis S. Soreide
Ashley R. Corbin	Charles Mager, Jr.	Carl T. Sullins, Jr.
Clarence H. Drown	Richard O. Mahan	George N. Tex
Wm. C. Elledge	Thos. J. Murphy, Jr.	Wright R. Todd
Samuel D. Farmer	Anton Navratil (2)	Ray Wittemore
George W. Garland	Clarence Norris	William F. Zens (2)

Honor Awards

The Act of May 23, 1920, provided for the retirement of civil employees of the United States when they reached a certain age and had the requisite years of service, but it was not until August 2, 1946, that the Interior Department granted Honor Awards for distinguished services, or when separated from the Department. Topographic Branch personnel who received Honor Awards are shown in table 53.

Table 53. Personnel of the Topographic Branch who have received Honor Awards.

Name	Date of separation	Type of award
Gerald FitzGerald	5/6/49	Distinguished
Ralph W. Berry	2/23/49 (Death)	Commendable
John E. Blackburn	3/31/47	Commendable
Raymond G. Clinite	10/31/53	Meritorious
Leona T. Dowdy	11/7/53 (Death)	Commendable
Jesse A. Duck	1/31/49	Meritorious
James R. Ellis	7/11/47	Commendable
Earl J. Essick	7/31/52	Commendable
Richard T. Evans	7/31/51	Meritorious
Archie A. Farrell	6/30/52	Meritorious
Charles R. Fisher	4/30/59	Meritorious
William A. Fisher	5/31/53	Commendable
Alvah T. Fowler	11/30/48	Meritorious
J. Irving Gayetty	3/31/47	Commendable
Wilmer S. Gehres	8/31/53	Meritorious
Elmer L. Hain	6/30/50	Meritorious
Fred M. Hart	7/31/51	Meritorious
Frank W. Hughes	6/30/49	Meritorious
Hattie B. Ingersoll	10/15/48 (Death)	Meritorious
Hoyt L. Johnston	3/31/49	Commendable
Lawrence A. Kelley	5/23/53 (Death)	Meritorious
Harry T. Kelsh	7/31/52	Distinguished
Reuben A. Kiger	6/30/48	Meritorious
Edward C. Kruse	7/31/53	Commendable
Herbert S. Lewis	2/28/50	Meritorious
John L. Lewis	8/31/53	Meritorious
Edward I. Loud, Jr.	4/1/54	Distinguished
Sigurd G. Lunde	4/28/50	Commendable
Robert L. McCammon	1/31/48	Commendable
Clarence P. McKinley	1/27/50	Meritorious
W. Klett McKinley	9/30/32	Meritorious
Frank J. MacMaugh	12/31/46	Commendable
Harry S. Milsted	8/11/52	Commendable
William H.S. Morey	3/31/47	Commendable
Lee Morrison	3/31/47	Commendable
Laurie O. Newsome	4/30/48	Commendable
Clarence Norris	9/10/53	Meritorious
Alexander J. Ogle	6/30/53	Meritorious
Thomas P. Pendleton	4/1/47	Distinguished
Leonard Powell	4/18/54 (Death)	Commendable
Robert H. Reineck	4/2/48	Meritorious
Carl L. Sadler	6/30/48	Distinguished
James L. Sanders	8/31/53	Meritorious
Rufus H. Sargent	3/31/47	Distinguished
Raymond C. Seitz	11/30/53	Commendable
Ernest A. Shuster, Jr.	11/30/51	Meritorious
George E. Sisson	7/4/53 (Death)	Meritorious
John G. Staack	6/30/47	Distinguished
Rhea B. Steele	7/13/51 (Death)	Meritorious
William C. Thompson	12/31/52	Meritorious
Frank L. Whaley	7/31/53	Meritorious
Ralph F. Wilcoxon	3/19/49 (Death)	Meritorious
Wilbur Williford	5/21/50 (Death)	Commendable

Contract Mapping

In order to meet the immediate demands of a largely expanded defense mapping program, contracts were placed with commercial map-making companies during fiscal year 1952 for the performance of separate operations involved in the mapping of areas aggregating 16,000 square miles, at a total cost of $1,500,000. Because this was a departure from usual Survey operations, it required writing detailed specifications of the work to be performed, and establishing procedures for inspecting and testing the completed work for compliance with prescribed standards.

Following is an excerpt from the detailed specifications:

"1. <u>General Statement to Bidders</u>.

"a. These specifications cover the execution of basic horizontal and vertical control, procurement of aerial photography, execution of necessary supplemental control, and stereocompilation of topographic maps, insofar as each of these operations is required to produce topographic map sheets that are as complete as it is practicable to compile from aerial photographs without field completion, and that are standard in accuracy, content, and appearance. All operations required to be performed must compare in quality and content with corresponding work accomplished by the Geological Survey in accordance with current practice on such work, so that the results may be made available in the customary manner for public use. Bidders are advised that the accuracy and completeness of the work called for in these specifications are considered the minimum essential to the purposes for which the products are intended.

"b. The Geological Survey will furnish the contractor whatever basic information and material are available in the way of photography and control, as enumerated in the Bidding Schedule, together with standard metal tablets, field notebooks, and computation forms, as needed. The contractor shall furnish all other materials, superintendence, labor, equipment, and transportation, shall procure the necessary aerial photography and execute the surveys called for in the Invitation and Bidding Schedule, shall establish permanent and intermediate survey marks as called for in Sections 24, 25, 34, and 35, shall perform all computations and adjustments necessary to determine positions and/or elevations for all such points, shall execute and finish the map compilations of the areas specified in the Bidding Schedule, and shall deliver to the designated consignee all information and material called for in these specifications and in the Bidding Schedule. All work shall be executed expeditiously and in conformity with good engineering practice, to the satisfaction and acceptance of the contracting officer, in complete accord with these specifications and other conditions in the Invitation.

"c. The Geological Survey will inspect and test the field work and computations of the basic control surveys and will inspect, test and field complete the map sheets. Bidders should not underestimate the difficulties of meeting the requirements of the specifications and must assure themselves, by thorough examination of available information, or by investigation of the area involved, that their facilities and proposed processes are adequate to insure completed compliance with all the terms of these specifications * * *."

Production Schedule Requirements

In order to meet the Geological Survey schedule for completion and publication of the maps to be compiled under this contract, the following delivery schedule is required for all items:

"Deliveries of completed and inspected map manuscripts shall begin on or before June 1, 1953, and shall continue at a rate as nearly uniform as practicable of one third of the total number of quadrangles in each item, in each succeeding 3- month period. Delivery of all manuscripts must be completed by March 1, 1954, in order that the Geological Survey may test the maps for accuracy; that faulty work, if any, may be corrected by the contractor; and that final delivery and acceptance of all materials be made and all vouchers approved and paid prior to June 30, 1954 * * *."

In fiscal year 1952, contracts for 15,821 square miles of stereocompilation services, including some control, were awarded to commercial mapping organizations. Of this total, about 6,000 square miles were completed in 1953 and 9,300 in fiscal year 1954, leaving about 500 square miles to be done to complete the contracts.

American Society of Civil Engineers

The American Society of Civil Engineers (ASCE) was founded on November 5, 1852, and its first annual convention was held in New York City in June, 1869. Its constitution stated that its objects shall be the advancement of engineering knowledge and practice and the maintenance of a high professional standard among its members. Among the means to be employed for this purpose shall be meetings for the presentation and discussion of appropriate papers and for social and professional intercourse; the publication of such papers and discussions as may be deemed expedient; the maintenance of a library, the collection of maps, drawings and models; and the establishment of facilities for their use. The Society now issues regularly three types of technical publications: Proceedings-Separates, Transactions, and the magazine, Civil Engineering; members of the Topographic Branch have contributed to these publications. The first members of the Topographic Branch admitted to the ASCE are listed in table 54 and members who served on committees of the Surveying and Mapping Division are listed in table 55.

Table 54.   First members of the Topographic Branch admitted to membership in the American Society of Civil Engineers.

Name	Date admitted	Name	Date admitted
S.H. Bodfish	1889	R.B. Marshall	1904
H.M. Wilson	1890	C.L. Nelson	1906
F.E. Matthes	1901	Frank Sutton	1906
E.M. Douglas	1901	G.R. Davis	1907
Hersey Munroe	1902	C.L. Sadler	1908
E.C. Barnard	1902	C.H. Birdseye	1912

Additional members			
J.K. Bailey	R.K. Bean	P.F. Bermel	C.A. Biever
R.E. Bruce	J.G. Collins	E.J. Coon	T.V. Cummins
R.O. Davis	M.F. Denault	G.S. Druhot	C.A. Ecklund
R.T. Evans	E.J. Fennell	C.T. Galloway	M.J. Gleissner
R.S. Gould	Peter Gribok	G.B. Grunwell	E.L. Hain
L.D. Hankins	M.J. Harden	F.M. Hart	R.E. Isto
C.C. Juhre	Albert Kenig	Daniel Kennedy	W.V. Kennedy
J.B. Leachman	R.H. Lyddan	A.C. McCutchen	R.R. Monbeck
C.N. Mortenson	R.F. Murphy	R.S. Pearse	T.P. Pendleton
D.H. Rutledge	J.P. Rydeen	S.W. Smith	J.L. Speert
J.G. Staack	R.G. Stevenson	A.C. Stiefel	F.K. Van Zandt
G.D. Whitmore	R.M. Wilson	F.L Witkege	

The Region Engineers and other members have taken an active part in their local sections.

Table 55.   Topographic Branch members who served on various committees of the Surveying and Mapping Division of the American Society of Civil Engineers.

R.K. Bean	E.J. Fennell	T.P. Pendleton	G.D. Whitmore
C.H. Birdseye	R.H. Lyddan	J.G. Staack	R.M. Wilson

November 5, 1952, marked completion of a century of life for this Society. Celebration of the Centennial was the outstanding activity of the year. The Centennial of Engineering reached a climax in the great Convocation at Chicago during the period September 3 to 13, inclusive. The activities included a symposium program in 12 sections, each of which was devoted to a subject of broad import to the profession and to the public. More than a hundred speakers, each a person distinguished in the engineering profession or in the business and industrial life of the country, took part. Sixty-five engineering societies and allied organizations participated in the Convocation. Some scheduled only banquet or committee meetings, but 38 held technical sessions, many of which were of exceptional merit. Nearly 1,000 papers were presented on the programs of the societies. Robert H. Lyddan, Chief, Plans and Coordination Branch, Topographic Division, gave a very interesting talk on "A Century of Topographic Surveying and Mapping," which was later published as Paper No. 2632. George D. Whitmore, Chief, Research and Technical Control Branch, delivered a paper called "Improvements in Photogrammetry Through A Century," which was written by Mr. Whitmore and Morris M. Thompson, Staff Engineer, Photogrammetric Section. This was published as Paper No. 2633. With a total attendance of nearly 30,000 people, including 400 from other countries, the Convocation became the greatest and most significant gathering in the history of the profession.[47]

Survey Order No. 219

Survey Order No. 219, dated October 14, 1952, is as follows:

> In accordance with the Topographic Division policy to locate operational activities outside the Washington area in order to conserve critical space needed for staff activities the Special Map Projects Section, a section engaged in the preparation of the state base map series, the millionth scale map, the 1:250,000 series, and other small scale maps, has been abolished.
>
> The functions performed by this Section are transferred to the four regional offices of the Topographic Division. A Small-Scale Compilation Unit has been established in the Cartography Section of each region to accommodate this work, and the drafting and editing operations have been integrated with the existing Drafting and Editing Units.
>
> /s/ W.E. Wrather
> Director

Survey Order No. 220

Survey Order No. 220, Establishment of Special Maps Branch, dated December 5, 1952, is as follows:

---

[47] See American Society of Civil Engineering, Official Register, published 1953.

The organizational unit of the Topographic Division, which has been responsible for the preparation of small-scale aeronautical charts for the U.S. Air Force and other special maps since 1946, has been designated as the "Trimetrogon Section." Since this office represents a first major functional breakdown of a Division and constitutes an organizational entity, as far as function is concerned, in accordance with the standard organizational nomenclature of the Department, it has been designated as a "Branch."

In addition to the original function of this office, in the past few years its duties and responsibilities have been expanded to include the following:

1. The compilation of large-scale planimetric and topographic maps of standard accuracy by means of vertical and oblique Kelsh and Multiplex stereophotogrammetric equipment.

2. The photogrammetric compilation of large-scale planimetric maps of standard accuracy by means of monocular and stereo equipment.

3. The processing of photographic reproductions up to the press plate stage for divisions of the Geological Survey and the Aeronautical Chart and Information Service.

4. The compilation of maps from radar scope photography and the preparation of radar prediction charts from photographic, cartographic, and other source materials.

5. The production of shaded relief plates, a monotone depicting hills, valleys, slopes, ridges, and other terrain features through judicious use of shading and highlights.

6. The extension of vertical and horizontal control, from existing control to areas which are remote, or where control is sparse or nonexistent, by means of the photo alidade.

7. The performance of operational research to determine methods and procedures for mapping areas where unusual conditions exist.

Accordingly, the Special Maps Branch is hereby established in the Topographic Division with headquarters at Washington, D.C. The Trimetrogon Section is abolished.

/s/ W.E. Wrather
Director

Memorandum from Director to Division Chiefs

Memorandum from the Director to Division Chief, dated June 23, 1953, "Objectives statements for the Geological Survey":

"The principal objective of the Topographic Division is the preparation of the topographic map series of the United States, its territories and possessions. Activities necessary to attain this objective include research in cartography; the development of more efficient equipment and techniques for the compilation of maps; review of the adequacy of existing or proposed map series to meet current needs; systematic revision of existing maps to maintain their usefulness; and the preparation of the results of mapping activities for publication.

"The Division schedules its mapping activities in relation to the investigations and development of natural resources; to the program for geologic mapping; to map requirements of national defense; and to civilian engineering requirements, such as highway, municipal, and industrial planning.

"The Division provides the Department of Defense and other Federal Agencies with mapping services.

"a. Plans and Coordination Branch—The objectives Coordination Branch are the formulation of broad mapping programs and the planning of operations for the Topographic Division; the accomplishment of overall management activities, such as scheduling, production and fiscal control, and cost analysis; and the coordination of Division activities with mapping programs and requirements of other organizations.

"The Branch provides a continuing service to major map users and the general public by disseminating map data.

"b. Research and Technical Control Branch—The objectives of the Research and Technical Control Branch are the improvement of technical operations of the Topographic Division by research and the development of new methods, procedures, and instruments; the establishment of standard specifications and procedures in the four major phases of topographic mapping, topographic surveys, geodetic operations, photogrammetry, and cartography; and the adherence to these standards in the accomplishment of technical operations.

"c. Special Maps Branch—The principal objective of the Special Maps Branch is the compilation of charts and radar predictions for the Aeronautical Chart and Information Center of the U.S. Air Force as an integral part of the overall program of the Air Force for the production and maintenance of adequate world-wide aeronautical charts, primarily in the interest of national defense. Activities necessary to obtain this objective involve the cartographic or photogrammetric compilation or revision of base charts preliminary to photolithographic reproduction.

"The Branch also compiled special planimetric and topographic maps for the Survey, and for other agencies, and carries out operational research on special problems.

"d. <u>Atlantic, Central, Rocky Mountain, and Pacific Regions</u>—The objective of each regional office of the Topographic Division is the accomplishment, within its assigned geographic area, of all operational functions involved is the production of topographic maps, except final reproduction. Activities necessary to attain this objective include the establishment of basic horizontal and vertical control, and supplemental control, the field of stereocompilation, field and office completion, final drafting and editing, as well as incidental cartographic photography."

Topographic Division Personnel Serving on Committees

Topographic Division personnel serving on committees are listed in table 56.

State Mapping Advisory Committees

Briefly, the system, as developed by the Bureau of the Budget and the Topographic Division, involves an annual expression of mapping needs, in terms of four priorities, on the part of about 20 Federal map using agencies. The Topographic Division receives these expressions through the Bureau of the Budget about March first of each year and makes an analysis of them for each State. Priority figures are reversed to form weights, which are in turn combined with other weights, evaluating the various uses that will be made of the maps, and totaled for each 15-minute quadrangle. These totals are entered on the proper quadrangle blocks of an administrative index map of the State concerned. Other requests received from various sources, including individuals, the State Mapping

Table 56. Topographic Division personnel serving on committees.

Name	Affiliation
Altenhofen, R.E.	American Society of Photogrammetry
Bailey, J.K.	Atlantic Region Security Officer
	Safety Committee, Chairman (Atlantic Division)
Baird, W.E.	Map Appraisal Committee
Bean, H.A.	Map Appraisal Committee (Advisory)
Belton, R.J.	Atlantic Region Editor, Topographic Division Bulletin
Bean, R.K.	Photogrammetric Techniques in Geology
	Standardization of Technical Nomenclature
	Standards, Costs, and Procedures
Birdseye, C.W.	Safety Committee (Atlantic Region)
Buckmaster, J.L.	Shop Policy and Requirements
Cummins, T.V.	Security Office (Rocky Mountain Region)
	Safety Committee (Rocky Mountain Region)
	Rocky Mountain Survey Committee
	(When acting as Rocky Mountain Region Engineer)
Davey, C.H.	Map Appraisal Committee
	Publications Committee–Subcommittee for Recommendations for U.S. Base Maps
	Standardization of Technical Nomenclature (Vice Chairman)
	Standards, Costs, and Procedures
Davidson, J.I.	Security Committee (Special Maps Branch)
	Survey Building Proposed
Davis, R.O.	Rocky Mountain Survey Committee
	Missouri River Basin Field (Alternate)
Druhot, G.S.	Distribution Advisory
	Foreign Activities

Table 56.   Topographic Division personnel serving on committees.—Continued

Name	Affiliation
Echardt, C.V.	Subsidence Committee–San Joaquin Valley
Ecklund, C.A.	Pacific Coast Survey
Eller, R.C.	American Congress on Surveying and Mapping
Fennell, E.J.	Army Map Service, Liaison Officer
	Program Planning
	General Staff (Alternate)
	American Congress on Surveying and Mapping
Fischer, G.A.	Security Office–Alternate (Rocky Mountain Region)
FitzGerald, Gerald	Executive Committee
	U.S. Advisory Committee on Cartography of the Pan American Institute of Geography and History
	Cadastral Surveys–Topography–Coordination Committee (ad hoc)
Frye, H.M.	Special Committee
Fuechsel, C.F.	Negative Scribing (Interagency)
	Standardization of Technical Nomenclature
	Symbolization for Topographic Maps (Subcommittee of Interagency Committee on Map and Chart Symbols)
	Special Maps (Subcommittee of Interagency Committee on Map and Chart Symbols)
Galloway, C.T.	Arkansas, White, Red Basin Interagency (Mapping Work Group)
	Security Officer (Central Region)
Hall, R.D.	Central Region Editor, Topographic Division Bulletin
Harman, W.E.	Security (Aerial Photo Procurement)
	American Society of Photogrammetry
Harmel, M.A.	Visual Aids Subcommittee for Geological Survey in Denver
Higginson, W.S.	Safety Committee Photogrammetry Section (Atlantic Region)
Holmes, M.E.	Security Officer–Alternate (Central Region)
Isto, R.E.	Alaska Survey Committee
	Quarters Evaluation Board–Palmer, Alaska
Jones, A.B.	Security Officer–Alternate (Pacific Region)
Kalbfleisch, A.C.	Map Appraisal Secretary
Kellogg, L.D.	Security Officer–Alternate (Atlantic Region)
Kilmartin, J.O.	Map Appraisal
	Procurement Coordination (Department of State)
	Exhibits
Kimble, Delar	Special Committee
Korin, Stephen	Safety Committee (Atlantic Region)
Landen, David	Photogrammetric Techniques in Geology–Secretary
Law, J.A.	Safety Committee (Atlantic Region–Cartography Section)
Linck, M.K.	Safety Committee (Topographic Division)
Lyddan, R.H.	Map Appraisal Committee
	General Staff
	Standards, Costs, and Procedures
	Study Committee on American Society of Civil Engineers
McDonald, A.L.	Incentive Awards
	Security Committee (Alternate and Personnel)
McMillen, H.J.	Survey Building, Proposed
Maltby, C.S.	Map Appraisal Committee (Chairman)
	Meteorological Instruments Committee
Mann, F.M.	Budget Committee, G.S.
	Survey Building, Proposed (ad hoc)
Marsden, L.E.	Standardization of Technical Nomenclature (Secretary)
	Topographic Instructions

Table 56. Topographic Division personnel serving on committees.—Continued

Name	Affiliation
Monbeck, R.R.	Pacific Region Editor, Topographic Division Bulletin
Moore, L.C.	Symbolization for Topographic Maps
	(Subcommittee of Interagency Committee on Map and Chart Symbols)
	Special Maps (Subcommittee on Map and Chart Symbols)
	Interagency Committee on Negative Scribing (Alternate)
Mundine, J.E.	Alaska Survey Committee (Alternate)
Nelson, O.H.	Special Committee
Overstreet, W.B.	Special Committee
Peckinpaugh, C.L.	Technical Committee on Surveying Instruments of Federal Specification Board
Rutledge, D.H.	Security Office (Pacific Division)
Schlatter, E.J.	Reporting for Publication in Military Engineer
	Security (Aerial Photo Information and Sales)
	Editor-in-chief, Topographic Division Bulletin
Speert, J.L.	Standardization of Technical Nomenclature
Stevenson, R.G.	Pacific Southwest Field Committee–Subcommittee on Mapping (Chairman)
Striker, A.F.	Publications Committee (Alternate)
	Subcommittee for Recommendations for U.S. Base Maps
	Security (Map Security)–Alternate
	American Congress on Surveying and Mapping
Thompson, M.M.	American Congress on Surveying and Mapping
Thurston, R.F.	Map Appraisal Committee
	Program Analysis Committee
	Standards, Costs, and Procedures
	Standardization of Technical Nomenclature
Upton, W.B., Jr.	Rocky Mountain Region Editor, Topographic Division Bulletin
Van Camp, C.P.	Management Services Committee
	Performance Rating Review Committee
	Property Board
	Records Liaison Committee
	Survey Building, Proposed (ad hoc)
Watkins, J.L.	Safety Committee (Atlantic Region–Field Surveys Section)
Whitmore, G.D.	Study Committee on Standards, Costs, and Procedures
	Standardization of Technical Nomenclature (Chairman)
	Topographic Maps and Aerophotogrammetry of Commission on Cartography (PAIGH)
Wilson, R.M.	Lake Mead Sedimentation Survey
Witkege, F.L.	Security Officer (Topographic Division)
	Alaska Survey
Zens, W.F.	Building Wing Warden (First Aid)

Advisory Committee if one exists, and Interior Field Committees, are then weighed with the Federal Agency requests to identify the areas where lie the greatest concentration of map needs. "New starts" for topographic mapping are selected in these areas after special priority needs, such as the present defense mapping requirements, have been given due consideration.

Early in 1952, the Bureau of the Budget, with the assistance of the Topographic Division and other Federal map-making agencies, began a review of the procedures involved; when the new instructions came out as Circular A–16 on January 16, 1953, they included wording to the effect that the priority statements would "be utilized, along with the expressed needs of the States, as the basis for developing its (meaning the Geological Survey) annual mapping program designed to meet Nation-wide needs for mapping."

Now the State Legislature of Minnesota had, in 1949, created a Mapping Advisory Board to take over the duties of a similar group, appointed by the Governor a few years earlier to study map needs in connection with a proposed State cooperative mapping program. Other states had made studies of their map needs; Pennsylvania made a detailed study in 1945. Even earlier than this, California had begun a study, and several reports were published. Colorado established a numerical priority for every quadrangle in 1947. Undoubtedly, there are others; but the effort of the Minnesota Board represents the first attempt to coordinate State mapping needs for use in developing the Federal mapping program along the lines suggested in Bureau of the Budget Circular A–16.

With an active committee in Minnesota as an example, other states became interested, and in about 2 years some 14 states (Minnesota, Kansas, Washington, Nevada, Mississippi, Florida, Montana, New Mexico, Wyoming, Iowa, Maine, New Hampshire, Vermont, and Indiana), had organized committees of this kind. Eight of these have actually submitted priority expressions of their map needs to the Topographic Division.

The State Mapping Advisory Committees have a value to the state, to the Geological Survey, and to the general public, that goes far beyond the coordination of requests for mapping. They can, in effect, be clearing houses for information related to surveying and mapping, serve as distribution centers, and help state map users solve their mapping problems. Surely the maps would be used to good advantage by many times the number of people now supplied, if the general public could be better educated in the use, value, and appreciation of these maps.[48]

State Cooperation

Since fiscal year 1885, a considerable portion of the funds available to the Topographic Branch for mapping has come from States and Territories (or subdivisions thereof) through cooperative mapping projects (table 57). Such projects are now based on an equal sharing of the cost between the local agency and the Geological Survey, although other proportions have occasionally been used in the past.

The controlling factor in the Survey's decision to sign an agreement for cooperative mapping is whether or not the mapping requested is in line, both as to location and type of mapping, with the Survey's broad objective to complete, as soon as possible, the National Topographic Map Series covering the entire country. The areas to be surveyed, the scale, and the contour interval, usually are determined by mutual agreement

---

[48] G.S. Druhot, Topographic Division Bulletin, June 1954.

Table 57.   Cooperators with the Topographic Branch.

State	Date of cooperation (fiscal years)	Cooperating agencies
Alabama	1900–05; 1911; 1915; 1924–32	State Geologist
Arizona	1920–21;	State Bureau of Mines; Governor and State Land Department
	1926–27;	Governor and State Water Commission
	1954.	Colorado River Boundary Commission
Arkansas	1928–29; 1937–45; 1946–54	State Geologist
		State resources and Development Commission
California	1904–07;	State Board of Examiners
	1908–18; 1920–22;	State Department of Engineering
	1923–54;	State Department of Public Works
	1954;	Colorado River Boundary Commission
	1914–15;	University of California
	1915;	Los Angeles Board of Public Service Commissioners
	1922–23;	Klamath–Shasta Valley Irrigation District
	1925;	East Bay Municipal Utility District
	1923–38.	Los Angeles County Surveyor
Colorado	1920–21;	State Geologist
	1924–25;	Governor and State Engineer
	1926;	Governor and State Inspector of Oils
	1927;	State School of Mines
	1928–33; 1952–54;	State Metal Mining Fund Board
	1936–37;	City and County of Denver
	1938–40;	State Highway Department and Denver Board of Water Commissioners
	1941–47.	Denver Board of Water Commissioners
Connecticut	1890–91;	State Board of Commissioners
	1923–25;	State Highway Commission
	1928–29;	State Attorney General
	1949–54.	State Highway Department
Delaware	1926–27	State Highway Department

Table 57.   Cooperators with the Topographic Branch.—Continued

State	Date of cooperation (fiscal years)	Cooperating agencies
Georgia	1926–27;	State Geologist
	1952–54.	State Department of Mines, Mining, and Geology
Hawaii	1910–11;	Superintendent of Public Works
	1912–15	Governor and Commissioner of Public Lands
Idaho	1913;	State Engineer
	1919–23; 1926; 1930–31; 1934	State Bureau of Mines and Geology
Illinois	1906–18;	State Geological Commission
	1919–54.	State Department of Registration and Education
Indiana	1920;	State Geologist
	1938–54	State Conservation Department
Iowa	1908–17; 1919–31; 1950–54	State Geological Survey
Kansas	1938–54	State Geological Survey
Kentucky	1904–17;	State Geological Survey
	1919–20;	State Commissioner of Geology and Forestry
	1921–25; 1927–30;	State Geological Survey
	1949–50;	Louisville and Jefferson County Planning and Zoning Commission
	1950–54	State Agricultural and Industrial Development Board
Louisiana	1903–04;	Director of State Experiment Station
	1907;	State Geologist
	1910;	Fifth Louisiana Levee District Commissioners
	1923;	State Department of Conservation
	1933–41;	Board of State Engineers
	1942–54	State Department of Public Works
Maine	1900–10;	State Survey Commission
	1911–15; 1920–23;	State Water Power Commission
	1916–19; 1924–34;	State Public Utilities Commission
	1940–45; 1947–54	
Maryland	1897–1901; 1903–12;	State Geological Survey
	1946–49	State Department of geology, Mines, and Water Resources
Massachusetts	1885–87;	State Topographic Survey Commissioners
	1933–54	State Department of Public Works
Michigan	1902; 1904–21; 1926–28;	State Geologist
	1929–31; 1938–39;	State Department of Conservation and
	1932–37;	State Highway Commissioner
	1938–54	State Highway Department
Minnesota	1910–17;	State Drainage Commission
	1930;	State Department of Drainage and Waters
	1946–47;	State Geological Survey
	1950–54;	State Department of Conservation
	1950–54	Iron Range Resources and Rehabilitation
Mississippi	1902–03;	State Experiment Station
	1908–09;	Tallahatchie Drainage Commission
	1920–23.	State Geological Survey
Missouri	1908;	State Bureau of Geology and Mines
	1909–17; 1920–35;	State Geologist
	1936–54	State Highway Commission and State Geological Survey and Water Resources
Montana	1928–29;	State Engineer
	1931–33.	State Bureau of Mines and Geology
Nebraska	1912–14; 1916	Board of Regents, University of Nebraska
Nevada	1930; 1932–33; 1948–49; 1952–54	State Bureau of Mines
New Hampshire	1906;	Governor
	1925–33	Governor and State Highway Department
New Jersey	1885–87	State Geologist

Table 57.   Cooperators with the Topographic Branch.—Continued

State	Date of cooperation (fiscal years)	Cooperating agencies
New Mexico	1927–29;	State Engineer
	1928–29; 1931–32;	State Geologist
New York	1893–1917; 1920–27;	State Engineer and Surveyor
	1928–54;	State Department of Public Works
	1928;	Palisades Interstate Park Commission
	1931	Monroe County Regional Planning Board
North Carolina	1897; 1902–03;	Governor
	1906–07;	State Department of Agriculture
	1909–10;	State Geologist
	1946–47;	State Department of Conservation and Development
	1948–52	State Highway and Public Works Commission
North Dakota	1920–21; 1939;	State Geological Survey
	1926–30;	State Engineer
	1942–54	State Water Conservation Commission
Ohio	1902–16;	Governor
	1936;	Lucas County Commissioners
	1950–54	Mahoning County Engineer
Oklahoma	1906–08;	Territorial Survey Commission
	1912–15; 1928;	State Geological Survey
	1928–31;	State Highway Commission
	1948–49;	Division of Water Resources, State Planning Commission
	1952–54	Tulsa Metropolitan Area Planning Commission
Oregon	1906–17; 1922–33;	State Engineer
	1922	State Bureau of Mines and Geology
Pennsylvania	1900–15; 1919–23;	State Topographic and Geologic Survey
	1924–27;	Department of Forests and Waters
	1928–54	State Department of Internal Affairs
Puerto Rico	1937–50;	Commissioner of the Interior
	1915–54	Department of Public Works
Rhode Island	1889	State Commissioners
South Carolina	1933	State Relief Council
South Dakota	1922–23	State Geologist
Tennessee	1911; 1922–23; 1926–32;	State Geologist
	1925–26;	State Department of Highways and Public Works
	1950–54	State Department of Conservation
Texas	1902;	State of Texas
	1905;	University of Texas Mineral Survey
	1910;	State Levee and Drainage Board
	1915–17;	Harris County
	1920–24;	State Bureau of Economic Geology and Technology
	1920–31;	State Board of Water Engineers
	1925–32;	State Reclamation Engineer
	1929–30;	Cameron County
	1929–30;	Galveston County
	1953–54	City of Austin
Utah	1922–23;	Tooele County
	1922–26;	Utah County
	1922–28;	Salt Lake County
	1923–27;	Weber County
	1924;	Box Elder County
	1926–27;	Davis County
	1950–54	State Engineer
Vermont	1914–17;	Governor
	1919–54	State Geologist

Table 57.   Cooperators with the Topographic Branch.—Continued

State	Date of cooperation (fiscal years)	Cooperating agencies
Virginia	1909–18; 1920–40;	State Geologist
	1928–29;	State Conservation and Development Commission
	1941–48;	State Conservation Commission
	1949–54	State Department of Conservation and Development
Washington	1919–21;	State Board of Geological Survey
	1922–25; 1927–31; 1936–37;	State Department of Conservation and Development
	1939–54;	
	1948–50	City of Bremerton Water Department
West Virginia	1899; 1902–17; 1919–26;	State Geological Survey
	1928; 1930–31	State Geologist
Wisconsin	1915; 1917; 1919–31;	State Geologist
	1932; 1938–54;	State Highway Commission
	1945–1954;	State Geologist (Regents of Wisconsin University)
	1950–1954	State Conservation Department

between the Survey and the cooperator. In the case of long-term extensive operations, such as those now active in California and Kentucky, the selection of areas is determined to best satisfy the priority needs of the cooperator and allow efficient, well planned.operations. The current Puerto Rico agreement is unique in that it is the first to provide for systematic and continuous inspection and revision at a sufficiest rate to maintain the content of the maps at a state of practical completeness at all times. This is accomplished by a resident engineer.

Seventy-Fifth Anniversary

" * * * On March 3, 1879, 75 years ago today, the U.S. Geological Survey was officially established under the direction of Clarence King.

"A month ago, at the Waldorf-Astoria, the man who is Mr. King's current successor as Director of the Survey, Dr. William E. Wrather, was given, the John Fritz Medal, one of the top awards in American engineering circles. Since its foundation, the 'U.S.G.S.', as it is known to thousands of practicing prospectors and other thousands of map-minded amateurs, has moved from a somewhat academic agency for the collection and publication of accurate scientific data to an ultramodern center of up-to-the-minute facts on such fascinating subjects as volcanology, thermoluminescence and tectonophysics, meanwhile, plugging away at its perennial assignment of completing a uniform, large-scale map of the nation.

"The man who succeeded King, in 1881, was John Wesley Powell. Powell planned and personally led a party through the unexplored Grand Canyon in boats to observe the features of that river of mystery which terrified the Indians. The swashbuckling

days of Major Powell have given way to more prosaic types of reconnaissance. All the resources of scientific and human ingenuity have been drawn upon to secure engineering accuracy for topographic maps which you may purchase at insignificant prices. In the early days of the mapping project, distances were often reckoned by counting the paces of a horse, or computing the revolutions of a buggy wheel, and barometers were used to establish elevations where sighting equipment could not readily be utilized on account of the terrain. After 1896, permanent bench marks determined by spirit leveling were made the official control system for all mapping work. Aerial photography has become an important part of the Survey's activities since early experiments were tried in 1920. In 1925, legislation was passed to secure completion of the mapping of the country within 20 years, but appropriations have been allowed to remain at levels which have prevented achievement of this goal.

"The United States is one of the few nations in history that had ever had the obligation or the opportunity to explore and map from scratch the greater part of an entire continent, and the Geological Survey is still in a sense 'pioneering' as long as any part of that task remains unfinished. The magnificent work that has been done for 75 years by experts, from the days of King and Powell, right down to Dr. Wrather's present staff, is an impressive monument to skill and scholarship in the service of the national interest."[49]

---

[49] The New York Times, Wednesday, March 3, 1954.

U.S. Geological Survey 75th Anniversary Observance

The Washington Society of Engineers, in cooperation with the District of Columbia Council of Engineering and Architectural Societies, sponsored a meeting in the Interior Department Building auditorium in Washington, D.C. on April 21, 1954, commemorating the 75th anniversary of the Geological Survey. Present at the meeting were Douglas McKay, Secretary of the Interior; Felix Wormser, Assistant Secretary of the Interior, and W.C. Mendenhall, former Director of the Geological Survey (1931–43); each of whom were called upon for brief remarks.

The following program was presented by Director W.E. Wrather and his staff:

U.S. Geological Survey 1879–1954—75 Years
of Scientific Investigation

Dr. William E. Wrather, Director
U.S. Geological Survey

Geologic Investigations
Dr. Wilmot S. Bradley, Chief Geologist

Conservation Activities
Harold J. Duncan, Chief, Conservation Division

Water Resources Investigations
Carl G. Paulsen, Chief Hydraulic Engineer

Topographic Mapping
Gerald FitzGerald, Chief Topographic Engineer

Publications of the Geological Survey
Robert L. Moravetz, Chief, Publications Office

Topographic Mapping (1879–1954)

The areas mapped as of 1954 are listed in table 58, and the funds expended by fiscal year are listed in table 59.

Gerald FitzGerald, Chief Topographic Engineer

Topographic mapping has always been a major activity of the Geological Survey, which at the time it was established in 1879, was probably much more geographic than geologic. The Hayden, Wheeler, Powell, and King Surveys, which were in effect consolidated to form the new Survey under the Department of the Interior, were equipped and trained to conduct geographic exploration and to do topographic mapping. Clarence King, the first Director, after discussing functions of the new organization with members

Table 58.   Areas mapped in 1954.

State	Total area (square miles)	Area mapped at 1:62,500 or larger	
		Square miles	Percent of total
Alabama	51,078	18,563	36.3
Arizona	113,580	30,330	26.7
Arkansas	52,725	11,741	22.3
California	156,803	90,475	57.7
Colorado	103,967	20,036	19.3
Connecticut	4,899	4,223	86.3
Delaware	1,978	1,250	63.2
District of Columbia	61	61	100.0
Florida	54,262	20,854	38.4
Georgia	58,518	15,464	26.4
Idaho	82,808	10,351	12.6
Illinois	55,947	50,198	89.8
Indiana	36,205	11,695	32.3
Iowa	55,986	6,924	12.4
Kansas	82,113	10,949	13.3
Kentucky	40,109	31,165	77.7
Louisiana	45,177	19,672	43.5
Maine	31,040	25,344	81.6
Maryland	9,887	9,887	100.0
Massachusetts	7,907	7,907	100.0
Michigan	57,022	18,512	32.4
Minnesota	84,068	11,617	13.8
Mississippi	47,420	11,800	24.9
Missouri	69,270	50,918	73.5
Montana	146,316	18,474	12.6
Nebraska	76,653	21,085	27.5
Nevada	109,802	11,245	10.2
New Hampshire	9,024	7,924	87.8
New Jersey	7,522	5,556	73.8
New Mexico	121,511	26,615	21.9
New York	47,929	37,697	78.7
North Carolina	49,142	18,871	38.4
North Dakota	70,054	16,489	23.5
Ohio	41,122	37,988	92.2
Oklahoma	69,283	11,440	16.5
Oregon	96,350	12,336	12.8
Pennsylvania	45,045	34,442	76.5
Rhode Island	1,058	1,058	100.0
South Carolina	30,594	18,919	61.8
South Dakota	76,536	9,579	12.5
Tennessee	41,961	27,847	66.4
Texas	263,644	51,700	19.6
Utah	82,346	13,290	16.1
Vermont	9,278	7,062	76.2
Virginia	39,699	31,741	80.0
Washington	66,977	19,727	29.4
West Virginia	24,090	23,944	99.4
Wisconsin	54,715	13,094	23.9
Wyoming	97,506	17,579	18.0
Total	2,980,987	985,638	33.0

Table 59.   Funds expended annually in fiscal years.

[--, no funds expended]

State	1954		Totals through 1954		
	Federal	State	Federal	State	Federal and State
Alabama	--	--	$109,502	$93,897	$203,399
Arizona	$17,754	$17,754	75,530	74,833	150,363
Arkansas	1,314	3,777	52,787	52,529	105,316
California	317,754	317,754	3,103,410	3,310,754	6,414,164
Colorado	4,896	5,918	156,864	158,172	315,036
Connecticut	16,632	21,205	189,233	189,556	378,789
Delaware	--	--	9,808	15,486	25,294
Florida	--	--	--	--	--
Georgia	7,129	7,086	21,292	21,235	42,527
Hawaii	--	--	266,398	331,174	597,572
Idaho	--	--	31,574	18,544	50,118
Illinois	18,724	28,250	1,407,643	1,387,942	2,795,585
Indiana	48,396	50,000	658,713	654,323	1,313,036
Iowa	19,795	10,000	96,761	92,902	189,663
Kansas	16,371	14,000	170,811	167,816	338,627
Kentucky	689,290	689,290	3,283,623	3,270,895	6,554,518
Louisiana	33,480	30,000	639,282	647,048	1,286,330
Maine	28,759	20,000	496,687	504,985	1,003,672
Maryland	--	--	103,776	103,755	207,531
Massachusetts	11,614	9,341	600,636	572,467	1,173,103
Michigan	30,328	20,000	658,637	642,856	1,301,493
Minnesota	121,900	90,000	503,543	479,900	983,443
Mississippi	--	--	59,777	72,428	132,205
Missouri	56,431	49,282	1,221,389	1,218,750	2,440,139
Montana	--	--	8,066	25,929	33,995
Nebraska	--	--	6,971	7,250	14,221
Nevada	15,643	16,222	61,902	61,760	123,662
New Hampshire	--	--	150,204	150,419	300,623
New Jersey	--	--	35,074	19,671	54,745
New Mexico	--	--	33,113	30,579	63,692
New York	52,869	46,119	946,953	978,295	1,925,248
North Carolina	--	--	215,668	204,406	420,074
North Dakota	32,226	17,500	244,666	232,915	477,581
Ohio	1,998	1,997	339,826	437,387	777,213
Oklahoma	4,962	5,706	83,338	79,413	162,751
Oregon	--	--	140,311	133,776	274,087
Pennsylvania	5,473	9,372	824,090	827,992	1,652,082
Puerto Rico	12,471	5,691	339,774	417,528	757,302
Rhode Island	--	--	4,866	4,866	9,732
South Carolina	--	--	1,917	787	2,704
South Dakota	--	--	9,822	6,994	16,816
Tennessee	80,611	80,612	494,855	493,585	988,440
Texas	4,254	4,255	436,296	529,682	965,978
Utah	5,923	6,000	70,646	68,334	138,980
Vermont	4,548	4,548	115,507	113,734	229,241
Virginia	32,714	32,714	732,503	725,005	1,457,508
Washington	10,522	10,522	326,922	337,490	664,412
West Virginia	--	--	273,322	296,626	569,948
Wisconsin	41,509	35,000	626,725	621,905	1,248,330
Wyoming	--	--	--	--	--
Total	1,746,290	1,659,915	20,443,013	20,888,575	41,331,288

of Congress concluded in part "that the intention of Congress was to begin a rigid scientific classification of the lands of the national domain for the general information of the people of the country and to produce a series of land maps which would show all those features upon which the intelligent agriculturists, mining engineers, and timber men might hereafter base their operations and which obviously would be of the highest value to all students of the political economy and resources of the United States." Topographic mapping, therefore, was made a part of the work of the Survey and funds were allotted each year from 1879 to 1888 for mapping surveys. Thereafter, up to, and including the present year, Congress has made annual appropriations to the Geological Survey specifically for topographic surveys.

John Wesley Powell, the second Director of the Survey, often was criticized both in and out of Congress for spending such a large proportion of appropriated funds on topographic mapping. In spite of this criticism, Major Powell was far-sighted enough to propose to Congress that a 20-year mapping program be authorized and financed in order to provide a sound framework for scientific study and national resource development.

During the first 20 years of topographic mapping under the new Survey, most of the work was published in quadrangle form at a scale of 1:250,000 for 1-degree sheets, and 1:125,000 for 30-minute sheets. Gradually the scale of much of the mapping, both in eastern and western United States, was increased to 1:62,500 for 15-minute quadrangles.

Early work was carried on with rather crude methods, such as the tape and compass traverse and the use of the aneroid barometer for elevations. The improvement of the plane-table, the development of the telescopic alidade with vertical-angle arcs, greatly increased the accuracy of early topographic surveys. In 1896, Congress authorized the Geological Survey to execute spirit leveling and to set permanent bench marks for the control of mapping.

Because most of the skilled topographers in the United States were members of the Geological Survey during World War I, a large number were commissioned for duty with the Topographic Corps of the Army or as artillery orientation officers; many served overseas.

Aerial photography used in World War I principally for intelligence purposes created great interest in Survey engineers on duty with the Army. After the War, "photogrammetry" became the keynote for a revolutionary advance in the techniques of surveying and mapping, and ceased to be of interest only to the long-haired scientist in the laboratory. It was accepted as a full fledged partner with "field surveys" in modern topographic mapping. Since the late 1930's, practically all standard topographic maps produced by the Geological Survey have utilized aerial photography in one way or another.

In 1925, Congress passed the Temple Act, authorizing a 20-year mapping program. Although this act was strongly supported by engineering organizations, as well as other influential groups, funds were not appropriated by Congress to carry it out.

While the years between the two world wars saw many changes in techniques, the actual accomplishment in new mapping was disappointingly slow, especially to the engineers who had worked so hard for the passage of the Temple Act; we were, in fact, losing ground each year. The 45 percent considered adequately mapped at the end World War I became 20 percent when we reinventoried the situation a short time after the attack on Pearl Harbor.

World War II had a far-reaching effect on topographic mapping in the United States. Production of maps of both domestic and foreign areas was greatly increased to meet military needs and millions of people became "map conscious" as a result of the complex geographical problems involved in global warfare.

Beginning in 1946, there has been a marked increase in the amount of money made available to the Geological Survey for topographic mapping. This, of course, has been accompanied by a corresponding growth in our organization. Trained personnel of the Topographic Division has increased from less than 600 in 1945, to more than 2,000 at the present time, while during the same period, we have stepped up our mapping from 9,000 to 90,000 square miles a year. The four Region offices of the Division have been furnished with the most modern map-making equipment obtainable.

During the first 60 years of the Survey's existence, about 5,000 topographic quadrangles were produced, and during the past 15 years an additional 8,000 have been published. The present rate is about 1,500 quadrangles a year.

Under our current program, new mapping, or revision, is being carried on in every State of the Union, as well as in Alaska, the Hawaiian Islands, Puerto Rico, and the Virgin Islands. The National Topographic Map Series consists of 7.5-minute quadrangles published at a scale of 1:24,000; 15-minute quadrangles published at a scale of 1:62,500, and reconnaissance maps published at a scale of 1:250,000.

The Special Maps Branch of the Division, organized before Pearl Harbor for military mapping, continued to add hundreds of thousands of miles of new work each year to the astounding area covered during World War II for the production of aeronautical charts.

While we have made substantial progress, especially in the last few years in mapping, we realize that we still have a long way to go to complete the National Topographic Map Series. Although the methods and techniques we use today are producing maps faster and better and cheaper than ever before, we know improvement is not only possible, but probable. New developments in procedures and new instruments now being added to the production line will ensure faster progress toward our goal of providing and maintaining a complete series of modern quadrangle maps of the United States and its possessions.

We are proud of our Topographic Division and grateful for the support it has received these first 75 years, which we know were the hardest. We especially are proud of the cooperation between our Survey and other Federal and State agencies that work side by side with us to get this mapping job done.

The following articles were submitted by the Topographic Division as contributions to the Science Magazine in connection with the 75th Anniversary of the Survey:

"The Helicopter and the Walkie-Talkie in Field Surveys"

The helicopter and the portable radio-telephone have become partners in triangulation surveys for mapping projects in mountainous areas of Alaska and the western states. The helicopter provides quick and easy transportation for the triangulators, and the radio makes it possible to coordinate the activities of the large field crews employed.

A typical project, extending over 3,000 square miles, or more, of desert mountains, can be triangulated in one season by seven or eight engineers with helicopter and radios. The helicopters operate a shuttle service from a base camp or from roads, landing the men with their instruments on mountain peaks, and moving them from peak to peak as required. Each triangulation observing party carries a portable radio-telephone (Walkie-Talkie), and the operation as a whole is directed from a master transmitter mounted at the base camp or in a Jeep. The men can talk to each other and to the helicopters over line-of-sight distances through the base radio. If necessary, a group conference can be held with each participant on a separate mountain peak.

Both horizontal positions and elevations are determined by triangulation, using optical-reading theodolites. This kind of surveying frequently requires measuring vertical angles reciprocally and simultaneously between two stations to avoid errors from atmospheric refraction. To carry out this operation, the triangulators use a "skirt" of fluorescent cloth around the instrument tripod as a signal. Radio contacts between observers make it possible for them to relocate stations quickly when the line of sight is blocked by trees, or visibility is otherwise impaired. By these techniques, elevations have been extended as far as 50 miles across rugged mountainous terrain with an accuracy of about 2 feet.

Helicopters are operated in pairs, usually, so a means of rescue will be close at hand in case of accident and as insurance against prolonged delays from mechanical failure. Engineers working with helicopters are spared the arduous mountain climbing ordinarily involved in triangulation, but the work still has interesting aspects. Helicopters are not as maneuverable at higher altitudes as they are near sea level, and landing on high peaks where cross-winds are always blowing is a critical job, even for a skilled pilot. Taking off calls for still greater skill and sometimes a little luck. Pilots prefer a peak with a sheer drop-off where they can dive immediately after taking off to gain flying speed. There have been no serious accidents in 4 or 5 years of operation, but it is still far from a routine means of transportation.

The value of helicopters and radio-telephone has been effectively demonstrated during 5 years of use on surveys in Alaska, and the Geological Survey expects that the combination will do a great deal to accelerate the mapping of large areas in Alaska and in the western United States.

"Scribing"

A new technique of drafting with special tools on coated plastic, known as scribing, is rapidly supplanting pen-and-ink drafting (fig. 23) in the final stages of map production. An adaption and refinement of negative engraving used in photolithography, the new method produces a more legible map with neater and sharper line work, in a shorter time and at less cost.

In the older method, the printers copy, the drawings from which printing plates are made is the product of skilled freehand inking. The map detail is reproduced photographically on metal-mounted paper, and traced in ink by the draftsman. A separate drawing is required for each color, the line weight and registration must be very precise, and the appearance of the printed map depends largely on the talent and patience of the draftsman. Because of the shortage of competent draftsmen, the operation frequently is a bottleneck in the production line.

With scribing (fig. 24), more dependence is placed on mechanical aids; therefore, the result is more uniform. A clear plastic base sheet, to which the map copy is photographically transferred, is first coated with an opaque material, usually yellow paint. Using a metal stylus ground to the width of line desired, the draftsman traces the copy, cutting out the opaque coating along the lines that are to be printed. The product is a line negative that the photo-engraver can use directly to make the printing plate.

The principal advantage of scribing is that the width of a line is determined entirely by the scribing tool used, in contrast to pen-and-ink drafting, when line weight depends on the size of the pen point, the fluid qualities of the ink, the surface paper, and on the pressure on the pen. The map scriber has few critical factors to control, and he can devote his full attention to productive work. The training period for new employees is relatively short, and their output is superior both in quantity and quality to pen-and-ink drafting.

For the "gadgeteer", scribing has opened up a new field of activity; the design of special tools for the various symbols and types of lines on topographic maps. The usual tool is made from a phonograph needle ground to a chisel-shaped point and held in an ordinary penholder, but many others have been developed for particular purposes. A swivel head scriber cuts two parallel lines at the same time to trace double-line roads, a special templet is used to cut rectangular building symbols, and there is even a motor-driven scriber to produce small circles representing oil or water tanks. There is still an unfilled need for a device to scribe dotted lines conveniently and accurately.

The most critical problems in the introduction of scribing techniques concerned the materials—the base sheets and the opaque coatings. The base sheet must be transparent, dimensionally stable, and have a smooth surface hard enough to prevent scratching by the scribing tool. Glass is the ideal material as far as these qualities are concerned, but it is fragile and heavy, and the problems of transportation and storage are formidable. After considerable experimenting, a type of vinyl plastic sheeting was selected as the most satisfactory, although there is still room for improvement.

"Convergent Aerial Photography"

Convergent photography (fig. 25), used in conjunction with the new map-plotting instruments under development by the Geological Survey, is expected to bring about big improvements in the accuracy and economy of maps prepared from aerial photographs. The essential geometry of convergent photography is illustrated in the diagrams. Photographs are taken simultaneously in pairs by two cameras rigidly coupled so the optical axes lie in a common plane and form an angle of 20 degrees with the plumb line and 40 degrees with each other. Stereoscopic models are made by combining the forward oblique photograph at one camera station with the rear oblique of the succeeding station.

One advantage of convergent photography over the commonly used vertical pictures is the greater area covered by each pair of photographs, the ground coverage being more than twice as great for pictures taken at the same altitude. Another great advantage is the stronger intersection angle formed by conjugate image rays on the stereoscopic model. That is, the angle of intersection at any ground point of the image rays to the two different camera stations will ordinarily be close to 60 degrees, resulting in elevation determinations almost twice as accurate as those made with vertical photographs at the same altitude, which normally have an intersection angle of about 30 degrees.

Thus, ground elevations and contours as seen in the stereoscopic model formed by the photographs may be determined much more accurately from convergent photographs than from conventional verticals taken at the same altitude; or, more important, for any required elevation accuracy or contour interval, the convergent photographs can be taken at a much higher altitude, resulting in lower costs for mapping operations. For example, if the camera altitude is increased so that the accuracy is the same as that ordinarily obtained with vertical photography, the stereoscopic models from the convergent photographs will cover about six times as much ground area, with a corresponding saving in cost of field surveys for control points, reduced cost of orienting the fewer models in the photogrammetric laboratory, and so on.

Photogrammetrists have known for many years about the theoretical possibility of convergent photography, but practical use could not be made of it without suitable map-plotting instruments. In addition to a new map-plotting

instrument of special design, that also permits photogrammetric triangulation with the convergent photography, a projector is required capable of transmitting the long image rays to the corners of the large convergent model. The newly developed photogrammetric triangulation instrument for handling convergent photography is known as the Twinplex, a prototype model of which has already undergone successful operational tests.

The companion photogrammetric projector, also newly developed and tested in operational trials, utilizes an ellipsoidal reflector instead of the conventional condensing lens system. The production trials have been quite satisfactory, and the Geological Survey has already contracted for several hundred square miles of convergent aerial photography.

"The Special Maps Branch"

The Special Maps Branch, organized prior to Pearl Harbor for the purpose of charting Alaskan air lanes for the Air Force, occupies a unique position in the Topographic Division of the Survey. With the outbreak of hostilities, the program of this branch was expanded to include all inadequately mapped areas over which planes would be ferried to combat areas. It has, since the war's end, continued to compile small-scale air navigation charts. This work is neither humdrum nor routine. The source material, maps, control, and photographs available for the compilation of maps and charts in many parts of the world often are not of the type that would be requested by an organization efficiently planning a new mapping operation.

The Branch was instrumental in the development of trimetrogon mapping, a system that utilizes three photographs having ground coverage from horizon to horizon, and taken simultaneously from the same air camera station. Advantages lay mainly in the extensive photographic coverage made available from each airplane flight. The system was designed to permit the use of a large number of operators of limited experience, and simple and easily procurable instruments. The operations are analogous to a factory production line wherein each person performs a small unit of work contributing to the finished product.

Since the end of World War II, numerous changes in the Branch's operations have been made. Installation of varied stereoplotting equipment has enabled it to exploit the advantages of different types of aerial photography, and to broaden the scope of the Branch's mapping capabilities.

Mapping assignments, presenting challenging photogrammetric problems, are processed by the use of a combination of the various instruments. Each photogrammetric instrument is assigned that phase at which it is most proficient. Successful completion of a project often requires the introduction of novel photogrammetric techniques designed to overcome the limitations of the given material. The procedures also are governed by the accuracy requirements and funds available for the job.

Included in the Branch's activities are operational research problems. The compilation of limited areas serve as tests to determine the feasibility of suggested new methods and techniques. Operational research continually is directed toward the establishment of methods that will result in a better map product at a lower cost, as well as special maps to meet deadline dates.

At present, the Branch is compiling topographic maps and aeronautical charts ranging in scale from 1:8,000 to 1:1,000,000. Since its organization, the Branch has compiled approximately 9,000,000 square miles of maps and charts. This included areas of considerable civilian and military interest, such as North Africa, northern Alaska, Labrador, and Alcan Highway, the Dakar–Khartoum Air Ferry Route, and areas within the Arctic and Antarctic Circles. Whereas the major part of its program is confined to the preparation of maps and charts for the U.S. Air Force, the Special Maps Branch compiles many maps urgently needed for strategic mineral investigation.

"The National Topographic Atlas"

The Geological Survey has been at work since 1879, mapping the interior of our country. It has responsibility for completing the National Topographic Atlas, turning out maps at a scale of 1 inch to the mile for most of the country, and at a larger scale for thickly settled areas, or where greater detail is needed for special purposes. It is altogether proper that this function should be a Federal responsibility. It is a task requiring centralized direction, precise accuracy, standardization, and authority, and ability to shift work from one part of the country to another, as national requirements may dictate.

During the 75 years since the Geological Survey was established by Congress, topographic maps have been published covering about one-half of the Nation. Many of these are now obsolete. Much more detail and higher standards of accuracy are required. Only two states are now considered adequately mapped—Massachusetts and Rhode Island. Several others like Connecticut, New York, California, and Illinois, however, are not far behind. By 1956, Kentucky will join the top-notchers when its 5-million dollar, 5-year cooperative program is finished.

Largely as a result of World War II, this Nation has never been so map conscious as it is now. In 1940, 704,345 topographic maps were sold by the Geological Survey. Today the figure approaches 3,000,000 annually.

In the early 30's, Delaware, Connecticut, Maryland, New Jersey, New York, Ohio, West Virginia, the District of Columbia, and the Hawaiian Islands, were completely covered by maps considered adequate at that time. These maps already have become obsolete. Fortunately, we can be sure that this will not happen to maps we are making today, for they are being turned out in greater detail and at a larger scale. We also are much better equipped today than we were earlier, not only to estimate future needs, but to keep our modern maps up to

date. Gerald FitzGerald, Chief Topographic Engineer, said that it will take almost 30 years to complete the job at our present rate of progress.

"Rate of Production of Topographic Maps"

Before 1948, Geological Survey topographers struggled to approach a goal of 300 maps a year. In the fiscal year 1952, they produced 897. In 1953, they turned out 1,452. And by the end of fiscal year 1954, it is estimated that more than 1,900 will have come off the presses.

Since two Geological Survey topographers, Sumner H. Bodfish and John H. Renshawe, first started mapping in Arizona on July 16, 1879, more than 15,000 topographic maps have been turned out by that agency. Today, work on approximately 7,000 new or revised quadrangles is in progress. These will be published as time and funds permit. Altogether, the American public already has access to a stock of more than 25,000,000 sheets.

Our long range goal is to complete the National Topographic Atlas and to produce more maps and better maps, faster, and at lower cost. At the same time, it is essential to conduct an educational program to let the public know the wealth of information provided by maps that are already available.[50]

Summary of Current (1952) Trends In Topographic Mapping Procedures[51]

Some of the changes underway, in the offing, or proposed for trial, are outlined below:

Aerial Photography

Ten new KC–1 cameras (formerly called T–12) owned by the Survey, are to be delivered in 1954. No direct reduction in cost of photography is anticipated, but, of greater importance, the quality of photography, which has a direct effect on compilation and field completion costs, should be improved considerably.

Convergent photography is being procured in 1954, enough to keep the Twinplex bridging instruments (one in each Region to be delivered this year) continuously occupied with bridging work. Here again, there may be no direct reduction in the cost of procurement of photography, but later operations, especially supplemental control, should cost less.

---

[50] From talk on "Surveys and Maps Prepare the Way for National Growth," by Secretary McKay, presented at the 14th Annual Meeting of the American Congress on Surveying and Mapping, Washington, D.C., March 22–24, 1954, and published in Surveying and Mapping Journal, April–June 1954.

[51] George D. Whitmore, Chief, Reserch and Technical Control Branch, from notes prepared by Section chiefs for staff meeting, published in Topographic Division Bulletin, March 1954.

Camera calibrating will be better, more frequent, and quicker, because of a radically different method of camera calibration by use of our new multicollimator calibrating instrument. This should mean fewer of the costly troubles that have resulted from contractor's cameras getting out of calibration.

Experiments in Regions will continue as to value of cross flights in aerotriangulation, and lower-altitude flights for planimetry, along with regular high-altitude flights.

Experiments are already underway to see what advantages might result from using the new high- altitude (scale of 1:70,000 plus or minus) photography being procured by AMS for 1:250,000 compilations. This may prove particularly useful for 15-minute sheets with 40- or 60-foot contours.

Basic Horizontal and Vertical Control

Best chance for getting basic control costs down lies in doing less of it, and/or adopting different accuracy standards. A study is underway in RT–1 as to advantages vs. disadvantages cost wise, and otherwise, in changing accuracy standards of basic control, to say plus or minus 10 feet in horizontal position, and plus or minus 0.5 foot or 1.0 foot in elevations.

If we adopt the 100-foot tolerance for well defined points on the 15-minute maps, instead of the present 40-foot tolerance, we probably could do with longer horizontal bridges, and make greater use of control-saving techniques, such as stereotemplets for aerotriangulation of project areas. Thus, a direct result of such change in specifications for 15-minute maps might be less new traverse and triangulation.

Supplemental Vertical Control (Fourth-Order)

We may be able to reduce greatly the number of V points as a result of the larger-area models of convergent photography and high-altitude verticals, and/or by doing vertical bridging over one or two intervening models, in the case of vertical photography. (The possibilities of vertical bridging with the stereoplanigraph, and other heavy plotters, will soon be thoroughly explored in a carefully planned research experiment).

The photoalidade, with specially-flown low-altitude oblique photographs, may be increasingly used for determining picture-point elevation of 40-foot and 80-foot interval maps, in an effort to reduce the amount of field surveys for supplemental vertical control in areas of difficult access.

The fourth-order classification, now a catch all, may soon be subdivided into fourth and fifth orders, if proposals now being considered by several Government agencies are approved.

Stereocompilation

Costs should start coming down somewhat, because:

- The more efficient ER–55 reflector-type projectors (fig. 26) are scheduled to replace, as rapidly as feasible, the B and L condenser-lens projectors on Multiplex bars.

- Increasingly greater use will be made of convergent photography and higher altitude vertical photography.

- Better diapositives, with clearer imagery and less distortion, will be available as soon as the new-type printers (now on order) are put to use.

- The new variable-ratio precise pantography will probably ultimately be used on all double-projection instruments, which should facilitate inter-changeability of compilation between different type instruments, and permit use of a smaller-scale on original compilation, with a full quadrangle on one sheet.

Field Completion

It is hoped that the amount of work required for field completion will be reduced as a result of a more complete and better stereocompilation. This is to be expected because of better photography, more helpful advance field interpretation by supplemental control parties, and more complete stereo-contouring due to convenient inter-changeability (because of variable-ratio pantographs) of compilation among all types of stereoplotters.

Map Finishing

The elimination of several indirect operations are feasible under the scribing process. E.g., if the original compilation is scribed on scale-stable material, there is no need for a film mosaicking operation for the original color separation negative. If the original operation is scribed at reproduction scale, which may be feasible for 1:24,000 maps, there is no need for a copy camera at this stage. Also, as the scribed color separation sheets themselves are negatives, there is no need for camera work in preparing the composite prints for editing. Later, during reproduction, there is no need for preparing wet-plate camera negatives, as press plates can be made directly from the scribed sheets (assuming scribing is done at reproduction scale) with some chance of better press plates.

Also, if the photogrammetrists can successfully scribe contours on a separate sheet, as is hoped, the color-separation scribers may be able to salvage the greater part of this work. Thus, the cost of the color separation work for the brown plate may be reduced ultimately by a considerable amount.

Tribute to Survey Wives

In closing, a tribute is extended to the Survey wives for their sympathetic understanding and forbearance.

"How Soon can you Pack?"[52]

When I became the blushing bride of a civil engineer,
How little did I think my life to others might seem queer,
    To those who settle down for life with fixings that are new,
    A bed, a lamp, a dinette set, to mention but a few.
    Who polish up their wedding gifts, and shine up pots and kettles,
    And hang up drapes of gay cretonne, and chase the dust that settles.
They hope to do this always in their own artistic home,
And look with pity on their friends whose lot it is to roam.
How we whose lives have taken us to many different places,
Have met folks north, south, east, and west, and even alien faces.
    We've lived in nipa shacks and huts, apartments and in flats,
    We've answered ads for "lovely rooms" and found them full of rats.
    We've taken rooms in "mansions"—that's the landlord's wily phrasing,
    (We've learned from much experience to do our own appraising.)
We've eaten hearty dinners on a rolling ocean liner,
We've eaten just as hearty in Al's Kozy Korner Diner.

We've camped in pleasant valleys, and we've scaled the mountain peaks,
And we've sun-tanned by the ocean, and fished in sparkling creeks.
    We've rustled food from markets, and from a grocery shelf,
    And lived on farms where food that's fresh seemed just to grow itself,
    We've gathered luscious fruit from trees, in Texas or Cebu,
    And picnicked on a glacier when the sun shone at Taku.
In fact, what most folks think of as an elegant vacation,
Is standard stuff for map makers, in spite of transportation.

The lucky wives can go along; and for those who like to roam,
Its a challenge to see what it takes, to make a place called home.
    You do not need a davenport, or whatnot—just a broom—
    You can get the spirit nicely in a little furnished room.
    You can learn to be a gypsy who loves to pioneer,
    You do not have to settle down when mapping's your career.
For it's "Here, my love, and there, my love, and how soon can you pack?"
For it's "Here, my love, and there, my love, across the world and back!"

—N.C.B.

Selected Biographies

Marcus Baker

Marcus Baker was born at Kalamazoo, Mich. He spent 2 years in Kalamazoo College and two at the University of Michigan, from which he was graduated in 1870. The degree of LL. B. was received from Columbia University in 1896.

In the summer after graduation, he assisted Professor Watson, of Ann Arbor, in computations for the Nautical Almanac. Then, for a year, he held the chair of mathematics in Albion College, and for 2 years was instructor in mathematics at the University of Michigan. For 13 years, he was a member of the U.S. Coast and Geodetic Survey. He assisted Dr. W.H. Dall in surveys of the coast of Alaska, having for his special function the astronomical determination of latitude and longitude.

He resigned, and on April 15, 1888, he accepted a position as geographer with the U.S. Geological Survey. For a number of years, he had charge of the Northeastern Topographic Division, supervising the mapping of Massachusetts, Rhode Island, Connecticut, and a part of Pennsylvania. He was afterwards an editor of topographic maps. For more than 10 years, Mr. Baker was secretary of the Board on Geographic Names. He also was editor of its bulletins. His Geographic Dictionary of Alaska was published in 1902 as Geological Survey Bulletin No. 187.

When a commission was appointed by our Government to investigate the matter of the Venezuelan boundary, Mr. Baker was employed as a geographic expert, taking leave of absence from the Geological Survey for that purpose.

At the time of his death on September 12, 1903, he was filling the position of Assistant Secretary of the recently established Carnegie Institution of Washington. "Yearbook No. 2, 1903" of that institution stated that Mr. Baker was a scholarly man, of broad culture, and talented in many fields; he was a conscientious, painstaking, accurate man, doing thoroughly and well that which he undertook; and he was an optimistic and affable man, who delighted to be of service to others. His activities were so varied and his responsibilities so numerous that the Carnegie Institution is but one of several organizations to suffer by his loss. He was a man of science, occupied chiefly with geographic and bibliographic researches, but a contributor also to history, and a lifelong student of mathematics.[53]

Edward C. Barnard

Edward Chester Barnard, a New Yorker, graduated in 1884 from Columbia University with a degree in mechanical engineering. On July 23, 1884, he started work with the Geological Survey as an assistant topographer, and mapped sections of four states and of the three Pacific Coast States. He had charge of a field party sent to Alaska to map the Forty-Mile District in 1898, and the Nome District in 1900. He was chief topographer of the United States–Canadian Boundary Survey in 1903, 1904, and 1905, locating the boundary along the 49th parallel west of the summit of the Rocky Mountains. On June 4, 1907, he was appointed Geographer-in-Charge of the Rocky Mountain Division, with summer field headquarters

---

at Denver, Colo. Of medium size, he was extremely agile, a fast walker, and was seldom seen without a cigar. On June 10, 1910, he was detailed to the State Department for Canadian Boundary work and he resigned May 2, 1915. He died February 6, 1921.

William M. Beaman

William M. Beaman, a native of Annapolis, Md., and a graduate of Massachusetts Institute of Technology, was appointed to the Survey as topographic aid in 1889. He was promoted through the topographer grades to topographic engineer and made topographic surveys in many States. He invented the Beaman Arc, which was used on telescopic alidades. From 1907 to 1917, he was an inspector of field topography. During World War I, he served as a major in the Army Engineer Corps, and was engaged in mapping the country for military purposes.

In 1920, Mr. Beaman was designated Topographic Engineer-in-Charge of the Section of Inspection and Editing of Topographic Maps, and he held this position until his death on March 1, 1937.

In 1922–23, he was furloughed for an assignment in Brazil in connection with a Government map exhibit. Mr. Beaman also was author of the chapter on topographic mapping in Survey Bulletin No. 788, "Topographic Instructions."

Claude H. Birdseye

Claude Hale Birdseye was born in Syracuse, N.Y. His family moved to Oberlin, Ohio, and in 1901, he received a bachelor of arts degree from Oberlin College. He was an instructor in physics at the University of Cincinnati in the spring of 1901, and later in that year he began work with the Geological Survey as a levelman with a topographic field party. That fall, he enrolled at Ohio State University for post graduate study in engineering, and in June 1902, again joined a field party of the Survey. On October 19, 1904, he was appointed topographic aid and continued through the various topographic grades.

Mr. Birdseye's experience and skill in topographic mapping was gained by hard work of actual practice in mountains and forests. His early assignments were not easy ones, but his natural resourcefulness and ability became so apparent that he was frequently chosen for special assignments of a kind that gave scope to his venturesome spirit. He was in charge of the topographic mapping in Hawaii, from 1909 to 1912, a task that involved the mapping of the active volcano, Kilauea, with its lake of fire. In the following year (1915), he supervised the mapping of Mount Rainier, Wash. On the summit of that mountain, 14,408 feet above sea level, his party was caught in a blizzard and was forced to seek precarious refuge in a steam vent. In 1916, he was appointed Geographer-in-Charge of all topographic mapping in the Rocky Mountain Division.

His World War I record was outstanding. He began as a captain of engineers and, about a month later, was promoted to the rank of major. He sailed for France on August 16, 1917, for duty as orienteur officer with the 1st Brigade of Coast Artillery, and a year later was promoted to the rank of lieutenant colonel and assigned to the Office of the Chief of Coast Artillery, American Expeditionary Forces, in charge of artillery ranging and map information. These rapid promotions, and the fact that instruction manuals written by him were published by the War Department, show that his abilities were productive and appreciated. His achievements were further acknowledged in 1919 when the French Orders of University Palms, grade of Officier de l'Instruction Publique, was conferred upon him.

He returned to the Geological Survey in June 1919, and was designated Acting Chief Geographer during the Chief Geographer's absence from the office. He was appointed Chief Topographic Engineer on October 1, 1919, and in this new and responsible position he directed all field and office work of making topographic maps. The art of mapping was entering a critical period, for the use of airplanes and aerial photography, which had developed so rapidly during the War, was beginning to alter the technique of mapping through the introduction of photogrammetric methods.

Colonel Birdseye's adaptability, experience, and initiative qualified him to understand the value of these methods, and to guide the Geological Survey into its modern mapping practices. His studies and administrative direction in the uses of aerial photography soon made him one of the foremost photogrammetrists in the United States. His character counted for as much as his ability; he was so unfailingly considerate and understanding that he won the wholehearted loyalty of associates and inspired in them a high degree of cooperation. It is fortunate indeed that so able a leader was in charge of the Topographic Branch during this period of transition.

In 1923, Colonel Birdseye's untamed spirit of adventure again broke forth, and he took time from his administrative duties to lead an expedition by boat down the rough waters of the Colorado River through the Grand Canyon. The reconnaissance surveys made by the expedition were highly useful in choosing the site for Hoover Dam. For this, and other contributions to American geography, he was awarded the Charles P. Daly Medal by the American Geographical Society on April 22, 1924.

Continuing as Chief Topographic Engineer, he found time during his efficient administration to compile, revise, and enlarge the "Topographic Instructions," published in 1928 as Geological Survey Bulletin No. 788, and prepared "Formulas and Tables for the Construction of Polyconic Projections," which were published as Geological Survey Bulletin No. 809 in 1929. Colonel Birdseye also wrote many shorter articles that were published in various technical periodicals. He took an active part in many scientific organizations and was never too busy to help anyone.

In September 1929, he resigned from the Geological Survey to become president of the Aerotopograph Corporation of America, and directed the detailed photogrammetric surveys of the site for Hoover Dam.

Colonel Birdseye was reinstated in the Geological Survey on February 1, 1932, as Engineering Assistant to the Director, and on August 1, 1932, was appointed Chief of the Division of Engraving and Printing, where he was in charge of the reproduction of the maps published by the Survey. He continued in this work until his death on May 30, 1941.

Colonel Birdseye should be credited with other achievements of an even higher order than those he gave to science directly, achievements of a kind that will perpetuate his memory in the hearts of all who knew him. By friendly counsel and assistance, he was able to inspire many with enthusiasm for greater attainments; thus did his spirit generously pervade the lives of his associates.

Sumner H. Bodfish

Sumner Homer Bodfish was born August 16, 1844, at Chicopee, Mass. He attended the public schools of New Haven, Conn., and in the spring of 1863, he enlisted in the 46th Massachusetts Infantry, in which he served 9 months. He then reenlisted for the remainder of the war in the 2nd Massachusetts Heavy Artillery, and saw active service in the Battle of Kingston, N.C., and in the vicinity of New Berne. After his regiment returned to Massachusetts, he saw further service in the Boston draft riots. At this time, President Lincoln gave him an appointment to West Point. He graduated in the class of 1868, was commissioned a second lieutenant in the 6th Cavalry, and ordered to Fort Tyler, Tex. As a result of his high standing on graduation, and the excellence of his conduct as an officer, he was made regimental adjutant, though a junior officer. Later on, he was made provost marshall.

In January, 1871, he resigned from the Army to engage in hydraulic engineering work with his father at Langley, Ga.

In the spring of 1872, he went to Washington, D.C., where he was shortly afterwards appointed first assistant engineer of the Board of Public Works. In the fall of 1878, Mr. Bodfish was appointed a topographer on the Powell Survey, and his first work was measuring angles, in a scheme of triangulation planned by G.K. Gilbert, which connected the expansion between the Gunnison and Kanab bases.

On July 10, 1879, Mr. Bodfish was appointed a topographer with the U.S. Geological Survey and was assigned to the Division of the Colorado under Capt. C.E. Dutton. He was given charge of a field party in northern Arizona, in order to make topographic surveys of a portion of the Grand Canyon of the Colorado, and a great number of canyons subsidiary thereto. The Grand Canyon district was a labyrinth of deep gorges; through its center flowed the Colorado, more than 6,000 feet below the general level of the country, and the tributary canyons were profoundly carved, resulting in a district of deep gorges and towering cliffs. This survey was successfully completed.

During the winter months, Mr. Bodfish and assistants were engaged in the construction of the maps, materials for which had been collected during the field season. At the close of the fiscal year these maps were completed. In 1889, Mr. Bodfish proceeded to the Kaibab Plateau, Utah, for the purpose of continuing the topographic work essential to the completion of the special map of that plateau, and of the Grand Canyon, and also to establish monuments and make the observations required to connect the salient points of the plateaus with the primary triangulation of this survey. Captain Dutton inspected Mr. Bodfish's previous work and reported:

> "I am very glad to be able to report that it is exceedingly successful, and that Mr. Bodfish is, in my opinion, entitled to high praise and warm congratulation for the exceedingly able and accurate manner in which he has performed one of the most difficult pieces of detail topography which can be encountered in the West. He will continue his work this season in the southern and southeastern portions of the plateau front, in those localities which are of greatest interest and moment to the geologist."

In 1882, he was engaged in detailed topographic work in the quicksilver mining region of California. The following six seasons were spent, until the fall of 1888, on map work at a scale of 1 mile to 1 inch, and 20-foot contours, in the neighborhood of Washington, D.C., and in various portions of Massachusetts.

Shortly after the creation of the Irrigation Survey, Mr. Bodfish was appointed Irrigation Engineer in the spring of 1889, and spent that season and the early portion of 1890, until the Irrigation Survey was discontinued by Congress, in charge of the Colorado Division. In the course of this work, he made general surveys for the discovery of reservoir sites and to ascertain their capacities, and the modes by which water could be conducted from them to the surrounding arid lands. He also made detailed surveys of the 11 reservoir sites in the upper valley of the Arkansas, these surveys being at very large scales, from 200 to 400 feet to 1 inch, and 2-foot and 4-foot contours. The winter of 1889 was occupied in computing the cost of the dams and the capacities of the reservoirs, and in making detailed plans for their construction.

Upon the discontinuance of the Irrigation Survey, Mr. Bodfish resigned and returned to private engineering practice, being occupied in the fall of 1890 in mapping for the War Department, the fortifications on the Potomac River. His last work was a careful survey of the Great Falls of the Potomac for B.F. Butler's Boston company, and a careful determination of the quantity and value of the water power that could be derived from these falls. In October, 1893, he became an invalid and died on May 17, 1894. He was buried at Arlington, the funeral being conducted by the Loyal Legion, of which he was an honored member.

Mr. Bodfish was always a careful, painstaking, and conscientious worker who was at his best in designing and conducting careful surveys or engineering works, which called for care and thoroughness in the execution of their details.[54]

George R. Davis

George Robert Davis, a native of Riverside, Calif., was a high school graduate with later studies in a business college and engineering school. He joined the Geological Survey as recorder in October 1898, was appointed assistant topographer on November 20, 1903, and promoted through the various grades to Geographer-in-Charge of the Pacific Division on July 1, 1912. Included in his work of topographic surveying was the mapping of the Mount Whitney, Mount Goddard, Bakersfield, and McKittrick Quadrangles, in California, as well as mapping in the Yosemite National Park, and Kings River Canyon, Calif., Mount Rainier National Park, Washington, and many other areas. In 1915, he was a delegate to the International Engineering Congress in San Francisco, and in 1920, a delegate to the Panama–Pacific Scientific Congress in Honolulu. In 1916, he married Miss Adelena Marie Fontaine, who was Mr. Marshall's very efficient chief clerk. They had a daughter named Anna. He died March 13, 1922.

Edward M. Douglas

Edward Moorehouse Douglas, a native of Saratoga, N.Y., graduated in civil engineering from Columbia University in 1881, and then joined the firm of C.L. Berger & Sons, makers of surveying instruments, at Troy, N.Y.

He was appointed topographer with the U.S. Geological Survey in 1882, and conducted topographic and geodetic surveys in nearly every state in the West. On July 1, 1897, he was appointed Geographer-in-Charge of the Rocky Mountain Division, and on July 1, 1903, Geographer-in-Charge of the Division of Western Topography. In 1911, he became Chief of the Computing Division and served in this capacity until his transfer to the War Department on September 1, 1925, which service continued until his retirement on August 31, 1930, after 48 years of Government service. He joined the Cosmos Club on March 7, 1887.

Mr. Douglas's special accomplishments also covered boundary surveys, among these being the survey and marking of the Idaho–Washington boundary line, and the survey and marking of the boundary of the Loquillo Forest Reserve in Puerto Rico. He was the author of many Government reports; the most widely used being Survey Bulletin No. 817, "Boundaries, Areas, Geographic Centers, and Altitudes of the United States and the Several States." He died July 1, 1932.

Conrad A. Ecklund

When H.H. Hodgeson retired voluntarily on November 30, 1942, Conrad A. Ecklund was appointed Pacific Division Engineer. Mr. Ecklund, a native of Illinois, and graduate of the Armour Institute of Technology, Chicago, with a bachelor of science degree in civil engineering, was appointed a Junior Topographer with the Geological Survey on June 3, 1909. He worked in the Rocky Mountain Division until 1916 when he was detailed to the Pacific Division. During World War I, he was a 2d Lt., U.S. Engineers, from January 1, 1918, to June 27, 1919, with overseas duty. He again performed topographic mapping in the Pacific and Rocky Mountain Divisions, and from 1930 to 1931, he was in charge of the Rocky Mountain Section. In 1932, he was transferred to the Pacific Division and placed in charge of the South Pacific Section. From 1933 to 1938, he was supervisor in charge of the Los Angeles County Cooperative Project, California, and since December 1, 1942, has continued in charge of the Pacific Division.

Gerald FitzGerald

Gerald FitzGerald was born in Burns, Oreg., and was educated in public and private schools in Washington and Oregon, including Seattle College and the University of Washington. He joined a topographic mapping party of the Geological Survey on August 23, 1917. He worked field seasons of 1918 and 1919 in Washington, Oregon, and California; accompanied a Geological Survey party to Santo Domingo, West Indies, in 1919, for a year's work and, upon his return, looked to Alaska as an outlet for his enthusiasm for the great outdoors, lore of adventure, and his great capacity for gruelling physical labor. For 22 seasons, he mapped areas in Alaska, chiefly remote Alaska, north of the Yukon towards the Arctic Ocean, where it was necessary to start in February with dog sleds while the country was snow covered and frozen, and establish caches of food to be found later after the thaw, when the topographic and geologic mapping could be done. Then, transportation was by pack train, canoe, raft, and by foot.

In 1938, Mr. FitzGerald, recognized as a specialist in reconnaissance mapping, was placed in charge of topographic surveys in Alaska. In this assignment, he organized and assisted in the original development of "Trimetrogon mapping," a cooperative program undertaken by the Geological Survey and the Army Air Forces to provide worldwide map coverage for military operations. In 1942, Mr. FitzGerald was commissioned in the Air Corps and assigned to the Aeronautical Chart Service, which he commanded with the rank of colonel, from 1943 to 1945. Due to his sound plans and brilliant supervision, the development of the Trimetrogon method of aerial photographic compilation on worldwide basis was universally recognized as the outstanding contribution to reconnaissance mapping and charting during World War II. In recognition of his work, he received the Photogrammetric Award of 1944 for "outstanding achievement in photogrammetry." He

---

[54] Memoir prepared by H.M. Wilson, Chief Geographer, and published in proceedings of the American Society of Civil Engineers, v. XX, 1894, p. 96.

was designated a delegate of the United States, by the Department of State, to the Second Pan American Consultation on Cartography at Rio de Janeiro, Brazil, August 14 to September 2, 1944. In 1945, he was awarded the Legion of Merit for his exceptionally meritorious conduct as Chief of the Aeronautical Chart Service.

At the request of the Secretary of the Interior, Mr. FitzGerald was relieved from active military duty on November 1, 1945, to return to the Geological Survey as Staff Topographic Engineer. In 1946, he was elected President of the Society of Photogrammetry. On May 22, 1947, he was appointed Chief of the Topographic Division which, under his direction, has continued its reputation of being one of the best equipped and best staffed mapping organizations in the world. Mr. FitzGerald was designated Chief of the American Delegation to the International Mapping Conference in London, England, in August 1947. This also was known as the Conference of Commonwealth Survey Officers. He was selected by the Director of the Geological Survey for a lecturer for the American Association of Petroleum Geologists, and he gave a series of talks before Geological Societies and University groups throughout the United States from October 4, to November 2, 1948. In recognition of his outstanding leadership in directing mapping operations for the Federal government in time of war and peace, Mr. FitzGerald received the Distinguished Service Award of the Department of the Interior in 1949, while still in office.

Henry Gannett

Henry Gannett, topographer with the Hayden Surveys, had been appointed to the Geological Survey on October 8, 1879, and immediately transferred to be the geographer of the Tenth Census. His Census work completed, the Director appointed Mr. Gannett as Chief Geographer-in-Charge of topographic mapping on July 1, 1882.

Mr. Gannett was born in Maine on August 24, 1846, received a bachelor of science degree at the Lawrence Scientific School of Harvard in 1869, and earned his degree as mining engineer at Hooper Mining School in 1870. In 1871, he accompanied a Harvard scientific expedition to Spain to observe the total eclipse of the sun. Upon his return, he declined an offer to join a North Pole expedition as astronomer and, instead, joined the Hayden Survey as topographer for the western division. Gannett was trained in topographic mapping at Cambridge by J.D. Whitney and Mr. Charles F. Hoffman, who influenced him to leave Wheeler and go to Hayden.[55] He wrote a section of Dr. Hayden's annual reports beginning with the Seventh (1873) to the final one (1878). He was a pioneer explorer, possessed great endurance, and was one of the first to ascend Mount Whitney in California. He was honored by having the highest peak in Wyoming's Wind River Range named for him.

He was a voluminous writer on geographical, statistical, and interrelated subjects, and did much to enrich the geographical literature of the world. In addition to many texts used in official publications, he was the author of "Manual of Topographic Surveying" (1893); Statistical Atlases, 10th and 11th Censuses; Dictionary of Altitudes (3d edition, 1899); Commercial Geography (1895); The Building of a Nation (1895); Gazetteer of Cuba (1902); Gazetteer of Texas (1902); Origin of Certain Place Names in the United States (1902); and List of the Mountains of the United States (1910–12).

As Chief Geographer of the Geological Survey and leading Geographer of the United States, Mr. Gannett was foremost in many geographical activities. As geographer of the Tenth, Eleventh, and Twelfth Censuses, he developed to a high degree of effectiveness in the graphic method of presenting statistical inquiries. He participated as assistant director in the censuses of Cuba and Puerto Rico in 1899, of the Philippines in 1902, and of Cuba in 1906.

U.S. Board of Geographic Names was another illustration of Mr. Gannett's skillful adaptation of the science of geography to the purposes of government. Originally an unofficial organization brought together by Mr. Gannett and Dr. Thomas C. Mendenhall, Superintendent of the U.S. Coast and Geodetic Survey, for the purpose of eliminating confusion and contradiction in geographic names constantly appearing in publications, the Board officially was established by an executive order issued by President Harrison on September 4, 1890. Mr. Gannett was chairman of the Board from 1894 to his death in 1914.

Mr. Gannett's participation in the forestry and conservation movements included two important assignments.[56] When in 1897, the Geological Survey, by act of Congress, was given the task of examining and surveying the forest reserves, which President Cleveland had set aside earlier in the year, Mr. Gannett was given charge of the examination work. In 1908, when President Theodore Roosevelt's Conservation Commission undertook to assemble authoritative data concerning the Nation's resources, Mr. Gannett was selected to edit the report.

Though actually in charge of the Topographic Branch, which did not attain the status of a Branch until 1889, Mr. Gannett's board activities and the adjustments to varying appropriations during the years 1882 to 1914, required his absence from Washington and changes in his title. He was Chief Geographer from July 1, 1882, to September 9, 1885; geologist-in-charge from September 10, 1885, to June 30, 1889; Chief Topographer from July 1, 1889, to April 19, 1896; and Geographer from July 1, 1896, to November 5, 1914.

At the time of his death on November 5, 1914, Mr. Gannett was co-founder and president of the Twenty-year Service Topographers; co-founder and president of the National Geographic Society; and co-founder and past president of the Cosmos Club. He was often called "the father of American Map-

---

[55] This is according to letters received from Hayden.

[56] "Henry Gannett, President of the National Geographic Society," by Simon Newton North, 1910–14. The National Geographic Society, 1915.

making" and will long be remembered for his unselfishness and devotion to his chosen work.

Publications of Henry Gannett

"A Dictionary of Altitudes in the United States" was compiled by Henry Gannett, and was published in 1884, as Geological Survey Bulletin No. 5. A second edition was published in 1891, as Geological Survey Bulletin No. 76; a third edition in 1899, as Geological Survey Bulletin No. 160; and a fourth edition in 1906, as Geological Survey Bulletin No. 274.

"'Boundaries of the United States and of the Several States and Territories, with a Historical Sketch of the Territorial Changes," by Henry Gannett, was published in 1885 as Geological Survey Bulletin No. 13. A second edition was published in 1900, as Geological Survey Bulletin No. 171; and a third edition in 1904, as Geological Survey Bulletin No. 226.

Mr. Gannett's letter of transmittal, dated October 16, 1884, states:

"I have the honor to submit herewith a sketch of the boundaries of the United States, the several States, and the Territories as defined by treaty, charter, or statute.

"Besides giving the present status of these boundaries, I have endeavored to present an outline of the history of all important changes of territory, with the laws appertaining thereto.

"This matter was in great part prepared under the direction of the Superintendent of the Census, and it is herewith presented for publication with his full concurrence.

"I have been greatly assisted in this work by Mr. Franklin G. Butterfield, who was formerly connected with the Census Office, by whose labors most of the material relating to the boundaries of the States upon the Atlantic borders has been compiled."

Some excerpts from "Boundaries of the United States and of the Several States and Territories, with a Historical Sketch of the Territorial Changes," by Henry Gannett are given herewith:

The Public Domain and an Outline of the History of Changes Made Therein

"Cessions by the States"

At the time the Constitution was adopted by the Original Thirteen States, many of them possessed unoccupied territory, in some cases entirely detached and lying west of the Appalachian Mountains. Thus, Georgia included the territory from its present eastern limits westward to the Mississippi River. North Carolina possessed a narrow strip extending from latitude 35 degrees to 36 degrees, 30 minutes, approximately, and running

westward to the Mississippi, including besides its own present area, that of the present state of Tennessee. In like manner, Virginia possessed what is now Kentucky, while a number of states, such as Pennsylvania, New York, Massachusetts, and Connecticut, laid claim to areas in what was afterward known as the Territory Northwest of the River Ohio, a region which is now comprised mainly in the States of Ohio, Indiana, Illinois, Michigan, and Wisconsin. These claims were to a greater or less extent conflicting. In some cases several states claimed authority over the same area, while the boundary lines were in most cases very ill-defined.

The ownership of these western lands by individual states was opposed by those states that did not share in their possession, mainly on the grounds that the resources of the General Government, to which all contributed, should not be taxed for the protection and development of this region, while its advantages would inure to the benefit of but a favored few. On these grounds, several of the states refused to ratify the Constitution until this matter had been settled by the cession of these tracts to the General Government.

Moved by these arguments, as well as by the consideration of the conflicting character of the claims, which must inevitably lead to trouble among the states, Congress passed, on October 30, 1779, the following act:

"Whereas the appropriation of the vacant lands by the several states during the present war will, in the opinion of Congress, be attended with great mischiefs: Therefore,

Resolved, "That it be earnestly recommended to the state of Virginia to reconsider their late act of assembly for opening their land office; and that it be recommended to the said State, and all other States similarly circumstanced, to forbear settling or issuing warrants for unappropriated lands, or granting the same during the continuance of the present war."

This resolution was transmitted to the different states. The first to respond to it by the transfer of her territory to the General Government was New York, whose example was followed by the other states.

"Washington Territory"

This was organized March 2, 1853, from a part of Oregon Territory. Its limits, as originally constituted, were as given in the following clause from the act of Congress creating it:

That from and after the passage of this act, all that portion of Oregon Territory lying and being south of the 49th degree of north latitude, and north of the middle of the main channel of the Columbia River, from its mouth to where the 46th degree of north latitude crosses said river, near Fort Walla Walla; thence with said 46th degree of latitude to the summit of the Rocky Mountains, be organized into and constitute a temporary government by the name of the Territory of Washington. (32d Cong., 2d sess.)

In 1859, on the formation of the state of Oregon, the residue of the Territory of Oregon, being the portion lying east of the present limits of the State, extending thence to the crest of the Rocky Mountains, was added to Washington. This area, with the part of Washington lying east of its present limits, was included in Idaho on the formation of that Territory in 1863.

The present boundaries of Washington Territory are as follows: Beginning on the coast at the mouth of the Columbia River; following up the main channel of the Columbia River to its point of intersection with the 46th parallel of latitude; thence east on the 46th parallel to the Snake River; thence down the main channel of the Snake River to the mouth of the Clearwater; thence north on the meridian that passes through the mouth of the Clearwater to the boundary line between the United States and the British possessions; thence west with that boundary line to the Pacific * * *.

"A Manual of Topographic Methods"

"A Manual of Topographic Methods," 300 pages, by Henry Gannett, Chief Topographer, was published by the Geological Survey in 1893, as Monograph XXII. Revised edition was published in 1906 as Geological Survey Bulletin No. 307.

In his letter of transmittal to Director Powell, Mr. Gannett states:

"I have the honor to submit herewith for publication a manual of the topographic methods in use by the Geological Survey, accompanied by a collection of constants and tables used in the reduction of astronomical observations for position, of triangulation, of height measurements, and other operations connected with the making of topographic maps. It must be understood that the methods are not fixed, but are subject to change and development, and that this manual describes the stage of development reached at present.

"In the preparation of this work, I have to acknowledge the aid of many of my associates, notably Mr. H.M. Wilson and Mr. S.S. Gannett. To Mr. R.S. Woodward, now connected with the U.S. Coast and Geodetic Survey, I am indebted for the "Instructions for the Measurement of Horizontal Angles" in Chapter III. These instructions, which were drawn up by Mr. Woodward several years ago for the guidance of field parties engaged in primary triangulation, have resulted in a great increase in accuracy and considerable economy of time and labor. To Messrs. G.K. Gilbert and W.J. McGee, I am indebted for their kindly criticism, especially concerning the chapter upon the "Origin of Topographic Features.

"Some of the tables have been prepared in this office; others have been compiled from various sources, notably from appendices to reports of the U.S. Coast and Geodetic Survey and 'Lee's Tables and Formulae.'"

Interesting excerpts from this Manual are as follows:

"Plan of the Map of the United States"

The field upon which the Geological Survey is at work is diversified. It comprises broad plains, some of which are densely covered with forests, while upon others trees are entirely absent. It contains high and rugged mountains, plateaus, and low, rolling hills. In some regions its topographic forms are upon a grand scale, while in others the entire surface is made up of an infinity of minute detail. Some parts of the country are densely populated, as much so as almost any region upon the surface of the globe, while great areas in other parts of the country are almost without settlement. Geologically, portions of the country are extremely complex, requiring, for the elucidation of geologic problems, maps in great detail, while other areas are simple in the extreme.

It is obvious that from this diversity of conditions, both natural and material, maps of different areas should differ in scale, and that with the difference in natural conditions and the difference in scale there must come differences in the methods of work employed. The system which is found to work to advantage in the high mountain regions of the west is more or less inapplicable to the low forested plains of the Mississippi Valley and the Atlantic Plain.

"Scale"

The scales that finally have been adopted for the publication of the map are 1:62,500, or very nearly 1 mile to 1 inch, and 1:125,000, or very nearly 2 miles to 1 inch.

When this work was commenced in 1882, three different scales were used for different parts of the country, depending upon the degree of complexity of the topography and the geological phenomena, upon the density of population, and the importance of the region from an industrial point of view. These scales were 1:62,500, 1:125,000, and 1:250,000. The maps, as fast as produced, have found extended use, not only among geologists, but in all sorts of industrial enterprises with which the surface of the ground is concerned, and already have become well nigh indispensable in the projection of railroads, water works, drainage works, systems of irrigation, and other similar industrial enterprises. Their extended use has developed a requirement for better maps; i.e., maps upon a larger scale and in greater detail. At one stage of its development this requirement was met by discontinuing all mapping upon the scale of 1:250,000, which it was recognized at that time was inadequate to the requirements. Since then, the standard of the requirements has continued to rise and, consequently, the plan of the map has been enlarged by the extension of the areas mapped upon the scale of 1:62,500, and a corresponding reduction of the areas to be mapped upon the scale of 1:125,000. Meantime, however, large areas have been mapped upon the discarded scale, and the maps have been published and widely distributed. Such areas will be remapped for the larger scales only as special needs may arise * * *.

"Contour Interval"

The relief of the earth's surface is now represented upon maps almost entirely by contour lines or lines of equal elevation. Until a comparatively recent date this element, secondary in importance only to the horizontal element, or the plan, has been ignored.

The contour intervals that have been adopted for the map of the United States are as follows:

For the scale of 1:62,500, the intervals range from 5 to 50 feet; for the scale of 1:125,000, 10 to 100 feet; and, for the scale of 1:250,000, the interval is 200 or 250 feet.

"Size of Sheets"

Atlas sheets are designed to be approximately of the same size, 17.5 inches in length by from 12 to 15 in breadth, depending upon the latitude, and all those of the same scale cover equal areas, expressed in units of latitude and longitude; that is, each sheet upon the 4-mile scale covers 1 degree of latitude by 1 degree of longitude; each sheet upon the 2-mile scale, 30 minutes of latitude and longitude; and each sheet upon the 1-mile scale, 15 minutes of latitude and longitude. The sheets are thus small enough to be conveniently handled, and, if bound, form an atlas of convenient size. From the fact that each sheet is either a full degree or a regular integral part of a degree, its position with relation to the adjacent sheets and to the area of the country is easy to discover.

"Geometric Control"

From the constructive point of view, a map is a sketch, corrected by locations. The work of making locations is geometric, that of sketching is artistic. This definition is applicable to all maps, whatever their quality or character. However, numerous the locations may be, they form no part of the map itself, but serve only to correct the sketch, while the sketch supplies all the material of the map. The correctness of the map depends upon four elements: first, the accuracy of location; second, the number of locations per square inch of the map; third, their distribution; and, fourth, the quality of the sketching.

It is in connection with the first of these elements that it seems desirable to define what constitutes accuracy. The greatest accuracy attainable is not always desirable, because it is not economical. The highest economy is in properly subordinating means of ends, and it is not economical to execute triangulation of geodetic refinement for the control of maps upon small scales. The quality of the work should be such as to ensure against errors of sufficient magnitude to appear upon the scale of publication.

While the tendency of errors in triangulation is to balance one another, still they are liable to accumulate and this liability must be guarded against by maintaining a somewhat higher degree of accuracy than would be required for the location of any one point. It is not difficult to meet this first condition of accuracy of the maps. The maximum allowable error of location may be set at one-hundredth of an inch upon the scale publication. This admits of an error upon the ground not greater on a scale of 1:62,500, than 50 feet * * *.

The education of the topographer, therefore, consists of two parts, the mathematical and the artistic. The first may be acquired largely from books, and this book knowledge must be supplemented by practice in the field. The second, if not inherited, can be acquired only by long experience in the field, and by many can be acquired only imperfectly. In fact, the sketching makes the map, and, therefore, the sketching upon the Geological Survey is executed by the best topographer in the party, usually its chief, whenever it is practicable to do so.

"Classification of Work"

The work involved in making a map usually comprises several operations, which may in practice be more or less distinct from one another. They are enumerated as follows:

First.—The location of the map upon the earth's surface, by means of astronomic observations.

Second.—The horizontal location of points.

This usually is of three grades of accuracy; primary triangulation, or primary traverse, in cases where triangulation is not feasible; secondary triangulation for the location of numerous points within the primary triangulation; and ordinary traverse, for the location of details.

Third.—The measurement of heights, which usually accompanies the horizontal location, and which may, similarly, be divided into three classes, in accordance with the degree of accuracy.

Fourth.—The sketching of the map.

Nearly all of the geometric work of the Survey, the work of location, is executed by five instruments.

- Theodolites, of powerful and compact form, used in the primary control.

- Planetable, with telescopic alidades of the best type, used for secondary triangulation and height measurements.

- Planetable, of crude, simple form, with ruler alidades, used for traversing and minor triangulation.

- Odometers, for measuring distance.

- Aneroids, for the measurement of details of heights.

With these instruments, nine-tenths of the work is done, and these instruments will be described in their proper places with much fullness of detail as seems necessary * * *.

As an aid in the interpretation of the various topographic forms that present themselves, the following brief discussion is appended.

"Topographic features originate from a variety of causes and are modified by many agencies. They are formed by the uplift from beneath, of great or small extent. They are formed by deposition from volcanoes, glaciers, water, and the atmosphere. They are formed or modified by aqueous and ice erosion. They are modified by gravity.

"These are the principal agencies in producing topographic forms as we see them today. These forms are only in rare cases the work of a single one of the above agencies; generally two or more have taken part in producing the present condition. Of all these, aqueous agencies are by far the most potent. Their work is seen in nearly all topographic forms, while, in those of great age, their action has been so extensive as to mask or obliterate all superficial traces of the action of any other agency * * *."

In 1929, the Map Information Office of the Federal Board of Surveys and Maps, J.H. Wheat, Secretary, published a pamphlet called "The Mountains of the United States," and the first page is as follows:

"The following pages give descriptions of the principal mountain ranges of the United States and indicate relationships between them. This list was compiled from the most reliable maps obtainable, preference being given to the Geological Survey quadrangle maps.

"The compilation was made in 1910–12 by Henry Gannett, formerly geographer of the Geological Survey, and chairman of the U.S. Geographic Board, and probably, at that time, the most competent person in the United States to prepare such a list.

"No changes of importance have been made in the original manuscript except to add an index."

Tribute to Henry Gannett

At a meeting at the office of the Geological Survey on Monday afternoon, November 9, 1914, to express sympathy at the death of Henry Gannett, some of his old associates, F.W. Clarke, Arthur P. Davis, Morris Bien, George H. Ashley, and Alfred H. Brooks, spoke of Mr. Gannett's personal character, and of his contributions to science, and the following resolutions were adopted:

"We have met today to express our sorrow at the death of Henry Gannett, the first Chief Geographer of the U.S. Geological Survey, and the leading geographer of the United States, and to pay tribute to his character and his work. When Mr. Gannett, in 1882, took charge of the work of the Survey's topographic branch, he established the methods and fixed the standards that have since been followed in the great task of making the maps for a Government atlas of the United States, a task that will for many years continue to employ a large part of the energies of the Geological Survey. The general recognition of the high quality of the Survey's maps is itself a testimonial of appreciation of Mr. Gannett's work. These maps are in large part the products of field methods devised by him. He first used the method of sketching traverse work on the board graphically in the field, and he established the present method of primary traverse. His "Manual of Topographic Methods" has been published in two forms by the Survey, and is still a standard book of reference.

"Mr. Gannett was a geographer in no narrow sense of the word; his interest in geography included such diverse subjects as forestry, rainfall, the profiles of rivers, the origin of the names of places, and the historical detail of our national geographic expansion, and of the determination of the areas and boundaries of the States, on all of which subjects he prepared reports that were published by the Survey.

"His government work outside of the Survey, chiefly for the Census of the United States, the censuses of Cuba, Puerto Rico, and the Philippines, and his services as chairman of the U.S. Geographic Board, and as Geographer of the National Conservation Commission, covered a wide field and showed large administrative power.

"In Mr. Gannett's death, the Survey has lost a man of broad view, wise initiative, and great service, and the science of geography has lost a master spirit. We tender to his family our heartfelt sympathy in their bereavement, and feel that they may find consolation in the knowledge that his work for mankind was faithfully and well done.[57]

Thomas G. Gerdine

Thomas Golding Gerdine took over Mr. Marshall's duties at Sacramento, Calif., though the effective date of his appointment as Geographer-in-Charge of the Pacific Division was 6 months later, January 23, 1908. He was born at West Point, Miss., in 1872, and graduated from the University of Georgia in 1891, with the degree of bachelor of engineering. On July 1, 1893, Mr. Gerdine joined the U.S. Geological

---

[57] "Henry Gannett, President of the National Geographic Society," by Simon Newton North, 1910–14, The National Geographic Society, 1915.

Survey as recorder in a topographic field party, working on the Santa Monica, Calif., Quadrangle. On May 3, 1894, he was appointed assistant topographer.

He was transferred to the General Land Office on April 22, 1895, and served for 2 years as Examiner of Surveys, acting as Federal representative in the examination and approval of land surveys under the old contract system. On July 12, 1897, he returned to the Geological Survey, served as party chief in the extreme western part of the United States and in Alaska, and worked on difficult topographic and geodetic survey projects. From May 1, 1899, to January 17, 1908, he was assigned to the Alaskan Division, and throughout his entire Alaskan service he showed not only marked professional ability, but all those other qualities, physical and personal, that are demanded of leaders of parties engaged in fieldwork under difficult and unusual pioneer conditions. He was personally popular for his extremely good nature and resounding laugh.

Mr. Gerdine was Geographer-in-Charge of the Pacific Division from January 23, 1908, to June 30, 1912, when he became Geographer-in-Charge of the Northwestern Division, consisting of the States of Idaho, Oregon, and Washington. His headquarters were still at Sacramento, Calif., but his summer field office was at Portland, Oreg. He was commissioned a major in the Engineer Officers Reserve Corps on September 16, 1917, and was assigned to administration and supervision of military mapping, first in New Mexico and Texas, and later along the Atlantic Coast of Virginia, and at some of the eastern artillery training camps. He was discharged on March 31, 1919, and reinstated as Geographer-in-Charge of the Northwestern Division. From 1912 to 1921, he was Geographer-in-Charge of the Northwestern Division; 1918–19, also in charge of the Atlantic Division; 1918–21, also in charge of the Rocky Mountain Division; and from 1922, to his death on October 31, 1930, Topographic Engineer-in-Charge of the Pacific Division.

Richard U. Goode

Richard Urquhart Goode was born in Bedford, Va., on December 8, 1858. He was educated at Hanover Academy, Norfolk, Va., and attended the University of Virginia for several terms. In 1877 to 1878, he was an assistant in the Engineer Corps of the Army. On July 12, 1879, he received an appointment as topographer with the U.S. Geological Survey. He was assigned to the Division of the Colorado, and conducted topographic surveys, under the direction of Sumner H. Bodfish, in Utah and Arizona. In 1880, he was given charge of the topographic work, under Capt. C.E. Dutton, in connection with surveys of the plateau region in the neighborhood of the Grand Canyon of the Colorado, in Arizona. In 1881, he reported to Professor J.H. Gore, in charge of the measurement of a base line from which to extend primary triangulation near Fort Wingate N. Mex., on which work he was engaged until his resignation on May 1, 1882. From 1882 to 1884, Mr. Goode was employed as topographer on the Northern

Transcontinental Survey, under Professor Raphael Pumpelly and A.D. Wilson, chief topographer, and conducted extensive surveys in Montana, Washington, and elsewhere in the Northwest.

Mr. Goode returned to the Geological Survey on July 23, 1884. He was placed in supervisory charge of a group of parties engaged in surveys in Missouri and Kansas, and he continued on this work until 1886, when he was given charge of the section of topographic surveys in Texas. This assignment continued until 1889, except that he was granted leave of absence in the spring of 1888, to assist the engineers of the Panama Canal Company in important topographic and land surveys covering property rights on the Isthmus of Darien. In April, 1889, he was given charge of the Southern Central Division of Topography and was promoted to geographer. In September, 1890, on the reorganization of the Topographic Branch into two principal divisions, eastern and western, he was transferred to the western branch, under the general direction of Professor A.H. Thompson, and placed in charge of the Kansas–Texas Division, which assignment he retained until August, 1894, when he was transferred to the more important charge of the Pacific Section. His management of that important work, to which were later added surveys in Alaska and forest-reserve boundary surveys, gave eminent satisfaction during the next 7 years. Early in June, 1903, the four topographic sections were consolidated into two, and Mr. Goode was placed in charge of the Western Section, including all the states west of the Mississippi Valley, approximately all of the United States west of the 100th meridian.

The large amount of administrative work that developed upon him as a result of these added duties, and a generally run-down condition, rendered Mr. Goode incapable of resisting an attack of pneumonia which caused his death on June 9, 1903.

Fred Graff, Jr.

Claude H. Birdseye, Chief, Section of Engraving and Printing, and former Chief Topographic Engineer, died on May 30, 1941, and Fred Graff, Jr., long-time topographic engineer, was appointed Chief, Division of Map Reproduction, formerly known as the Section of Engraving and Printing, on August 1. Mr. Graff, a native of New York State and graduate of Syracuse University, began work with the Geological Survey on June 18, 1901, as a rodman. He became a topographic aid in 1905 and was loaned for almost a year to the Navy Department to make a topographic survey of Guantanamo Naval Station, Cuba. Three years later he became assistant topographer and worked on complex field assignments in extremely difficult terrain. He progressed as supervisor of one party and then of several parties and finally, in 1936, became Chief of the Rocky Mountain Section of the Central Division and Project Director of the large WPA Project at Denver, Colo., where he demonstrated outstanding technical competence in training men and handled difficult administrative problems with efficiency. He continued as Chief of the Map Reproduction Branch until his retirement on March 31, 1953, and under his direction the number of new maps printed annually increased from approximately 100 to more than 1,200 without significant increase in personnel or space. He received the Distinguished Service Award of the Department of the Interior.

William H. Herron

William Harrison Herron, a native of Monticello, Ill., attended the State Normal College at Eastern Illinois, receiving an A.B. degree. He began work with the Geological Survey on July 1, 1885, as a field assistant and was appointed an assistant topographer in February 1886. Like many other Survey appointees, during the winter office season in Washington, he attended Corcoran Scientific School (later a part of George Washington University), beginning in 1886. He mapped areas in most of the Central States and in Montana, Oregon, and Wyoming. On June 1, 1907, he was appointed Geographer-in-Charge of the Central Division. His summer field office was at the University of Illinois, Urbana, Ill., and he made many friends for the Survey by his friendliness, wealth of good stories, and personable appearance. He was Acting Chief Geographer from January 22 to December 30, 1916, while Mr. Marshall was assigned to the National Park Service. During World War I, he served from June 25, 1917 to March 25, 1919, as major, Engineer Officers Reserve Corps, and mapped strategic areas in the United States. While Major Sutton was overseas, during 1918–19, Majors Herron and Gerdine were also in charge of mapping in the Atlantic Division. He was reinstated on March 26, 1919, and continued as Geographer-in-Charge of the Central Division and then as Central Division Engineer until his death on October 9, 1929.

Herbert H. Hodgeson

On October 17, 1929, Herbert H. Hodgeson was appointed Central Division Engineer after William H. Herron's sudden death on October 9. His work with the Survey began in 1899 as a field assistant during the summer seasons. He had graduated from South Dakota State College in 1898, taught two terms in a country school and been appointed assistant topographer in 1902. He specialized in both triangulation and topography. During World War I, he commanded Company B, 29th Engineers, at Fort Devens, Mass., and across the Atlantic, where he was assigned many important duties. In June 1919, as major, he commanded a battalion homeward bound and was honorably discharged on July 12.

Back with the Survey he continued geodetic and topographic field surveys and after his appointment to the Central Division in 1929 directed the field and office work pertaining to topographic mapping of areas in the midwestern and some of the Rocky Mountain States. His excellent character, able leadership, and kind consideration for his subordinates won for him their hearty cooperation and loyalty.

Thomas G. Gerdine, Pacific Region Engineer, died October 31, 1930, and Mr. Hodgeson was chosen to take his place and Glenn S. Smith was appointed Central Region Engineer. December 9, 1930, Mr. Hodgeson proceeded to his new headquarters at Sacramento, Calif., where his adaptability, experience and initiative enabled him to understand and keep abreast of the rapid development in aerial photo mapping and the modern practices in geodetic and topographic surveying, until his voluntary retirement on November 30, 1942.

Willard Drake Johnson

In October 1879, W.D. Johnson began work with the U.S. Geological Survey on old Lake Bonneville, under G.K. Gilbert. He was about 19 years of age at that time. He continued fieldwork until some time in January, 1880, and afterwards continued in Gilbert's Division of the Great Basin.

Mr. Johnson was appointed an assistant topographer with the Topographic Branch in 1882 and continued in that Branch until 1896. He was possessed of a most charming personality and had unusual ability as a teacher of the science of topographic engineering and untiring energy in every work he undertook. While connected with the Topographic Branch he worked perhaps more than any other topographer to improve the methods of topographic mapmaking. His only thought was the advancement of the scientific work of the Geological Survey, sacrificing his health, pleasures, and means to that end. He invented and patented on May 3, 1887, a new style of planetable tripod, known as the "Johnson Planetable Movement;" the advantages of which were so evident that it was at once adopted by the Geological Survey and Mr. Johnson waived his royalty to the Government for all tripods needed. Another invention which he gave to the Government was the paper holding thumb screws for use in planetable boards. Thousands of these have been made for the Survey and they are standard equipment for such purposes all over the United States. Other devices and methods proposed by him have had their effect on the topographic work of the Geological Survey for many years.

In 1888 he was placed in charge of topographic surveys of the Arkansas River in Colorado, continuing in charge of that work until its completion in the fall of 1890. In 1891 he was given charge of the California office, which position he held for 3 years. He was one of the charter members and director of the Sierra Club in California in 1892. Mr. Johnson made maps for Professor I.C. Russell in the high Sierra around Mono Lake in 1883; then went to Massachusetts for topographic work in cooperation with that State.

He was in Oklahoma in 1897, working under F.H. Newell, hydrograper of the Geological Survey, on the Underflow Water of Arkansas River and Allied Problems of the Great Plains; and as a result of his work with the Water Resources Branch, Mr. Johnson published a report on "High Plains and Their Utilization:" the first part of which appeared in Part IV of the 21st Annual Report and the conclusion in Part IV of the 22d Annual Report.

In 1901 he went to Utah with G.K. Gilbert. One map of the Fish Spring or House Range, was copied for Charles D. Walcott and published in the Smithsonian Miscellaneous Collection, probably volume 12, about 1903. He worked on the completion of his maps at Salt Lake City and afterwards at Provo. About 1905 Mr. Johnson changed from topography to geology and worked under Professor T.C. Chamberlin, with headquarters at Bridgeport, Calif., for at least two seasons, 1905–06, studying glacial geology. He went to Bishop, Calif., a little farther south, in 1907–08 on the same work. In 1907, he was in Lone Pine, in the Owens River Valley, east of Mount Whitney, Calif., studying the displacements of the earthquake of 1872. Mr. Johnson devoted considerable time to work on the Pleistocene problem in the Sierra Nevada with Mr. Gilbert and on the "Faulting in Owens Valley, of recent date—1872, and earlier, in several stages," in conjunction with Professor Hobbs of Michigan, but his efforts were hampered by illness.

During 1912, 1913, and a part of 1914 he was roaming around more or less. On May 23, 1912, he wrote from Carmel, Monterey County, Calif., as follows:

"The 20-years-Service Banquet announcement found me a little too far away to attend. I am glad to find myself associated with such a proud organization, and I hope I may be so fortunate as to find a place card for me at some future banquet board. Or do "Associates" feed at a second table?

"I hadn't fully realized that I dated so far back—so far beyond 20 years. It used to be a reproach that the U.S. Geological Survey, as compared with the Coast and Geodetic Survey, was made up of youngsters!"

In the spring of 1914 he came to Washington and accepted a position as Geographer with the Forest Service at Portland, Oreg. In 1916, he returned to Washington and was confined in St. Elizabeth's Hospital for several months. He was released from the hospital during the summer of 1916 and was working on a model of the Grand Canyon for the American Geographic Society at the time of his death in 1917.

Daniel Kennedy

Daniel O'Madigan Kennedy was born in St. Louis, Mo., on March 5, 1900, and, at the age of 17, enlisted in the U.S. Army during World War I. During this time he served with the 138th Infantry and was wounded while engaged in a battle in France.

Following the war, in 1921, he came to Rolla, Mo., and enrolled in a Topographic Vocational Program that was being offered at the Missouri School of Mines and Metallurgy (MSM) for veterans of World War I.

In 1926, he graduated from MSM with a bachelor of science degree in civil engineering and began a career with the U.S. Geological Survey. He continued his studies at MSM, and earned what was then a civil engineering (Professional) degree in 1935.

Kennedy assisted in the mapping of the Roseau River Valley, Minn., in 1929, a project of approximately 600 square miles, on a scale of 1:31,680 and contour interval of 2 feet. In 1937, he and Lee Walker designed a portable projector, which greatly facilitated the transfer of map data to the field sheets.

Following the attack on Pearl Harbor on December 7, 1941, he returned to active duty with the U.S. Army with the rank of captain. He served under Gen. George Patton, in the 3d Army as the General's chief map maker, in charge of all mapping, map supply, and other engineering intelligence. Early in World War II, he became commander of the 1681st Engineer Survey Liaison Detachment of the U.S. Army Corps of Engineers.

His outstanding work throughout five campaigns in the European Theater earned him a promotion to lieutenant colonel, and for his ingenuity in translating the German map grid system to the American standard, he was decorated with the Bronze Star by General Patton. Other military decorations included the Legion of Merit, the Purple Heart with clusters, and eight battle stars.

Lt. Col. Kennedy was appointed Chief, Operations and Plans in 1947, and subsequently, Chief Topographic Engineer of the Army Map Service in Washington, D.C. In this capacity he had wide planning responsibilities involving the technical supervision of 3,000 people.

He was later called back to active duty by the Army as the officer best qualified to take charge of the survey and triangulation arrangements for the atomic tests at Eniewetok. For his accomplishments in these endeavors, the Army awarded him the Commendation for Meritorious Civilian Service and a promotion to the rank of colonel.

A change in Topographic Branch organizational structure on December 15, 1948, created four mapping regions, and Daniel Kennedy was assigned as Chief of the Central Region. He returned to the Rolla community in 1948 to become Central Region Engineer, succeeding the retired Capt. C.L. Sadler. A year later, his alma mater awarded him the degree of Doctor of Engineering in recognition of his many outstanding contributions to the engineering profession.

In 1961, he received the Distinguished Service Award of the Department of the Interior; and more recently he received recognition for his more than 40 years of Government service.

The Central Region underwent a steady and consistent growth under his management. He continued as its chief until his retirement on February 13, 1970. By this time another name change had occurred and the Central Region had become the Mid-Continent Mapping Center.

Mr. Kennedy's active participation and leadership have been recognized by the many professional and technical organizations he has been affiliated with during his career. He held numerous elective offices in these organizations and served on various committees of responsibility.

These organizations included: American Society of Civil Engineers (having served as President, Mid-Missouri Section, and Director, District 14); National Society of Professional Engineers; Missouri Society of Professional Engineers;

Society of American Military Engineers; American Society of Photogrammetry, where he served as a director; American Congress on Surveying and Mapping; Sigma XI; Chi Epsilon; Theta Tau; Tau Beta Pi; MSM Alumni Association; and the University of Missouri–Rolla Booster Club. His association with the ASCE continued from the time he was MSM Student Chapter President in 1925.

He became the donor for "Daniel Kennedy Essay Awards," a competition established in 1949 for senior civil engineering students enrolled at the MSM who were members of the ASCE Student Chapter. Mr. Kennedy was a registered Professional Engineer in the state of Missouri, and served as a member of the Professional Conduct Committee of the Missouri Society of Professional Engineers.

Always an active leader in civic affairs, Mr. Kennedy was ever ready to give of his personal time and effort. He has served on the Board of Directors of the Rolla Rotary Club as Vice President, President, and Program Chairman. In 1952, he was President of the Rolla Chamber of Commerce. He served in many other key positions, including: Co-chairman of the Rolla Planning and Zoning Commission; Board of Directors, Rolla United Fund; member of the Rolla Park Board; member of the Centennial Committee, University of Missouri–Rolla; and chairman for the fund drives for Cancer and Cerebral Palsy.

The Mid-Continent Mapping Center of the Topographic Division, U.S. Geological Survey, honored Daniel and Margaret Kennedy in a tribute to his 47 years of Government service upon his retirement as the Central Region Engineer on February 13, 1970. He also was honored by the Rolla Chamber of Commerce with a community appreciation luncheon. The guest speaker was Mr. Robert H. Lyddan, Chief Topographic Engineer. Following Mr. Lyddan, Rolla's Mayor Curtis W. Logan presented him with a commemorative plaque. The highlight of the evening was a personal letter to Mr. Kennedy from President Nixon, read aloud by William Radlinski, commending Mr. Kennedy's long service to Government through his vital role in the military and in our country's topographic mapping effort.

Several gifts were presented to the Kennedys, including an unusual, airbrush contoured portrait of Mr. Kennedy prepared by photogrammetrists of the Rocky Mountain Region; as well as the traditional Division scroll and a check, contributed by Survey personnel; a memento of his field days; a "road-runner" desk ornament; and an honorary American Congress on Surveying and Mapping lifetime membership, which were presented by Mr. Earle J. Fennell, Executive Director, American Congress on Surveying and Mapping, Washington, D.C.

Dr. William C. Hayes, Missouri State Geologist, presented Mr. Kennedy with a composite map of a St. Louis Quadrangle, illustrating the high quality of mapping maintained through the years. An All Sportsman trophy was presented as a token of appreciation for his continuing support of all local Survey-sponsored athletic activities. A present from the Washington staff was a handcrafted table of Missouri walnut with a top of an inlaid copper press plate of the Big Clifty, Ky., Quadrangle, the first map that Mr. Kennedy compiled in the field as a junior engineer with the Survey.

In a celebration ceremony on his 99th birthday, at Fort Leonard Wood, Mo., he was awarded the Legion of Honor for service in France during World War I. Then, in honor of his 100th birthday, he was presented a letter wishing birthday greetings and commending him for his contributions to the United States, the U.S. Geological Survey, and the U.S. Army for distinguished service in both World Wars.

Before his death on March 7, 2003, Mr. Kennedy authored a book recounting his life entitled, "SURVEYING THE CENTURY a Soldier of the Two World Wars, and Topographic Engineer, Remembers."

Robert B. Marshall

Robert Bradford Marshall, a native of Virginia, with sparkling black eyes and a most dynamic personality, began work with the Geological Survey on March 16, 1889, as a field assistant. On January 1, 1890, he was appointed a topographic aid, and a topographer on November 1, 1890. He advanced through the various grades until he was appointed Chief Geographer on January 23, 1908. During 1916, he was Superintendent of National Parks, and in 1917, was commissioned a major, Engineer Officers Reserve Corps, and later advanced to lieutenant colonel. During this period he supervised military mapping along with his topographic duties. He was honorably discharged and on April 1, 1919, was reinstated as Chief Geographer. At his own request, he was relieved of these duties on September 30, 1919, and received an appointment as topographic engineer when actually employed. He did not again work for the Geological Survey and his services were terminated without prejudice June 30, 1925.

Mr. Marshall was a lifelong crusader for the physical and economic integration of water, power, and reforestation. He once said, "Land and water are eternal. Credit and prices are ephemeral. No real progress will ever be built on waste." In 1891, he came to California as a topographer and his technically trained eyes saw the paradox of millions of parched and burning acres in the summer, and millions of acre feet of water pouring wastefully into the Pacific Ocean during the winter and spring. So he dreamed a dream of storing the rain waters of the Sacramento, San Joaquin, and Santa Clara Valleys, and of Southern California, and during the next three decades he sought the engineering facts to pound his dream into reality. In 1921, he went before the California Legislature and although called crazy, emerged with an appropriation of $200,000 for the State engineers to investigate his plan. That was the real start of the Central Valleys Project, which in those days was known as the Marshall Plan. Practically all of the bigger units of the project, such as the Shasta and Friant Dams, are basically the same as Mr. Marshall planned them.[58] He died on June 21, 1949.

---

[58] The Sacramento Bee, June 21, 1949.

T.P. Pendleton

Thomas P. Pendleton, a native of California, joined the Geological Survey in 1905 as a recorder, and received an appointment as Junior Topographer in 1909. From 1909 through 1917, he mapped many areas in the western United States with the planetable. In 1918, he was commissioned a second lieutenant in the U.S. Engineers, and worked principally with the early development of the first tri-lens aerial camera, the training of personnel, and the development of radial line mapping methods. After the war, he rejoined the Survey and used the panoramic camera to map a part of Mount McKinley National Park in Alaska. In 1920, he was furloughed for a special mapping assignment in Turkey, and upon his return in 1921, was appointed Chief of the newly-formed Section of Photomapping. In 1922, he again requested furlough for topographic mapping in Palestine and the Balkan States, returning to head the Survey's Photomapping Section in 1923.

In 1926, Mr. Pendleton resigned to accept a position as Chief Engineer of Brock and Weymouth, Inc., and for 5 years worked with the development of the Brock process and the production of maps with this method. In 1931, he was Chief Engineer of the Aerotopograph Corporation of America, with headquarters in Washington, D.C. In 1932, he reentered Government service as Chief Photocompiler for the Coast and Geodetic Survey, and in 1934, returned to the Geological Survey and directed a Tennessee Valley Authority mapping project on which about 500 engineers were employed. Later, he directed tests that resulted in the development of the Multiplex, an instrument for the projection of aerial maps in mapmaking, which is now in wide use. In 1942, he was Chief, Section of Photomapping, with headquarters at Washington, D.C.

From his earliest interest in the use of photography as an important tool in the production of maps more than 30 years ago, Mr. Pendleton followed through in the development of this idea. His perseverance and unswerving application under adverse circumstances in pioneering in the field of photogrammetry have been instrumental in the development of new mapping methods. He provided the leadership and initiative applying these new principles to the art of mapmaking. He was author of numerous technical articles on photogrammetry and mapping, and co-author of Bulletin No. 788, "Topographic Instructions of the U.S. Geological Survey."

Mr. Pendleton received the following letter from Secretary Ickes, dated February 29, 1944:

"Enclosed is a formal notice of a salary increase which I have been glad to approve upon the recommendation of the Board of Awards of the Interior Department's Suggestions System in recognition of your especially meritorious services in promoting the advancement of the stereoscopic method of mapping from aerial photographs.

"The notice carries its own story of material reward for your ingenuity and diligence, but I wish to add a personal word to express my own appreciation of, and thanks for, the splendid service that you have rendered. Never has there been greater need than now for all that intelligence and devotion can offer, and in behalf of the Department, and the public that it serves, I gratefully acknowledge your contribution."

He was the official delegate of the Department of the Interior to the Second Pan American Consultation on Geography and Cartography held in Rio de Janeiro, Brazil, August 14 to September 2, 1944.

Mr. Pendleton continued as Chief topographic engineer from March 18, 1943, to April 1, 1947, when he requested retirement following a long illness. He received the Distinguished Service Award from the Department of the Interior. His death occurred on May 28, 1954.

Albert Pike

Albert Pike, who succeeded Glenn S. Smith as Atlantic Division Engineer, was a native of Rockbridge County, Va. On March 1, 1890, he was appointed a messenger in the Interior Department. Three years later, while still in his teens, he was sent with a Geological Survey field party to New York State and traversed roads with a horse and buggy. On May 1, 1894, he was appointed assistant topographer and continued to map in whole, or in part, about 100 quadrangles in all parts of the country, establishing a reputation as a pace-maker in executing the ground surveys accurately, and in inking the field manuscripts for reproduction in the Washington office.

During World War I, he served as Captain, U.S. Engineers, overseas with the 42d Rainbow Division, as topographic officer. After his return to the Survey, he continued in charge of the Atlantic Division until he requested optional retirement on February 28, 1946, after 56 years of service with the Geological Survey. He died September 3, 1952.

On February 13, 1946, Mr. Pendleton wrote Mr. Pike as follows:

"Information came to me several days ago of your decision to retire in the near future, rather than continue the difficult duties you have carried for many years. Your record of service is so long and enviable that it cannot be enhanced by further attempts to extend it, so I believe your decision is a wise one.

"You are among the very few engineers who have personal knowledge of mapping methods originally employed in this country, as well as those that have come into use in recent years. Your familiarity with topography of this country is equally broad. It would seem to me that recollections of these things should compensate you in great measure for any regrets you might have in leaving the organization with which you have so long been associated.

"I fully concur with your belief that Dallas H. Watson is by experience, and by natural ability, the logical successor to your position as Division Engineer, and it pleases me to know that you recommend him so highly."

John H. Renshawe

John Henry Renshawe was born on October 11, 1851, in a log house at Pattensburg, Ill. He graduated from the Lacon, Ill., high school and taught school for 2 years. From 1872 to 1878, he was a member of the famous J.W. Powell exploring expeditions in Utah, Nevada, and Arizona. During these expeditions, he conducted operations on land, while Powell was making his two boat trips through the Grand Canyon. This was extremely difficult work, in country inhabited only by Indians, and hundreds of miles from the only transcontinental railroad.

Mr. Renshawe was appointed a topographer on the first staff of the U.S. Geological Survey, on July 10, 1879, and he was assigned to the Division of the Colorado. He did the survey of the Uinkaret District, Ariz., an area of 1,475 square miles, in an arid region with but one known living spring and scattered water pockets. During 1880 and 1881, he supervised the publication of topographic maps at New York City and the Washington office. From 1882 to 1885, he was in charge of the mapping of Yellowstone National Park. His party was struck by lightning while on Yellowstone Lake and Mr. Renshawe has the distinction of bearing the indelible marks of a lightning stroke, while still living. From 1896 to 1903, he was in charge of the Central Division of the Topographic Branch.

Coincident with the reorganization of the Topographic Branch on July 1, 1903, provision was made for the inspection of topographic surveying and mapping, and Mr. Renshawe, geographer, was placed in charge of that section. His duties were not administrative, but general, in character, and were carried on through consultation and collaboration with the chiefs of the eastern and western sections of topography. He kept in touch with the topographers in the field, criticized their expressions of topographic forms and their choice of essential details, and eliminated, as far as possible, their mannerisms and individualities in contour sketching, with a view of gaining uniformity in style and expression for similar features throughout the country. When Mr. Renshawe returned from the field late in October, he began work on a complete and systematic revision of the three-sheet wall map of the United States. The map was on a scale of 40 miles to 1 inch, with a contour interval of 1,000 feet, and was based on the detailed topographic maps of the Geological Survey. This map showed all state and county boundary lines, the names and location of county seats, all railroads, and principal drainage. The relief was treated somewhat broadly, as prescribed by the horizon-

tal and vertical scale, but no effort was made to preserve the character of all topographic features.

From 1903 to 1914, Mr. Renshawe's work as Inspector took him into every portion of the United States, and made him familiar with the various topographic types encountered between Canada and Mexico, and from the Atlantic to the Pacific, so that with his natural artistic ability, he perfected, through a deft manipulation of light and dark tints, a system of relief maps, and his general relief maps of the United States, Alaska, and various states have placed him foremost in this line of work. On July 1, 1920, he was placed in charge of the Section of Relief Maps. He also was an artist of note and specialized in water color landscapes.

Mr. Renshawe's kindly unassuming manner endeared him to all of his associates and it became the custom, when passing his room, to enter for a few minutes, just to watch his easy manipulation of crayon or brush and the resulting symphony of color in maps or landscapes, and the members of the Topographic Branch who are fortunate enough to have "Renshawe Landscapes" will always treasure them.

On August 1, 1921, Philip S. Smith, Acting Director, wrote Mr. Renshawe as follows:

"I note from the personnel records that you will come within that provision of the act for retirement of employees in the Classified Civil Service, on October 11, 1921, which states that employees in the Classified Civil Service who have reached the age of 70 years and rendered at least 15 years of service shall be eligible for retirement.

"I am, however going to take advantage of that provision contained in the bill which permits your further employment for 2-year periods and ask for your extension, but I can give you no assurance as to when action will be taken on my recommendations, or whether they will be approved by the Department. This action is prompted because of my intimate knowledge of your present active work in developing the unique art of portraying vividly the topography of our country. The shaded maps of the United States, and of parts of the desert regions that have been published are masterpieces, and the maps on which you are now engaged of California and other parts of the West can only be brought to successful completion by yourself. The Geological Survey needs you, and I hope that it meets with your approval to have the necessary steps taken which will keep you in our official family. Won't you advise me in writing your wishes in this matter?"

And now Mr. Renshawe's reply:

"Please allow me to express my appreciation of your generous reference to my work, and to thank you for your offer to take the initiative in recommending the extension of my active service beyond the time limit fixed by the retirement act for employees in the civil service.

"Aside from my personal inclination as to retention in active service, I would like to finish several important pieces of work that I have in hand, and to assist in a fuller development and application of the system of mapmaking that I have introduced in the Survey.

"So, considering everything, I shall be glad to benefit by the provision in the retirement act, allowing for limited periods of retention in active service."

On August 4, 1921, the Topographic Branch prepared the following memorandum for the Acting Director:

"Mr. J.H. Renshawe, Geographer-in-Charge of Section of Relief Maps, will reach the retirement age of 70 years on October 11, 1921, and under the law, will be separated from the Topographic Branch of the Geological Survey, unless action is taken at once for his retention on active duty.

"Mr. Renshawe has served the U.S. Government 49 years, of this time, he has spent in the Topographic Branch of the Geological Survey, 42 years, holding various positions from that of assistant to geographer-in-charge of a division, and later that of chief inspector of topography. During the past 10 years, he has initiated and perfected the method of map shading in producing relief maps, to such point as to interest geographers throughout the world, and he is today considered the greatest expert in this line. The demand on the Geological Survey for this class of map work from various Governmental sources is so great that it is impossible to fulfill but a small percentage of the requests. Should the Geological Survey lose Mr. Renshawe's services through retirement in the near future, it would be impossible to replace him, thus a large number of maps now in progress of preparation by this method would have to be discontinued, or be completed by inexperienced personnel.

"In addition to being actively engaged in the numerous maps, Mr. Renshawe is instructing an understudy who, it is hoped, will develop to such an extent that he can carry on this specialized work when Mr. Renshawe reaches an age when it will be necessary for him to abandon active duty.

"Mr. Renshawe's retention on active duty cannot be too strongly urged for the reasons above stated, especially as he is physically able to perform his duties to the fullest extent, and the loss of his services to the Government, especially to the Geological Survey in the near future, would be impossible to estimate."

Mr. Renshawe received a 2-year extension, and a further 2-year extension, and was finally retired on June 30, 1925. He died on October 25, 1934, after 83 years of creative living.

Carl L. Sadler

Carl L. "Cap" Sadler succeeded Glenn S. Smith as Central Division Engineer upon his retirement on April 30, 1940. Mr. Sadler was a native of Arkansas, a graduate of the Arkansas State University as a civil engineer, and had spent most of his years in the states of the Mississippi Valley since he began as a traverseman with the Topographic Branch in 1899. He served as Captain, U.S. Engineers, during World War I, and afterwards was retained in the Central Division. Mr. Sadler was already well established in Rolla, Mo., where his section had maintained a field office for years and, since September 1933, a permanent headquarters. He continued as Central Division Engineer until his retirement on June 30, 1948. His citation for the Distinguished Service Award was as follows:

"Early demonstration of his ability to make topographic surveys, and with a capacity for hard work, extending far beyond normal requirements, he was soon recognized as a natural leader who invited responsibilities. He was advanced through the ranks to be appointed Division Engineer-in-Charge of the Central Division, the position which he held at retirement. Mr. Sadler was untiring in his efforts to extend operations in his Division. Always impatient with slow processes, he set the pace for map production. He was intensely loyal and drove hard to accomplish results. He demanded and obtained great production effort from all of his staff, and was able to secure the maximum amount of mapping at a minimum of cost. Because of the care and energy with which he managed these activities, he was eminently successful in promoting and continuing state cooperative mapping projects. His accomplishment in that work have been praised by many State officials * * *."

Mr. Sadler died in Rolla, Mo., on August 19, 1982.

Rufus H. Sargent

On March 1, 1937, William M. Beaman, Chief, Section of Inspection and Editing, died, and on April 1, Rufus S. Sargent was appointed Chief of that Section. Mr. Sargent, a native of Maine, joined the Geological Survey in June 1898, as traverseman and was engaged on surveys in the Black Hills of South Dakota, and the northern Rocky Mountains near the Canadian Border. He was appointed an assistant topographer on June 19, 1900, and because of his skill in the use of the planetable and his ability to carry on accurate mapping, in spite of hardships and unexpected emergencies, he was selected for work in Alaska in 1906.

Mr. Sargent spent 28 seasons in Alaska conducting reconnaissance expeditions into the far reaches of this unexplored territory. His urge for geographical conquests led him to undertake several expeditions into foreign countries for which he was granted furloughs. In 1903 to 1904, he was with the Carnegie Institute Expedition into China, where he mapped unexplored reaches of the Yellow River. As topographer for the Bolivia–Argentine Exploration Corporation in 1920 to 1921, he mapped rugged wilderness areas in South America. In 1907, and again in 1926 to 1927, he explored and mapped difficult terrain in Mexico for various mining companies.

He was Chief, Section of Inspection and Editing from April 1, 1937, to his retirement on March 31, 1947, having been retained for 16 months on extensions beyond statutory retirement age. In that position, he kept pace with the rapidly developing new science of photogrammetry, as applied to map making, and inaugurated many changes in the practices of cartography and in the processes of map reproduction. When it became necessary in 1942 to relieve the congestion in the Washington office, he selected quarters in Clarendon, Va., and established a branch drafting office.

In every sense a geographer, Mr. Sargent sought to pass on to others his knowledge of foreign lands through lectures and written articles, and he brought to all of his work and activity a zeal, enthusiasm, and ability that marks him as true explorer and pioneer. His love of nature and keen sense have added a human touch that makes his accomplishments stand out with all who knew him. He received the Distinguished Service Award from the Department of the Interior. His death occurred December 27, 1951.

Glenn S. Smith

Glenn Shepard Smith was with the Geological Survey as a field assistant during 1888 and 1889, and was appointed an assistant topographer on January 1, 1890. He was employed until December 31, 1892. In 1895, he worked as a field assistant at Cowee, N.C. and in 1896 and part of 1897, as a field assistant on the Indian Territory Survey. He was appointed an assistant topographer on July 6, 1897, and advanced through the various grades to Topographic Engineer. During 1916, he was in charge of the Central Division while W.H. Herron was Acting Chief Geographer. January 11, 1917, by Survey Order No. 78, he was given charge of the new Division of Military Surveys. June 22, 1917, he was commissioned major in the Engineer Officers Reserve Corps, and later was given the rank of lieutenant colonel. Overseas, he was Assistant to the Chief of Topographical Division, General Staff, G.H.Q., in charge of all field topographic mapping, and organization of the Base Printing Plant. Colonel Smith also was Director of the Base Printing Plant, 29th Engineers, in charge of topographic mapping. On July 1, 1919, Colonel Smith was relieved of his duties in connection with military surveys, and designated Topographic Engineer-in-Charge of the Division of West

Indian Surveys. He was honorably discharged on December 28, 1920, and reinstated in the Geological Survey.

He continued in charge of the Santo Domingo Surveys, and also was acting chief topographic engineer when the chief topographic engineer was absent. On March 8, 1924, he became Division Engineer-in-Charge of the Atlantic Division, and also in that year, Secretary to the Southern Appalachian Park Commission. He resigned his position as Atlantic Division Engineer on September 9, 1929, and was kept on the rolls as chief engineer, when actually employed. He continued as Interior Department representative with the Southern Appalachian Park Commission. In December 1930, he was appointed Topographic Engineer-in-Charge of the Central Division and continued in this position, with 1-year's extension, until his compulsory retirement on April 30, 1940. He died on June 2, 1951.

John G. Staack

John George Staack, a native of Wisconsin, graduated from the University of Wisconsin with a bachelor of science degree in civil engineering in 1904. During the summer of 1902 and 1903, he worked with topographic field parties and received a topographic aid appointment with the U.S. Geological Survey on July 1, 1904. During the period 1904–07, he was assigned to quadrangles possessing many types of land forms in states of the Mississippi Valley and as far west as the Pacific Coast.

During World War I, he served nearly 2 years as Captain with the Corps of Engineers on assignments mapping strategic areas in the United States. After the War, for a number of years, he obtained furloughs from the Geological Survey during July to instruct students in the Surveying Camp of the University of Wisconsin at Devil's Lake. For this instruction in planetable methods as practiced by the Topographic Branch, he received expressions of gratitude from the University authorities who appreciated his pleasing personality, teaching methods, and ability to make good topographic engineers from raw material.

His expert knowledge of the intricate topography of the glaciated and the unglaciated areas of the Upper Mississippi Valley led to his assignment, in 1924, as Chief of the Great Lakes Section, comprising the States of Minnesota, Iowa, Wisconsin, Michigan, and Illinois. His duties in these cooperating states included the supervision of field surveys and office drafting, which resulted in many outstanding surveys of this region.

On September 10, 1929, Mr. Staack was appointed chief topographic engineer, and for the following 13.5 years he guided the Topographic Branch through trying depression years and into the expanding activities required by the Armed Forces during World War II. The administration of a large Public Works Administration mapping program from 1933 to 1940 resulted in the employment and training of many engineers, and the maintenance of extensive topographic surveys. In

1933, a Branch Office was established at Chattanooga, Tenn., in cooperation with the Tennessee Valley Authority to map the Tennessee River Basin. During this period, the application of aerial photography and photogrammetry to topographic mapping was advanced through several stages in development.

With the increased demand of the Armed Services for maps of strategic areas, a large expansion of survey activity was begun in 1941. The increasing administrative problems and the necessity of maintaining maximum production of all mapping projects placed an increasing burden on the chief topographic engineer, as the number of employees and amount of appropriations increased two and one-half fold. In 1942, Mr. Staack was relieved of some of these duties by the appointment of Mr. Thomas P. Pendleton, outstanding photogrammetrist, as chief topographic engineer, and Mr. Staack continued his valuable services to the Survey as the assistant chief topographic engineer until his voluntary retirement on June 30, 1947.

The Distinguished Service Award was bestowed upon Mr. Staack by the Department of the Interior in recognition of services rendered through his ability and steadfast loyalty in the administration of the Topographic Branch during some of the most trying years of its history, and the maintenance and improvement of the high quality and standards of the topographic surveys.

Frank Sutton

Frank Sutton graduated, with a degree in civil engineering, from the Pennsylvania Military College in 1879. Between 1879 and 1886, he was employed in railroad location and construction. On April 3, 1886, he was appointed a draftsman in the Geological Survey, and on October 15, 1888, became a topographer. His natural ability as a topographer, combined with his training as an engineer, attained for him rapid promotions, and in 1906 he reached the grade of Geographer-in-Charge of the Atlantic Division. He was commissioned a major in the Engineer Officers Reserve Corps on January 23, 1917, and was called to active duty on June 22, 1917. Major Sutton was first assigned to special topographic mapping duty with the Geological Survey, and was later assigned to duty with the 25th, 29th, and 304th Engineers, having command of the latter. He served in France from August 1918 to April 1919, returned to the United States, and was honorably discharged on April 7, 1919. He was reinstated in his former position of geographer in the Geological Survey, which position he held until his retirement on account of ill health on March 7, 1924. His stability of character, devotion to duty, shrewd wisdom, and fine sense of humor made for him a host of friends. He died on July 22, 1926.

Sledge Tatum

Mr. Tatum was appointed surveyor in the U.S. Geological Survey in May 1895, and spent several years surveying land lines in Indian Territory. In 1899, he was appointed topographer and extended triangulation in various states. In 1903, he surveyed forest boundaries in Washington and Idaho, and the following year executed triangulation along the International Boundary. On June 10, 1905, he was transferred to the Department of Engineering, Isthmian Canal Commission. He began work in the Canal Zone as instrumentman on the Chagres River surveys, and advanced rapidly, at one time having charge of all the survey parties in the field, both on the Chagres River Division and the Canal Zone boundary. On November 1, 1908, Mr. Tatum became superintendent of construction and was employed on the building of the Gatum Dam.

He resigned on February 8, 1909, to accept a transfer to the U.S. Geological Survey, and was assigned to special investigations until his appointment as Geographer-in-Charge of the Rocky Mountain Division. On June 10, 1910, E.C. Barnard was detailed to the State Department for work on the United States and Canada Boundary Survey, and Sledge Tatum was placed in charge of the Rocky Mountain Division. He was one of those rare combinations of thorough administrative ability, good judgment, and consideration of the rights of others. A tall man with personal charm and a ready smile, he had the respect and friendship of his fellow workers. He was Acting Chief Geographer from December 10, 1915, to his death on January 18, 1916.

Almon H. Thompson

Almon Harris Thompson was born at Stoddard, N.H., on September 24, 1839. He graduated from Wheaton College, Ill., in 1861, and soon afterwards he married Ellen L. Powell, sister of his lifelong friend, Maj. J.W. Powell. During the Civil War, he served as first lieutenant, 139th Illinois Volunteer Infantry. After the war, he returned to Illinois where he served as Superintendent of Schools at Lacon (1865–67) and Bloomington (1867–68), and as Acting Curator of the Illinois Natural History Society (1869). His career as professional geographer began at the age of 31, when Powell selected him as scientific associate on a reconnaissance traverse from Salt Lake, southern Utah, across the Colorado and through the Navajo country, which ended at Fort Defiance, December 5, 1870. The purpose of this traverse was to make preparations for the systematic exploration and mapping of the Colorado River and its adjoining lands, the task assumed by Thompson for the period 1871–78, part of which (1871–75) is covered in his diary. This diary has been published by the Utah State Historical Society[59] and is very interesting. A short outline states that Professor Thompson was astronomer and topographer, second Powell expedition, and Rocky Mountain exploration expedition. Following are some excerpts from his diary.

---

[59] Diary of Almon Harris Thompson, published by the Utah State Historical Society, v. VII, nos. 1, 2, and 3, January, April, and July 1939. Introduction by Herbert E. Gregory, Geologist, U.S.Geological Survey.

"May 1871 to November 28, 1871. Down the Green and Colorado Rivers. November 29, 1871, to August 13, 1872; Fieldwork. Kanab to mouth of Dirty Devil, overland. August 13 to September 1871. Through the Grand Canyon.

"September 1, 1872, to June 4, 1873. Topographic work, Kanab, St. George, Pipe Springs, Paria, etc. September 8 to December 24, 1873, ditto. 1875, Topographic work. Gunnison, Castle Valley, Kaiparowitz Plateau, Warm Creek, Paria."

Professor Thompson possessed that rare combination of qualities that brings success to the explorer: a rigid insistence on discipline and order of procedure, kindness toward his subordinates, and sympathetic interest in the native people with whom he came in contact. Though in no sense egotistical, he had the confidence that views difficulties merely as routine problems capable of solution. His peace of mind seems to have been little disturbed by the dangers and hardships of field work.

A diary entry for February 17, 1873, states: "Got map finished. Fred and Jack started for Panguitch or farther." In these few words are recorded the completion and dispatch to Washington of the first map ever made of southern Utah and of the canyon of the Green and the Colorado Rivers; a map resulting from nearly 2 years of arduous and skilful work in a region largely uninhabited. These maps of the Colorado drainage basin place Thompson in the front rank of geographic explorers. Entries in the diary and his paper on "The Irrigable Lands of Utah," that reveal keen insight into utilization of natural resources, add to his reputation.[60]

On July 15, 1882, Professor Thompson accepted a position with the U.S. Geological Survey, and he was permanently appointed a geographer on August 11, 1882, and placed in charge of the Wingate Division of the Topographic Branch. On July 1, 1884, he was ordered to take charge, besides the old Wingate Division, of the topographic fieldwork in Texas, Missouri, Kansas, and Arkansas; the whole to be known as the Southwestern Division. On May 1, 1885, the Southwestern Division was extended to include the California Division and the name was changed to the Western Division. Professor Thompson was given charge of this Western Division. He was geographer under Henry Gannett from September 11, 1885, to December 28, 1888, in charge of the Western Division and made western field trips each summer.

During 1889 to 1890, Professor Thompson was in charge of the topographic work of the Irrigation Survey. The Western Section was reorganized and its personnel, as well as some men from the Eastern Section, were transferred to the Irrigation Survey. From September 1, 1890, to June 9, 1894, he continued in charge of the surveys west of the 100th meridian. Professor Thompson extended triangulation in the States of Colorado, Kansas, Texas, Arizona, Montana, Pennsylvania,

Ohio, and New York from 1894 to 1902. He assisted in the preparation and installation of the Geological Survey exhibit at the Louisiana Purchase Exposition, and on May 1, 1904, proceeded to St. Louis to take charge of the exhibit. From 1903 to 1906, he worked in office computations and as assistant in charge of office work to both the Eastern and Western Sections. He aided Messrs. Wilson and Douglas, during their absence on field inspection, by approving vouchers and looking after routine correspondence. He died on July 31, 1906, and was buried in Arlington Cemetery.

Gilbert Thompson

Gilbert Thompson was born at Blackstone, Mass., in 1839. He had a common school education and was a printer by trade. Mr. Thompson was a soldier with a U.S. Engineers Battalion from November 22, 1861, to November 21, 1864; an assistant engineer with the Headquarters Army of the Potomac (1864–65); on western explorations and surveys, etc., (1886–79);[61] and topographer on the Wheeler Survey. He is also said to have been the first person to make official use of fingerprints in the United States.

When asked how he became a topographer, Mr. Thompson liked to tell the following story:

In 1861, when he enlisted in Boston, he was asked what he was engaged in and replied, "I am a printer." He was ordered to the front immediately, and when he reached the mobilization camp, was sent to a section in which there were only officers. He asked a young officer about it and he replied, "You are an artist, I believe." "No," said Mr. Thompson. "When you enlisted and were asked what business you were engaged in, did you not say you were a painter?" "No, I said I was a printer." "Well," said the officer, for heaven's sake don't say anything. You have been assigned to the engineer branch of the service." Then the officer took Mr. Thompson back of his tent and showed him how to use a compass, pace distances, sketch topography, etc., and thus he became a topographer. (One of the best topographers the Geological Survey ever had.)

On May 1, 1880, Mr. Thompson was appointed a topographer with the U.S. Geological Survey, and assigned as topographer with G.K. Gilbert, Division of the Great Basin, to locate the shores of Lake Bonneville, Utah. The maps were compiled at the Salt Lake City office and Mr. Gilbert in his report to the Director in the Second Annual Report, 1880 to 1881, states:

"The field and office labors of my assistants have, without exception, been characterized by zeal and efficiency. The mapped outline of the lake, which constitutes the most tangible of our results, is a joint work to which each has contributed his quota. Drawn upon a larger scale by Mr. Thompson, it will be issued with the monograph of Lake Bonneville,

---

[60] Introduction to published Diary by Herbert E. Gregory, Geologist, U.S.Geological Survey.

[61] Who's Who in America, 1906–07.

now in preparation. The same volume will contain numerous local maps by Messrs. Wheeler, Thompson, Webster, and Johnson, and a report on the hypsometric work by Mr. Webster. These will speak for themselves in due time and need no commendation here."

On August 3, 1881, he left Salt Lake City and proceeded to Fort Wingate, N. Mex., where he assumed charge of the Wingate Division. From January 10, 1882, to June 30, 1882, he was acting chief topographer. He was placed in charge of the District of the Pacific, which was later called the California Division, and from 1882 to 1883, he completed a detailed map of Mount Shasta and its immediate surroundings. A field map covering about 24,000 square miles in northern California was now ready for the use of the geologists. On July 1, 1884, he was given charge of the Appalachian Division, comprising the country south of the Mason and Dixon Line and the Ohio River and east of the Mississippi River. In 1887, the name was changed to the Southeastern Division. Mr. Thompson was Chief Geographer from July 10, 1890, to June 30, 1894, in charge of the Southeastern Division. From 1894 to 1895, he was in the field on revision work of the Bristol and Abingdon Quadrangles, Va., and in 1896–97, he extended triangulation and primary traverse in the state of Tennessee. Office work was the revision of the nine-sheet U.S. map. During 1898, he mapped 143 square miles on the Helena Special Quadrangle, Mont. From 1899 to 1909, he was in the Washington office. He was commander of an engineers' battalion, D.C. Militia, from 1890 to 1908.

Tribute to Gilbert Thompson

Maj. Gilbert Thompson, the oldest member of the Topographic Branch of the Geological Survey died suddenly on Tuesday June 8, 1909, and at a meeting of the members of the U.S. Geological Survey, held on June 10, the following resolution was adopted, which has been embossed and transmitted to Mrs. Thompson.

"The members of the U.S. Geological Survey have learned with deep sorrow of the sudden death of their friend and associate—Maj. Gilbert Thompson

"His long, honorable, and valuable service upon the Survey; his devotion to duty; and his genial, kindly disposition, make his departure a serious loss to us all.

"We wish to express to his wife and daughter the high appreciation in which he was held among us, and to extend heartfelt sympathy in their bereavement."

Dallas H. Watson

During Mr. Pike's absences from the Washington office, the direction of the Atlantic Division had been for several years in the hands of Dallas H. Watson, who continued as acting division engineer until his appointment as Atlantic Division Engineer on April 17, 1946.

Mr. Watson, a native of Oklahoma, was educated in the State's Agricultural and Mechanical College, emerging with a bachelor of science degree. in civil engineering. He received an appointment as topographic aid on June 1, 1911, and worked in all parts of the country, specializing in topographic mapping and triangulation.

During World War I, he served in France as Captain of Artillery, and after his return to the Survey, was detailed to Santo Domingo for topographic mapping during the period August 16, 1919, to August 15, 1921. In 1922, he was in charge of Puerto Rico topographic mapping; in 1931, Senior Topographic Engineer-in-Charge of the New England Section; in 1940, Supervisor at Large, Atlantic Division; and in 1946, Atlantic Division Engineer, which position he has held for 8 years.

Herbert M. Wilson

Herbert Michael Wilson was born at Glasgow, Scotland in 1860. He studied at Plainfield, N.J., and Cooper Union, N.Y., and received a degree in civil engineering in 1881 from the Columbia School of Mines. He was a railway engineer in Mexico from 1881 to 1882. On August 12, 1882, he was appointed a topographer with the U.S. Geological Survey, and from 1882 to 1885, was assigned to the Wingate Division under A.H. Thompson. He was then given charge of the Gold Belt subsection, and later of the California subsection field work. On May 25, 1889, he was appointed an engineer in the Irrigation Survey, and on December 1, 1889, he was sent to India, where irrigation had been practiced for centuries and was well developed, for the purpose of making investigations. He reached Bombay on January 13, 1890, and, after an exhaustive investigation in the interior of India, he returned to Bombay on March 3, and traveled via Cairo, Rome, and Paris, and arrived in Washington on April 15. His observations were embodied in a report published in Part II, Irrigation, of the Director's annual report for 1889 to 1890. He was appointed geographer on September 3, 1890, and from 1890 to 1894, he was in charge of the Northeastern Division.

From July 1, 1894, to June 25, 1896, he was a Chief Geographer, and at the same time he was in charge of the Atlantic Section, which comprised the former Northeastern and Southeastern Divisions. From July 1, 1896, to 1903, he was a member of the Topographic Committee, as well as in charge of the Atlantic Section. From 1903 to 1907, he was Geographer-in-Charge with E.M. Douglas, of the Topographic Branch, and in charge of the Eastern Section. On April 6, 1907, he was transferred to the new Section of Technology,

Geological Survey, as chief engineer. The Technology Section became the Bureau of Mines on July 1, 1910, and he was engineer-in-charge from 1910 to 1914. Mr. Wilson was Director of the Department of Inspection and Safety of the Associated Insurance Companies of Hartford, Conn., from 1915 to his death on November 25, 1920.

Mr. Wilson lectured extensively on irrigation, geography, fuel testing, smoke abatement, mine fires and rescue, fire proofing building construction, swamp reclamation, etc. He was the author of the "Manual of Irrigation Engineering," 1893; "Topographic Surveying", 1899; American Civil Engineers' Pocket Book; official reports; and numerous magazine articles on the above mentioned subjects.[62]

Robert S. Woodward

Robert Simpson Woodward was born at Rochester, Mich. He attended the Rochester Academy and received a degree in civil engineering from the University of Michigan in 1872. He received many honorary degrees from various universities during his lifetime.

From 1872 to 1882, Dr. Woodward was an assistant engineer with the U.S. Lake Survey. He was an astronomer with the U.S. Transit of Venus Commission from 1882 to 1884, and was appointed an astronomer with the Topographic Branch of the U.S. Geological Survey on July 17, 1884. In 1885, he was promoted to geographer and placed in charge of the new Astronomic and Computing Section. On January 1, 1889, he was promoted to Chief Geographer, and resigned on June 30, 1890.

In his 11th Annual Report, 1889–90, Part I, Geology, Director Powell states:

> "Mr. R.S. Woodward's resignation was accepted with regret, as this Survey can ill afford to lose his rare ability for mathematical research. Since his first association with the Survey in 1884, he has not only supervised the computations made in connection with the triangulation and astronomic determinations, conducted the computation of a series of tables for the use of the Topographic Branch, and given aid to geologists having occasion to treat their data by mathematical methods, but he has made important additions to geologic science by discussing and advancing, on several lines, the theories of terrestrial physics . . . ."

Following are excerpts from the "Memoirs of Robert Simpson Woodward," by F.E. Wright:

> "In 1890, Mr. Woodward resigned from the Geological Survey to accept a position with the U.S. Coast and Geodetic Survey. Because of his long experience in precise triangulation, he was called upon to 'devise means of testing in the most thorough way practicable, the efficiency of the various forms of base apparatus used by the Survey, especially the efficiency of long steel tapes or wires.' He devised the iced bar apparatus for measuring base lines and for calibrating steel tapes. He was also the first to measure primary base lines with long steel tapes, and to prove that these tapes furnish the required degree of accuracy, namely, one part in one million * * *.

> "In 1893, Dr. Woodward was called to Columbia University as Professor of Mechanics and Mathematical Physics. From that time on, his work was that of the teacher and administrator. In 1895, he became dean of the College of Pure Science, and was confronted with many problems that required tact and perseverance for their solution* * *."

Dr. Woodward was President of the Carnegie Institution from December 13, 1904, to January 1, 1921, and during his administration, the Institution grew and flourished, and contributed much to the development of science in this country. He received many honors from different scientific societies.

In closing Mr. Wright states:

> "In addition to his great attainments, Dr. Woodward possessed a simplicity and open friendliness of manner and character that endeared him to young and old alike. It was a privilege to know him, to feel his radiant enthusiasm, and to be uplifted by his hopeful outlook on this world and its many problems."[63]

Dr. Woodward died on June 29, 1924.

---

[62] Who's Who in America, 1919.

[63] National Academy of Sciences, U.S.A., v. XIX–XX, Biographical Memoirs—Robert Simpson Woodward, 1849–1924, by F.W. Wright.

Index

D

X

Y

Publishing support provided by:
    Rolla Publishing Service Center

For more information concerning this publication, contact:
    Director, National Geospatial Program
    Geography Discipline
    U.S. Geological Survey
    USGS National Center
    12201 Sunrise Valley Drive
    Reston, VA 20192–0002
    (703) 648–5569

Or visit the National Geospatial Program Web site at:
    http://usgs.gov/ngpo/